AMERICAN ORIENTAL SERIES

VOLUME 75

CUNEIFORM MATHEMATICAL TEXTS AS A REFLECTION OF EVERYDAY LIFE IN MESOPOTAMIA

AMERICAN ORIENTAL SERIES

VOLUME 75

AMERICAN ORIENTAL SOCIETY

NEW HAVEN, CONNECTICUT

1993

CUNEIFORM MATHEMATICAL TEXTS AS A REFLECTION OF EVERYDAY LIFE IN MESOPOTAMIA

By
KAREN RHEA NEMET-NEJAT

AMERICAN ORIENTAL SOCIETY
NEW HAVEN, CONNECTICUT
1993

ISBN 0-940490-75-7

To my beloved father,
Nathan Waldman, Chief Machinist, CWO2, United States Navy,
30 years of service to his country,
may he rest in peace.

TABLE OF CONTENTS

ABBREVIATIONS

All abbreviations of books, periodicals, etc. conform to the *Assyrian Dictionary* of the Oriental Institute of the University of Chicago, with the exceptions listed below.

HKL	Borger, R. K., *Handbuch der Keilschriftliteratur*, 3 vols. Berlin: Walter de Gruyter & Co, 1967–75.
Limet, *Liège*	Limet, H., *Étude de documents de la période d'Agadé appartenant à l'Université de Liège*. Paris, 1973.
TMS	Bruins, E. M., and Rutten, M., *Textes mathématiques de Suse* (MDP 34). Paris: Librairie orientaliste Paul Geuthner, 1961.
TSŠ	Jestin, R. R., *Tablettes sumériennes de Šuruppak conservées au Musée de Stamboul* (Mémoires de l'Institut français d'archéologie de Stamboul 3). Paris, 1937.

ACKNOWLEDGEMENTS

I first became acquainted with some of the intricacies of the mathematical texts while writing my dissertation, which was published as *Late Babylonian Field Plans in the British Museum*, Studia Pohl, Series Maior, 11 (Rome: Pontifical Biblical Institute, 1982). Several years later in conversation with my husband I mentioned some of the problems discussed in the mathematical texts, such as bricks and brickworks, excavations of canals, earthwork constructions, inheritance, and field plans. He suggested that a study of how the mathematical texts did (or did not) reflect everyday life in Mesopotamia might prove interesting. His suggestion intrigued me. My survey of the material began as an article, and an in depth investigation led to this book.

I would especially like to express my thanks to William W. Hallo, Curator of the Babylonian Collection at Yale University, for his encouragement and guidance at various stages of the manuscript; he has always been a generous friend and mentor to me through the years. In addition, I would like to thank Ulla Kasten, Museum Editor at the Babylonian Collection, for her advice and friendship. Christopher B.F. Walker, Assistant Keeper, in the Department of Western Asiatic Antiquities of the British Museum, was always a willing reader and advisor. Professor Anne D. Kilmer of University of California at Berkeley generously shared her unpublished manuscript (1985) and translation of Vaiman 1963. In addition, Asger Aaboe, Chairman of Mathematics and History of Science Department at Yale University, helped me greatly in private discussions.

Special thanks are due to Professor Marvin A. Powell, Jr., of Northern Illinois University, DeKalb, and Professor Jens Høyrup of the University of Roskilde, Denmark, for their patient reading, valuable comments and suggestions. Also, I would like to express my gratitude to Professor Stephen A. Kaufman, AOS Sectional Editor for the Ancient Near East, for his advice and encouragement throughout the editorial process. Professor Maynard Maidman of York University read the manuscript before printing, and Professor Stanley Insler of Yale University provided valuable assistance in the publication process.

Above all, I would like to thank my family for their devotion and encouragement, particularly, my husband, Mordechai, who has been my friend and companion for twenty years. Lastly, I would like to acknowledge a love of mathematics nurtured in me by my father, Nathan Waldman, Chief Machinist, CWO2, United States Navy, may he rest in peace.

I first presented the material in chapters 1-4 and the bibliography for mathematical cuneiform texts at 4. Workshop über Begriffsentwicklung in der Babylonischen Mathematik, Berlin, in May 1988. At this conference I presented a detailed analysis of all extant cuneiform mathematical texts through charts, descriptions of each problem and categorical summaries. I developed a system for categorizing the practical problems. This analysis and categorization had not been done in a systematic way prior to my research. My participation in this event was underwritten by a grant from the Max Planck University and the Government of the City of Berlin.

K.R. Nemet-Nejat

July, 1990

CHAPTER 1

Economic Dimensions of Cuneiform Mathematics

Economic necessity triggered the invention of writing (Damerow et al. 1988a; Hallo and Simpson 1971: 33; cf. Seidenberg 1962a). The discovery of clay tokens and the bullae that enclosed them and on which the symbols of the tokens were sometimes impressed confirmed this view (Schmandt-Besserat 1988: 28–29). The tokens and bullae may have located the beginning of writing at about 8500 B.C. (Schmandt-Besserat 1977, 1979a, 1979b, 1988; Brandes 1980).[1] Schmandt-Besserat (1980: 361; 1988: 3–4) traced the earliest tokens to the primitive farming communities and proposed a link between the need for recording and the beginning of agriculture. She suggested that the tokens "represent the elaboration of pebble counting ... (and) were a better memory aid as the various distinctive shapes and markings stood for specific commodities" (1980: 357–58). At first the tokens were a simple household and market bookkeeping system, but later they developed to satisfy the more complicated accounting needs of expanding businesses and state administration, which coordinated local and long-distance commerce (Schmandt-Besserat 1979: 24; 1980: 358; 1986: 365, 368–69; 1988: 5). The token system underwent profound changes with the emergence of cities at about 3500 B.C. The evolution was visible in a proliferation of markings on the faces of the tokens, apparently indicating a need to record an increasing variety of goods (Schmandt-Besserat 1980; 1988: 34–35).

In a critical evaluation of Schmandt-Besserat's research, Lieberman (1980: 351–58) distinguished between calculi and "small clay objects." He defined calculi as solid clay spheres and cones used as signs, as illustrated by the impressions on the outside surfaces of spheroid bullae and accompanied by their presence within. He used the term "small clay objects" to describe other clay objects that may or may not have been used as signs (1980: 340–41). Lieberman noted that Sumero-Akkadian lexicons record clay calculi among other bookkeeping devices and suggested the Sumerian word for 'clay calculus' was **imna** 'clay stone'. A Nuzi bulla contained 48 'stones', which were represented in the inscription on the outside of the bulla by the term **na$_4$-meš** 'stones, counters' (Oppenheim 1958).

In a valuable complement to the articles by Schmandt-Besserat (1977–79), Brandes (1980) provided an extensive review of the publications on bullae and clay tokens and also attempted to verify Schmandt-Besserat's identification of 15

1. Damerow et al. 1988b suggested an even earlier date of approximately 10,000 B.C. for the clay token system.

categories of clay tokens with proto-Sumerian logograms for numerals and commodities by referring to A. Falkenstein's sign list in *Archaische Texte aus Uruk* (1936). Most of the tokens were simple geometric shapes and could be viewed as a kind of three-dimensional writing (Brandes 1980: 10). Commodities were usually depicted abstractly; only containers and animals were portrayed realistically (Schmandt-Besserat 1979a: 21). However, scholars were able to recognize signs for numbers and objects from everyday life in the earliest forms of writing. The numerical notations in the archaic proto-cuneiform and proto-Elamite documents acted as semantic indicators for a number of ideograms such as grain products, animals and humans (Damerow and Englund 1989). Archaeological evidence supported Diakonoff's theory that the different numerical signs used to represent quantities, capacity, area measures, etc., suggest an ancient tradition of concrete counting in proto- or prehistoric Mesopotamia (Schmandt-Besserat 1984).

The earliest written texts, which dated to the end of the fourth millennium, appeared to be records of daily matters such as grazing cattle entrusted to herdsmen by their owners. The script showed slightly conventionalized pictographic signs, which appeared to be logographic. Numerals were represented logographically as in most economic texts. The early texts appeared within the context of the political and economic changes of the fourth millennium, including the development of cities. The creation of cities implied a new type of social organization.[2] Tribal and familial loyalties were replaced by allegiance to a king, who headed a military machine. With new political and administrative centers came the accumulation of centralized capital, which probably spurred the growth of crafts and professional specialization (Hallo and Simpson 1971: 33–34; on the importance of political authority and social stratification, rather than urbanism, in the emergence of the state, see Adams 1966: 1–37).

Writing underwent further changes during the third millennium: the signs became stylized and lost their pictorial associations; they began to render individual sounds as well as separate words (Damerow et al. 1988a; Hallo and Simpson 1971: 151–52). The revised script was used by scribes trained in a standardized curriculum.

The numerous economic and administrative documents from the early periods attested to the development of the scribal craft and to a specialized scribal training in mathematics. The third millennium political and economic

2. Damerow and Englund 1989 indicated that proto-Elamite tablets recorded the allocation of products from agricultural activity, particularly grain rations to workers under the direction of household administrators and the probable disbursement of grain for the purpose of sowing "rather unimposing fields." The recorded numbers of animals, humans and the measures of grain are "without exception entirely within the range of expected local activity."

changes—expansion, centralized government and the development of a complex bureaucracy to handle the details of an empire—required the development of a new notation system in record keeping in order to handle both very large and very small numbers.

The Sargonic and the subsequent Ur III governments not only controlled a vast area politically but also placed large sections of the economy under the direct control of the state. An immense administrative bureaucracy of scribes and other civil servants managed and kept account of the flow of goods passing in and out of the state warehouses. Because the state was responsible for the subsistence of its dependents, it kept elaborate records. From about 2400 B.C., tablets from Southern Babylonia registered the distribution of large quantities of barley to dependents of the state. In this setting the place-value notation system was invented to replace the more cumbersome system of standard notation. Standard notation entailed separate symbols for each numerical power, that is, for 1, 10, 60, 6000 (60 x 10), 3600 (60^2), 36,000 (60^2 x 10). In the simplified place-value notation the position of the number in a sequence determined its value; that is, the higher powers of 60 were placed at the beginning, the lower powers at the end. The same symbols were also used for writing fractions. For example, the value of the symbols 𒁹 𒌋 𒐊 is 60 + 10 + 5 = 75 or 60^2 + 10 + 5 = 3615 or 1 + 15/60 = 1.25 (Walker 1987: 20–21). The system of place-value notation, also called the positional number system (Aaboe 1964: 17; van der Waerden 1961: 38), laid the groundwork for Old Babylonian mathematics (Powell 1976a: 422–23; 1976b: 97).

Lieberman argued that the system of place notation was actually developed in the fourth millennium. He contended that the use of different size and shape number symbols in the calculi laid the foundation for the development of place-value notation. What was once represented by the relative size of a sign was later represented by its relative place (Lieberman 1980: 342–45, 353; Neugebauer 1969: 19). Lieberman suggested that the clay objects could be viewed as one of the contributing factors to the advanced state of early Mesopotamian mathematics.

The recent work of Damerow, Englund and Nissen on archaic texts clarified the numerical sign systems of proto-cuneiform and proto-Elamite tablets. They noted three distinct systems: (1) sexagesimal, (2) bi-sexagesimal, and (3) decimal. Both proto-cuneiform and proto-Elamite texts used the sexagesimal system to record all discrete objects with the exception of rations, and they used a bi-sexagesimal system primarily to count rations of barley products, kinds of fish and possibly cakes of cheese. The decimal system used to register animals and humans was found only in the proto-Elamite texts (see Damerow and Englund 1987, 1989; Damerow, Englund and Nissen 1988b).[3]

The earliest well-defined group of extant cuneiform problem texts have been dated to the latter years of the Sargonic empire, around 2200 B.C. (the Old Akkadian period). Since the Old Akkadian tablets are 500 years older than their Old Babylonian counterparts, they were important in reconstructing the history of mathematical scribal training. Almost all of these texts contain errors. As often occurs, the errors were more instructive than the correct answers. These texts demonstrated that the solution was expressed in sexagesimal place notation at some point during the calculation and that the sign **GÍN** was used to indicate the position of sexagesimal fractions (see Powell 1976a, 1976b; Whiting 1984). This technique may already have been known in the Pre-Sargonic period.[4]

The vast majority of extant mathematical texts belonged to two distinct periods, separated by more than a thousand years: the Old Babylonian and the Seleucid periods.

3. Thureau-Dangin 1921: 123–27 discussed the decimal substratum and analyzed the system of Sumerian numeration. Powell 1972: 168–72 summarized studies by Neugebauer and Thureau-Dangin, while adding his own comments.

Neugebauer 1969: 17 noted:

> ...it is only in strictly mathematical or astronomical contexts that the sexagesimal system is applied. In all other matters (dates, measures of weight, areas, etc.), use was made of mixed systems which have their exact parallel in the chaos of 60-division, 24-division, 12-division, 10-division, 2-division which characterizes the units of our own civilization)...

4. Whiting 1984: 63–64 noted that, perhaps, the earliest indication of the use of sexagesimal place notation in conjunction with the metrological system to express fractions is found in *OIP* 14, 70 [Edzard 1968]. Powell 1984a argued in his analysis of metrological mathematics that "Old-Babylonian mathematics" must actually be dated at least to the Akkad period.

CHAPTER 2

Survey of Mathematical Training for Scribes

SURVEY OF TEXTS REFERRING TO MATHEMATICAL TRAINING

The organization and operation of schools in Mesopotamia are known from a series of cuneiform essays on school life, examination texts, royal hymns where kings refer to their education, lexical lists, and many student/teacher exercises (Kramer 1963: 229–248; Gordon 1960). This information has been discussed in general works and specialized articles (Falkenstein 1948, 1953; Gadd 1956; Gordon 1959: nos. 2.36–2.56, 1960; Curtis and Hallo 1959; Hallo and Simpson 1971; Jacobsen 1982; Kramer 1963; Laessøe 1953a; Landsberger 1954, 1960b; Oppenheim 1964: 235–39, 305–10; Roux 1964: 329–33; Saggs 1962: 420–27); however, a list of texts, specifically referring to mathematical training for scribes, has never been collected and analyzed.

From the lexical lists, we learn of various forms of mathematical instruction:

(1) In the Old Babylonian Lu-Series Recension A:[1]

464 **lú šumun-gi$_4$** = *[ša šu]-ma-ki-i*
465 **lú $^{[giš]}$dab$_4$ (=DUB)-dím** = š*[a d]a-ab-di-mi*
466 **lú ^{na_4}na** = *š[a ab-ne]-e*
467 **lú im-^{na_4}na** *š[a]* [...]

'man of (the) accounting board(?)'
'man of (the) abacus(?)'
'man of stone(s/weights)'
'man of clay stone(s)' (see Lieberman 1980: 346–47).

The group of profession(al)s in this passage of OB Lu is both preceded and followed by men who were officials in the school; therefore, we can assume the list cited above refers to teachers who trained students in the use of these devices. Lieberman (1980: 349) cautioned against assuming that an expert was limited to but one skill.

1. The ancient name of the text is unknown. There are three rescensions of the relevant passage, but only Rescension A is included here.

Turning to the literary texts, we find first in the *edubba* literature,[2] the following references to mathematical training:

(2) In Examination Text A,[3] line 27:

a-rá igi igi-ba igi-[gub-ba] nig-šid ku[ru]$_7$ šid-dù ga-lá à-dù-à
dù-a-bi dù-a ḫa-la ḫa-la-bi a-šà si-ge-dè ì-zu-ù
a-ra-a i-gi i-gi-ba-a i-gi-gu-ba-a nik-kas-si pi-qit-tam pa-qa-a-da a-de-e ka-la e-pe-šú zi-it-tam za-a-zu eq-la pa-la-ku ti-de-e

'Do you know multiplication, reciprocals, coefficients, balancing of accounts, administrative accounting, how to make all kinds of pay allotments, divide property, and delimit shares of fields?' (Sjöberg 1975a: 144; 1975b: 167).

(3) In Examination Text D l. 15:

na$_4$-rú-a ab-sar-e-d[è a-ša]-ga gíd-e-[dè] nig-šid sá-du$_{11}$-ge-[dè x x x] = *na-ra-a šá-ta-ri eq-la [šá-da-d]u nik-kàs-sa ka-šá-du*! [x x x]

'To write a stele, to draw a field, to settle accounts...' (Sjöberg 1972: 127, 129).

(4) In "Schooldays," ll. 59–62:

lú-tur-mu šu-ni i-ni-in-bad-du kù-zu i-ni-in-ku$_4$-re-en
nam-dub-sar-ra níg-galam-galam-ma-bi mu-ni-in-pàd-pàd-dè-en
šà-dub-ba-a šid-níg-šid ki-búr-búr-ra-bi igi mu-un-na-si-ga-aš
gu sum-ma-ab ki-dul-dul-a-bi dalla mu-ni-ne

'My young one has opened (wide) his hand, (and) you made wisdom enter there;
you showed him all the fine points of the scribal art;
you made him see the solutions of the mathematical tablets (and) of counting and accounting,
you (taught him how) to make deep(?) the cuneiform script(?)' (Kramer 1949: 203, 206; for translation, see Kramer 1963: 239).

2. Michalowski 1983: 226–27 noted that recently scholars have raised the question as to whether the *edubba* literature really reflected contemporary practices or merely portrayed an idealized version of how "it used to be."

3. Though Examination Text A is from the Neo-Assyrian period, the text reflects the curriculum of the Old Babylonian *edubba*.

(5) In the Old Babylonian *edubba* text, Dialogue 1: lines 57–58

a-rá ḫé-bí-šid zà-bi-šè nu-e-zu igi-diri ḫé-du$_8$ ki-ús nu-mu-ra-ab-dab$_5$

'you may recite the multiplication table, but you do not know it perfectly; you may solve ... *reciprocals*, (but) you cannot...for you' (Sjöberg 1975b: 167).

(6) In Dialogue 1: line 6

a-rá igi-diri nig-šid saḫar-gar-ra zà-bi-šè ì-zu

'you have learned perfectly multiplication, ... *reciprocals*, accounting and calculation of volume' (Sjöberg 1975b: 167).

A series of quotes from "The Disputation between Enkimansi and Girnishag"[4] also adds to our understanding of the mathematical training of a scribe:

(1) **gána ba-e-dè gin-na gána nu-mu-da-ba-e-en**
a-šà si-ge-dè gin-na éš-gána gi-DIŠ-ninda nu-mu-da-ḫa-za
giš-gag a-šà nu-mu-e-da-rú-e dím-ma nu-mu-e-da-an-ku$_4$ (UET VI/2, No. 150: 21–23)

'Go to divide a field but you won't be able to divide the field,
go to delimit a field, you won't be able to hold the measuring rope and the nindan (cubit) rod,
the peg of the field you won't drive in, you are not able to figure out the sense' (for translations see Sjöberg 1975b: 168; Kramer 1963: 241–42; Gadd 1956: 31–32)

4. Gadd 1956: 39–41 indicated that "The Disputation between Enkimansi and Girnishag" is an example of the literary genre known as contest literature; its central feature is a verbal contest of merit between two characters, and benefit is evaluated in terms of value to man. This literary genre was represented by the special sign (**adamin**) in writing of "man-man" (the second man being written upside down). In fact, the scribes boasted that the education they dispensed "created humanity" with the complementary sense of humanism. In the story of Oannes mankind had lived like animals before coming under the influence of civilization, as taught by Oannes and the *apkallu* or ancient sages, who were also called **um-mi-a** (*ummânu*), a term which came to mean 'craftsman' or 'headmaster of a school' (Gadd 1956: 13–14).

(2) **a-rá ḫé-bí-í[b...]**
igi-diri ḫé-bí [...]
saḫar-gar-ra x x [...]
ib-si$_8$ sag-dù x [...]
uš-uš í-kú gána bu-[...] (UET VI/2, No. 150: 34–38).[5]

'You may [recite] the multiplication table ...
You may ... *reciprocal* ...
Volume ...
A (side of a) square, a triangle ...
one side to another, an inclination, a field ...' (my translation; cf. Gadd 1956: 32)

All the texts indicate mathematical instruction includes "multiplication, reciprocals, balancing of accounts, administrative accounting, how to make all kinds of pay allotments and how to divide property and delimit shares of fields."[6] Many of these subjects occur, also, in the "problem texts"; however, the range of topics examined in the problem texts is much wider than this list (see Nemet-Nejat 1988).

Scribal education included both oral and written instruction. The teacher dictated a text which the student repeated and wrote down as indicated in the following proverb:

dub-sar šu ka-ta-sá-a e-ne-àm dub-sar-ra-àm

'A scribe (whose) hand rivals (his) mouth, he is indeed a scribe' (Sjöberg 1975b: 170–71; cf. Gordon 1959: 202: no. 2.40).

Part of the instruction was by question and answer as reflected in colophons noting *maš'alti ummāni* 'questioning by the teacher.' However, the primary source of information we have on instruction is the large number of

5. The substance of this passage is similar to the Old Babylonian *edubba* text, Dialogue 1: lines 57–58.

6. Høyrup 1985b: 9 suggested that during the Ur III period scribal schools and scribes alike associated their functions with state service; therefore, the mathematics curriculum became, in effect, "applied" mathematics.

school tablets excavated. Not only do we find many copies of the same texts but even multiple copies of the same text on one tablet.[7]

The value Mesopotamian society placed on scribal education in general and mathematical training in particular has been emphasized in the ancient literature. Even kings considered scribal knowledge to be a valuable asset. In two Sumerian hymns of self-glorification, King Shulgi emphasized his mathematical abilities in the following passages:[8]

(1) Šulgi B:16–17

nam-dub-sar-ra ki-nam-kù-zu-ba lú im-mi-re$_6$-re$_6$
zi-zi-i gá-gá šid níg-šid-dè zàg im-mi-til-til

'At the place where man goes to study the scribal art, (that is), in the place of learning the secret arts,
subtracting, adding, counting and accounting, he completes' (my translation; see Castellino 1972: 32–33, 85–90: B:16–17).

(2) Šulgi C:46–47

šid níg-šid giš-ḫur-kalam-ma-ka
igi-gál-sum-mu-bi á-bi-šè in-ga-zu

'of counting and accounting, the regulations of the nation,
their given wisdom I have also mastered' (my translation; see Castellino 1972: 252–53, 277–78: C:46–47).

Though Shulgi's hymns are replete with boasts, his claims to mathematical knowledge reflect the high regard for such accomplishments in the ancient Near East. A proverb differentiates between the educated and uneducated classes in terms of mathematical ability:

7. For example, *MCT*: 22 no. 99, 3 notes CBS 13567 + ...contains the same multiplication table copied three times.

8. Castellino 1972: 88 noted the use of **šid** = *manû* 'to count' > 'to reckon, to calculate' > 'to spell out a text rhythmically' (as is done with magical texts expressed by *manû*) > 'to recite, say aloud' (in an even tone, as children do when repeating their lesson). These several meanings occur in most of the edubba texts. Benno Landsberger 1954: 123 pointed out that the earliest Sumerian word for scribe is **umbisag**, which is written logographically ŠID.

šà níg-šid nu-zu [š]à igi!-gál-tuku

'A mind which does not know accounting, is it a mind that has intelligence?' (Castellino 1972: 88).

Other rulers also, such as Naram-Sin, Lipit-Ishtar, Assurbanipal and Darius, took credit for scribal knowledge, but their accomplishments are merely boastful declarations. Nonetheless, only a few kings claim literacy in the long history of Mesopotamia. This fact underscores the closed nature of the scribal guild and the inevitable dependence of the palace on the specialized skills of the scribes (Landsberger 1960b: 110–11). In fact, the scribal school may have been in royal service and composed hymns ordered by the royal court for the king; these hymns were then used by the *ummia* in his teaching in the *edubba* (Falkenstein 1948: 185). Three royal hymns actually indicate that they were composed in the school.

SURVEY OF CUNEIFORM MATHEMATICAL TEXTS

ALGEBRA AND NUMBER THEORY Babylonian mathematics shows a strong preference for algebra and number theory. Despite a considerable amount of geometrical knowledge, geometry often serves as a guise for essentially algebraic problems. Whenever a geometrical problem is posed, it is for the express purpose of computing some numerical quantity, be it length, area, or volume (Aaboe 1964: 29; Freudenthal 1977; Gandz 1929; *MCT*: 1; Neugebauer 1945: 11–13; van der Waerden 1961: 72).

In the series texts, which show systematic learning, no tablets exist formulating general rules without numbers. That is, the Babylonians never take the extra steps to prove statements and formulate theorems (Aaboe 1964: 29; *MKT* III: 79–80; Neugebauer 1963: 530). The scribes usually restrict themselves to performing the operations leading to the solution without defending them.

Babylonian mathematics does not have a graphic symbol for the unknown. Rather, the unknown is designated by 1 which is used provisionally to solve a 1st or 2nd degree equation by the method of false value (*sarrum*) in order to find the coefficients in these equations (see Thureau-Dangin 1937c, 1938b; *TMB*: xix–xx; Vogel 1936). Examples of 1st or 2nd degree equations in Babylonian mathematics show that the scribes possess all the necessary tools for algebraic technique: reduction of similar terms, elimination of the unknown by substitution, introduction of an auxiliary unknown, completing the square,[9] and, in second degree equations containing two unknowns, using $\pm$ in a single algebraic statement (*TMB*: xxxiv; Gandz 1937).[10] The purpose of many mathematical problems is to teach the student scribe the transformation of more complicated quadratic equations into "normal form" such that $xy=a$ and $x\pm y=b$ (Neugebauer 1969: 41–44).

Knowledge of reciprocals is a key in understanding the solution of mathematical problems. In the abstract sexagesimal system fractions are eliminated. The reciprocal or **igi** of a given number is another number which multiplied by it will give 60 (Thureau-Dangin 1939: 117–18, 140–41).[11] For example, the reciprocal of 5 is 0;12. Every number remains an integer because the system was intended to serve only as an instrument of calculation. The Babylonian number uses place value notation, that is, a very limited number of symbols whose magnitude is determined by position (Neugebauer 1945: 12). Division is not performed.[12] Numbers are multiplied by reciprocals. Even fractions are designated by a term implying an inverse relation, **igi-te-en**. For the Sumerians the fraction $1/n$ was conceived as the reciprocal of n. The Babylonian mathematicians conformed to the Sumerian tradition and treated the fraction m/n as m times the reciprocal of n (*TMB*: xiii; Bruins 1962: 281–83). In describing the education which he received, King Assurbanipal emphasized that he knew

9. To solve the quadratic formula $ax^2+bx+c=0$ by completing the square, we find that

$$x=\frac{-b\pm\sqrt{b^2-4ac}}{2a}$$

Thureau-Dangin 1939: 125–33 pointed out that completing the square presupposes the implicit knowledge of the two identities $(a\pm b)^2=a^2+b^2\pm 2ab$. Since Babylonian mathematicians were often practical rather than theoretical in their approach, they may have arrived at this information by arithmetical means. Also, see Gandz 1937; cf. van der Waerden 1961: 71–72; Vaiman 1961: 168–69.

10. Jens Høyrup noted in private communication that the Old Babylonian scribes were able to handle the two signs (or more specifically the two concepts) in parallel statements.

11. Thureau-Dangin 1936e: 55–56 revised his previous translation of **igi** as 'part' to 'opposite.' Gandz 1936 suggested that 'tables of reciprocals' was a misnomer and should be replaced by the more correct 'tables for the conversion of common fractions into sexagesimal fractions.' He also proposed an additional meaning for **igi** 'eye' as 'part.'

12. Only in one problem, BM 85196 #3, the scribes computed by dividing by 2 (indicated by the term *ḫepû*, 'to split in half') rather than multiplying by 0;30.

how "to solve complicated problems of *igiarû*," that is, solving *m*/*n* by "unravelling" the reciprocal (**igi**) of *n* and then multiplying it (**a-rá**) by *m*. Possibly two kinds of tables, reciprocal and multiplication, were consulted to solve the problem (Thureau-Dangin 1939: 121).

Algebraic problems in school exercises were constructed with an eye to their solution, so that ancient mathematicians rarely encountered an irregular number. A regular number is a number, *n*, for which its reciprocal, 1/*n*, can be obtained by division which ends after a finite number of steps. For this condition to be satisifed, *n* can contain no prime number which is not also contained in base 60; in other words, *n* must be of the form $n=2^x3^y5^z$ where x, y and z are integers or zero (*MCT*: 11). In instances in which the solution is expressed as a fraction, it is always a fraction that could be transformed into a sexagesimal representation. The solution is always expressed by a positive number. Probably Babylonian mathematicians never had the concept of a negative root (Vogel 1936: 717 contra Thureau-Dangin 1932e: 139–142).

Interest in approximations of reciprocals of irregular numbers is first attested in Old Babylonian times (Neugebauer 1969: 34–35). For example, M 10 deals with the reciprocals of 7, 11, 13, 14 and 17 in which two approximations, one greater and one less than the reciprocal of 7 are given in the first two lines of the text.

Babylonian mathematicians also knew the Heronic formula of approximating an irrational square root or $(h^2+w^2)^{1/2}$ as demonstrated in AO 6484 #8 and VAT 6598 #6.

In the Seleucid period the form of fractions in sums shows the scribes were sometimes guided by mercantile concerns rather than mathematical calculations. In computing prebends, the scribes often arrived at a long and complex total in order to convince the purchaser that he was buying a large fraction of the day.[13]

Cuneiform mathematical texts did not develop a specific sign for zero. Lack of the zero sign was probably tolerated because of the amplitude of base 60 (*TMB*: x; van der Waerden 1961: 38–39). Zero in initial or medial position is represented either by an empty space or the separation sign.[14] The intention of

13. Cocquerillat 1965: 240–41 suggested that the scribes were handicapped because they did not have the elementary rules for reducing fractions to the same denominator; rather their processes of reduction were empirical, varying according to each concrete problem.

14. Babylonian mathematics first indicates zero in medial position in the Old Babylonian period in *TMS*, Text 24 (see Friberg 1982: 90); however, medial zero occurs primarily in astronomy.

the "zero" sign is to avoid misreadings by separating the different units or places. Texts which use the "zero" sign employ it consistently; however, this usage is not adopted in all texts. Neugebauer (1941: 213–14) suggested the separation sign appears only in texts which either certainly or most probably come from Uruk. In fact, the colophons show that these texts were written by scribes belonging to one of the two schools which are represented in Uruk tablets, that is, the pupils (*liblibbu*) of *Ekur-zākir* or *Sîn-lēqe-unninni*. Neugebauer asserted that from the purely mathematical point of view the use of the "zero" sign is a step backward in the development of rational number notation, and therefore it was not generally accepted in astronomical texts.

GEOMETRY The invention of the geometry of angles must be credited to the Greeks (Gandz 1929) because Babylonian mathematics lacks basic geometric terms and concepts. It has no word for angle or slope. The concept of angle is described by **kú, šà-gal** = *ukullû* 'fodder, food,' or the fuller form of the expression **kú ì-kú** 'it ate fodder'; that is, for a given unit of height (1 **kùš** or ½ **kùš**), the profile of the figure in question (incline of a wall or embankment with a horizontal plane) recedes by a certain number of units (x **kùš** per x **kùš** ≈ %), this slope corresponding to an angle of 45° (see *TMB*: xvii and *MCT*: 81). Babylonian mathematics has no term to describe the idea of perpendicularity or parallelism either. The Babylonians refer to the perpendicular as 'descending' (*kī urīdu* or *warittu*) (*TMB*: xviii, 227–28; *TMS*: 4); for them, the perpendicular is a straight line which aids in calculating the area of a triangle that does not contain a right angle.

Babylonian mathematics shows familiarity with geometric shapes[15] and includes a list of coefficients for determining the area of polygons. However, the apparently chaotic order of the coefficient lists indicates that geometry was not a special mathematical discipline but was treated on the same level as any other kind of numerical relation between practical objects (Neugebauer 1969: 45). Triangles are, for the most part, right-angled or isosceles (Bruins 1953b: 412); for instance, Plimpton 322, a text dealing with Pythagorean triangles, may have been used as a 'teacher's aid' in formulating solvable problems involving right triangles (Friberg 1981a: 299–306). Areas of quadrilaterals are usually solved by

15. Seidenberg 1962b: 514 noted that the Babylonians were interested in rectangles, triangles, trapezoids, circles and pyramids, but, like the Egyptians, geometry (including theorems and proofs) was a secondary matter. They knew the Theorem of Pythagoras and Pythagorean triangles, that is, right triangles whose sides are integers (*MCT*: 38–41, 55 and 142). Seidenberg 1962b: 514 even proposed that they knew that in a triangle, a line parallel to the base divides the other sides proportionately. With the possible exception of the theorem that the angle in a semicircle is right, their basic knowledge of geometry did not exceed that of the Egyptians.

applying an approximate formula often used in calculating the area of a field: averaging opposite sides and multiplying them by each other, that is, the agrimensor formula

$$A = \frac{a+c}{2} \times \frac{b+d}{2}$$

TMS, Text 2, shows a hexagon and a heptagon for which the circumscribed circles, presumably used to facilitate the drawing, have been erased (Bruins 1959, 1962: 334). The Babylonians generally used a gross approximation of $\pi = 3$. However, the Susa coefficient list, *TMS*, Text 3: 30, may suggest that the Babylonians were aware of a more precise value of $\pi = 3\frac{1}{8}$ (Neugebauer 1969: 47; for discussion of the values of π for the Babylonians, Egyptians and Hebrews, see Seidenberg 1962b: 512–13).

METROLOGY The conversion of units of measurements involves the mastery of ratios between units, both through tables that are memorized or consulted and through simple calculations involving ratios. In a group of one-problem tablets, for instance, each tablet deals with finding the area of a square by squaring the number of **ninda** to obtain area-**sar** and converting the result into the customary smaller units, **gín** and **še**; in each tablet the problem is stated and the answer and the metrological conversion given in a short text marked off in the lower right corner of the tablet, the remaining surface being left uninscribed, except for numbers used in calculating the area of the square. These numbers never involve the unit conversion part of the problem, suggesting that the conversion in the lower right side derives from memorization or from referring to metrological tables (Neugebauer and Sachs 1984: 246–51; NBC 8082; Steele 1951: 25: pl. 7).

FORMULATION OF THE TEXTS The phrasing of the problem texts is highly condensed, especially when Sumerian words occur. Some scholars tend to regard the Sumerian logograms in Old Babylonian mathematical texts as pseudo-Sumerian, but there is no evidence in the writing system as a whole to indicate the development of new logograms of this kind in the Old Babylonian period. In fact, a review of Old Babylonian mathematical terminology demonstrates that all the basic mathematical procedures used in the texts—adding, subtracting, multiplying, raising to powers, finding roots, using reciprocals and coefficients—have Sumerian origins. In addition to these terms, a large group of nouns, adjectives, interrogative pronouns, pronouns, and adverbs, are of Sumerian origin (Powell 1976a: 429; Vogel 1958).

In the mathematical texts the Sumerian expressions do not always conform to the rules of Sumerian grammar, and suffixes are routinely omitted. For example, **zi** indicates 'subtract' (imperative) or 'I subtracted', as **daḫ** is used to indicate 'add' or 'I added' (*MCT*: 2–3).[16] Sometimes even Akkadian mathematical terms are abbreviated within a text; for example, *ta-ra-ḫi-ša* is sometimes written as *ta*, and *šuplum* or *šupulša* as *šu* (*MCT*: 88).

Modern translations of mathematical tablets are based on standard forms of Akkadian and on unabbreviated, more correct forms of Sumerian phrases when they occur.[17] According to this paradigm, the problem texts are translated as follows: the first person (singular preterite), that is, the teacher, is used in the description of the problem, and the second person (singular imperative or present-future), that is, the student, is used for operations performed in solving the problem (Foster 1983: 126 n. 16; *MCT*: 3; *TMB*: xxxix–xxx; Høyrup 1990: 59; 1985a: 47; Bottéro 1987: 18).

In problems that give details of the method of solution, intermediate results are sometimes signalled by the phrases *rēška līkil* or *ša rēška ukallu*, 'keep in your head' or 'that your head keeps', and the result may be announced by *illiakkum* (√*elû*) 'comes up for you,' *nadānum* 'to give,' and *tammar* 'you see'. In *TMS*, Text 9, a mathematical text from Susa, the instructor offers more than one method of solution, distinguishing between the two procedures by the term *ina Akkadî* (or *kīam Akkadî*), that is, the Akkadian versus the non-Akkadian procedure.[18]

SUMMARY

An examination of texts referring to the training of scribes and of mathematical texts strongly suggests that the mathematical curriculum trained scribes in "applied" or practical mathematics and demonstrates the level of mathematical knowledge the scribes acquired in order to perform these tasks.

16. Neugebauer 1932–1933 opposed replacing Sumerian ideograms with their Akkadian equivalents in transliterations of the text; in some cases, these ideograms may have functioned as mathematical symbols.

17. Cf. Høyrup (1990: 60–65) distinguished (by typefaces) in his translations among terms written in syllabic Akkadian, Sumerian phrases and logograms, and terms of unknown origin.

18. See also *TMS*, Text 7, in which two methods are used to solve systems of first degree equations in two variables in each of the two problems. For a discussion of this problem see Goetsch 1968–1969: 114–15.

CHAPTER 3

Cuneiform Mathematical Texts

The extant mathematical texts, for the most part, belong to two distinct periods, separated by more than a thousand years: the Old Babylonian and the Seleucid periods (*MCT*: 1; Oppenheim 1964: 306; Saggs 1962: 424–25; cf. Powell 1975: 185). In addition, a small group of mathematical exercise tablets from the Old Akkadian period (that is, the Sargonic Empire) form the oldest well-defined group of cuneiform documents of problem texts. Since these tablets are 500 years older than their Old Babylonian counterparts, they are important in reconstructing the history of mathematical scribal training.

The Old Akkadian texts will not be included in our discussion of the problem texts; they comprise a discrete group of limited range, which are adequately analyzed by Powell (1975; 1976) and Whiting (1984). Briefly, the exercises are two types: (1) the area and one side of a rectangle are given, and the other side must be determined; and (2) the side of a square is given, and the area must be determined. These texts involve a knowledge of reciprocals and elementary mathematics, and almost all contain errors (Powell 1975: 185, 1976a, 1976b; Whiting 1984).

The remaining cuneiform mathematical texts can be divided into three main categories: the table texts, the coefficient lists and the problem texts.

THE TABLE TEXTS

The table texts deal with the following topics: (1) multiplication in the form of single or combined tables (*MCT*: 19–33; *MKT* I: 32–67; *MKT* II: 36–37; *MKT* III: 49–50; *TMS*: 35: Text 4; Bruins 1954: 55: IM 54536, IM 54216, IM 52548, IM 55111, IM 55292, IM 52879; Neugebauer, 1931b; Soubeyran 1984: 19–28: Texts 1–3; Gurney 1973: nos. 182–89; Petschow 1974: no. 65; Pinches 1963: 41; cf. Aaboe 1969; for a discussion of computational aids and techniques, see Bruins 1962), (2) reciprocals (*LBAT*: nos. 1631–37; *MCT*: 11–16; *MKT* I: 8–32; *MKT* II: 36–37; Aaboe 1965; Hulin 1963; Delaporte 1911; Hunger 1973: no. 101 obv. ii and rev., which contains a multiplication table of 1,40; Neugebauer 1931c; Petschow 1974: no. 65 rev.; Pinches 1963: no. 42; Sachs 1947a, 1952; Ist 485 and STT 399, in which numbers and fractions are spelled out in syllabic Sumerian [Hulin 1963; Powell 1971: 54–72] ; see *LBAT*: nos. 1642–1645, texts analyzing reciprocals),[1] (3) operations with reciprocals including tables related to *maḫīru* or 'price equivalency' problems (*MCT*: 17–18), (4) squares and square roots and cubes and cube roots (Gurney 1973: nos. 155, 190–195: *LBAT*: no. 1636?, 1637 obv. 16–17, 1638?, 1639, 1640?, 1641; *MCT*: 33–35; *MKT* I: 65–77; *MKT* III:, 51–52; Soubeyran 1984: 28–30: Texts 4–5), (5) exponential tables (*MKT* I: 77–79; cf. Bruins 1970), (6) logarithms (*MCT*: 35–36),[2] (7) metrological tables of length, area, monetary conversion and weight units (*MKT* I: 88–94; Greengus 1979: no. 292, a table of area measures; Hunger 1973: no. 101 obv. i and no. 102; Gurney 1973: nos. 114 and 115; van der Meer 1938: no. 156; Sachs 1947b; Thureau-Dangin 1930b), and (8) varia and fragments (*MCT*: 36; *MKT* I: 79–82).

Two examples of multiplication tables are known from the pre-Sargonic period. VAT 12593, the oldest example of a Sumerian multiplication table, is from Shuruppak c. 2500 B.C. It is actually a table of squares in standard metrological notation (Deimel 1922: no. 82; *MKT* I: 91–92).[3] Another text similar to the Fara tablet is known from Bismaya (ancient Adab); it probably dates to 2400 B.C. and consists of a table of small units of length and their squares stated in standard metrological notation (Edzard 1968).

The table texts are most probably older than the problem texts (*MCT*: 1; on the Sumerian origin of the table texts, see von Soden 1965).[4] The systematic arrangement of information in the table texts makes them useful as a basic tool in learning mathematics and in solving problems. In fact, numerous standard reciprocal tables and corresponding multiplication tables, found at different sites, attest to the extensive use and homogeneous character of these learning devices (Sachs 1947a: 221). The table texts are for students at the elementary level of

1. The reciprocal tables can be classified as reciprocals of standard type, of larger extent and of regular and irregular numbers. Friberg 1982: 92–93 proposed that the table of reciprocals is a table of complementary numbers whose product is equal to one. Thureau-Dangin *TMB*: x noted that **igi** 'inverse' is an abbreviation from **igi-gál** 'that which is opposite.' For a discussion of 1-**da** **⅔-bi** 40-**àm** as the frequent beginning of reciprocal tables, see Steinkeller 1979.

2. Neugebauer and Sachs *MCT*: 36 and no. 96d indicated that the logarithmic tables are the logical extension of the exponential tables. Unfortunately, the Old Babylonian mathematicians omitted in their "logarithmic tables" the use of a common base and the tabulation for constant intervals, which would ultimately be needed for practical computations as in calculating interest.

3. Powell 1976a: 430–31 suggested that the reverse of the tablet which contains the product of smaller surface units listed in standard notation (in **bur**) was memorized by the student together with the equivalent number of **sar**.

4. Neugebauer 1969: 30–31 indicated, "There can be little doubt that the tables for multiplication and division were developed simultaneously with the economic texts."

scribal education and are probably used extensively outside the schools as well. The problem texts are exercises for more advanced students and often deal with matters of practical interest (Oppenheim 1964: 329; *MCT*: 1; Neugebauer 1969: 30–31). As expected, more table texts than problem texts have been found. The multiplication and reciprocal tables are prepared only for use in schools; they are not used by skilled mathematicians who are able to calculate by heart due to the flexibility of the sexagesimal system (Bruins 1971: 222, 240). Later the development(?) of reciprocal tables with ten or more places is motivated by the growth of theoretical astronomy in the Seleucid period (*MCT*: 1 n. 4; Neugebauer 1945: 12; 1969: 29, 34; Bruins 1970; see Thureau-Dangin 1913).

THE COEFFICIENT LISTS

The coefficient lists contain fixed values for a series of items, organized more or less by categories (for example, geometric shapes, standard measures, bricks, metals) for various types of mathematical problems. Like the table texts, they can be used to solve examples in the problem texts and, in fact, are cited in the calculations to the solutions of various mathematical problems. The coefficient, or *igigubbû*,[5] may be based on definition or formula as we know it, for example, 0;5, the circumference coefficient used to calculate the area of a circle from the circumference squared, is an approximation of $1/4\pi$ (for example, IM 49949: 1; IM 52301 rev. 23; Böhl Coll. 1821 left margin under drawing of concentric circles; see Kilmer: 1960: 284, 1985); or, the coefficient may represent an empirical constant, as expressed in the relation between the capacity in **bariga** calculated from the volume in **sar** by multiplying the latter by 6,0, the grain capacity coefficient (*TMS*: Text 3: 49; IM 49949: 9; IM 52301 rev. 21; in AO 6484 obv. 23; and Haddad 104 problems #1–5).

The lists display the following structure: on the left is the value of the coefficient and on the right the object to which it applies. Numbers in some lines are followed by a noun or nominal phrase while other lines show the pattern of number-*ša*-noun. The same value may appear in either format.[6] The coefficient

5. In *TMB*: xii **igi-gub-ba** (= *igigubbû*) is translated as 'fixed inverse.'

6. Neugebauer and Sachs *MCT*: 133 n. 303 suggested that when the noun directly follows the number, the number represents the actual value for the following noun; when the number is followed by *ša* and then the noun, the number represents a fixed metrological proportion between units. The appearance of seven additional coefficient lists since the publication of *MCT* offers many examples contrary to their analysis (see Glossary of Realia); a few examples will suffice:

7,12 **igi-gub**(?) BI *a*!-*ma-ri-im*: IM 49949: 6

lists are concerned with both non-mathematical operations and mathematical ones (Kilmer 1960: 273; Livingstone 1986: 17),[7] as, for example, the appearance of a list of string names and numbers in a musical instrument, comprising a catalogue of intervals (Kilmer 1960: 278, 298–300; Duchesne-Guillemin 1984; Wulstan 1968),[8] may suggest. That is to say, in the tradition of lexical lists, the coefficient lists may have been padded or expanded in order to achieve the idealized form of scientific organization (von Soden 1965). On the other hand, these lists may be viewed as underscoring the paucity of problem texts from which to draw conclusions (*MCT*: 132). In fact, the recently published Haddad 104 employs several coefficients, which were previously mysterious entries in the coefficient lists. For example, Haddad 104 uses the value of *našpaku* 'the grain capacity coefficient' (cited above) and a series of coefficients related to work with *allu* 'hoe' with reference to three kinds of brickmaking operations, namely, digging, mixing and molding.

Neugebauer and Sachs recognized only two main categories of cuneiform mathematical texts: the "table-texts" and the "problem-texts"; they did not indicate the "coefficient lists" in their classification of mathematical tablets (*MCT*: 1). However, only two coefficient lists, YBC 5022 and YBC 7243, were known at the time of publication of *MCT*, and only a few values were used from these coefficient lists in solving mathematical problems. Nearly twenty years later, Oppenheim (1964: 306) referred to the coefficient lists as a "third, small group of mathematical texts...which serve basically practical purposes."

The coefficient lists published to date include: (1) A 3553, (2) BM 47860, (3) CBS 10996, (4) IM 49949, (5) IM 52301 rev. 20–22, a brief list of 7

7,12 **igi-gub sig$_4$-anše**: A 3553: 24
7,12 *ša* **sig$_4$-anše**: YBC 5022: 16
7,12 **sig$_4$-[an]še**: YBC 7243: 40

3,45 *pí-ti-iq-tum*: IM 52301 rev. 21
3,45 *šà pí-ti-iq-ti*: *TMS*, Text 3: 43
3,45 **igi-gub gú**(?)**-gal**(?) *pí-*[*t*]*i-iq-tim*: IM 49949: 13
3,45 **igi-gub im-dù-a**: A 3553: 8
3,45 *ša* **im-dù-a**: YBC 5022: 17
3,45 **im-dù-a**: YBC 7243: 13

7. Neugebauer 1969: 45 suggested that the coefficient lists with their apparently chaotic order indicate that geometry was not a special mathematical discipline but was treated like other kinds of numerical relations between practical objects.

8. Kilmer 1990a: 759 indicated that Old Babylonian musicologists recorded their investigations into scales and intervals, string lengths and pitches, in order to bring 'law and logic, measure and number' to the art of music.

coefficients and found in the middle of a problem text, (6) IM 52916, IM 52685, IM 52304, (7) *TMS*, Text 3, (8) YBC 5022 and (9) YBC 7243.

THE PROBLEM TEXTS

The problem texts can be broadly divided into two categories: (1) algebraic/geometric problems; and (2) practical problems, that is, problems concerned with the realia of Mesopotamian life. The boundaries are not hard and fast (Høyrup 1985b: 12–17). Tablets may contain both categories of problems. The known corpus of problem texts to date includes 180 tablets; of these, 53% deal with purely algebraic/geometric problems, 40% with purely practical matters, and 7% are mixed. The tablets themselves may contain as few as one problem or as many as several hundred.[9] The majority of tablets are one-problem texts. These findings do not include Old Akkadian texts (for example, A 5443, A 5446, Const. no. 236, Limet, *Liège*, nos. 36–39, *MAD* 5: 112, *TMH* 5: 65, *TSŠ* 50, 77, 188, 671),[10] which form their own well-defined group, nor do they include the few assorted Ur III texts (AOT 304, AOT 305 and *UET* 3: no. 1386).

The type and content of the problem texts vary greatly. Some tablets contain only one example, which gives all the details of the method of solution. Other tablets contain the data and the solution only partially worked out. The

9. The distribution is as follows:

Algebraic/geometric texts: 95		
1 problem(s)	40	42%
2–10 "	26	28%
11–99 "	20	21%
100+ "	7	7%
? "	2	2%
Practical texts: 73		
1 problem(s)	27	37%
2–10 "	30	41%
11–99 "	14	19%
100+ "	0	0%
?	2	3%
Mixed texts: 12		
4–10 problems	8	67%
11–99 "	4	33%

10. Powell 1976a: 436 n. 19 listed additional mathematical texts which Vaiman recognized as belonging to the Fara period.

problem is sometimes accompanied by drawings (not to scale). At the opposite end of the spectrum are tablets that contain hundreds of problems in very condensed form giving either no answers or only some answers. Intermediate types include texts with two or more problems solved in great detail and well arranged according to degree of mathematical difficulty, texts with many varied problems arranged without regard to levels of mathematical skill, and texts that present a small collection of coordinated problems (*MCT*: 36; Neugebauer 1969: 43; Høyrup 1985b: 14).

Some tablets containing a great number of problems without answers have colophons with the serial number of the tablet, which suggests some attempt at systematic learning. Unfortunately, there is no evidence for a canonical mathematical series. The same tablet number can occur on several tablets with different contents. The numerical arrangement is merely a means to enable scribes to keep the various groups organized (see *MCT*: 36).

ANALYSIS OF THE ALGEBRAIC/GEOMETRIC PROBLEMS More than half of the algebraic/geometric tablets deal with plane geometric figures, such as squares, triangles, trapezoids, rectangles,[11] circles (Lewy 1947: 319; Bruins 1964a: 177–181), and partitioned triangles and trapezoids (on generating formulas, see Friberg 1981a); that is, the algebraic/geometric texts generally deal with problems of areas, not volumes. About one-quarter of these texts are accompanied by drawings (not to scale).[12] Although their terminology is geometrical, the method of solution is algebraic (Aaboe 1964: 29; *MCT*: 1, 37). The problems are solved, particularly in the case of rectangles and squares (for example, YBC 4612 and

11. Thureau-Dangin 1934b: 51–52 proposed that **uš sag** problems should perhaps be read **uš-sag** 'rectangle.' Gandz 1937: 410–11 suggested that the term rectangular equations, rather than quadratic equations, is historically more accurate because the problems of the rectangle gave rise to these equations. Seidenberg 1978: 314 suggested that the standard types of linear and quadratic equations appear to be problems of the rectangle, involving the area and the diagonal of the rectangle.

Seidenberg 1962b: 500 indicated:

> The Babylonians had a composite of algebra and geometry: they interpret a product as the area of a rectangle and a number multiplied by itself as a square. This composite is not geometric algebra, but rather "algebraic geometry," i.e., geometry done with numbers together with associated algebra.

12. Vaiman 1960 examined six drawings of three-dimensional figures in cuneiform mathematical texts. He analyzed three illustrations as three-dimensional figures which appear to be two-dimensional when seen from above in the ring wall problem in BM 85194 #4, the triangular fundament in BM 85196 #1, and the water reservoir in Erm 15073 #5. He suggested that three illustrations are "symbolic" drawings with no clear perspective; these are the dams with trapezoidal cross sections in Erm 15073 #6 and #7 and the siege ramp in BM 85196 #17.

BM 80209) and in one case of triangles (YBC 4696), by a series of equations with up to six unknowns (YBC 4714; Bruins 1962: 293) and even to the eighth degree (*TMS*, Text 19). In some tablets parenthetical expressions are compounded to be added and subtracted for 25 or more problems (for example, VAT 7537, YBC 4697 and YBC 4715). A handful of tablets deals with such miscellaneous subjects as cube roots (YBC 6295 and VAT 8547), metrology (see Neugebauer and Sachs 1984: 246–48), and numbers and their reciprocals (M 10, VAT 6505, VAT 8520 and YBC 6967). Finally, only 5 tablets deal with solid geometric figures (CBM 12648, NBC 7934, IM 54478, VAT 672, and 2N-T 174), of which one set of three rectangular prisms may, in fact, represent the foundation for a wall and a pillar (NBC 7934) and two figures appear in broken context (CBM 12648 and VAT 672). In the mixed category 5 tablets deal with planar figures but 2 (of 12 tablets) deal with solid geometric figures.[13] Note that tabulation of information presents many problems since tablets may deal with more than one type of geometric figure. In addition, it is difficult to determine whether to categorize geometric figures such as squares and rectangles as planar geometric forms or as (series of) equations that are linear, quadratic, cubic, bi-quadratic, etc., based on the method of solution.[14] For the most part, solid geometric figures are only referred to in the practical problems where they are indexed by names dealing with bricks and brickworks, excavations of canals and/or earth, (earthwork) constructions such as walls, dams, ramps, etc. That is, there is no systematic terminology to describe solid geometric figures; rather, the shapes of three-dimensional figures are usually deduced from the descriptions and measurements in the problem texts. Thureau-Dangin's comment (*TMB*: xix) that Babylonian geometry did not rise above the level of everyday needs is apt in this regard.[15]

ANALYSIS OF THE PRACTICAL PROBLEMS Practical problems, selected because they contain key words or concepts that reflect the realities and concerns of everyday

13. In AO 6484, problems #9–#13 involve a cube; in VAT 7848 #5, the example seems to treat a cube but the final value indicates the problem concerns area.

14. Friberg 1982: 50 argued that VAT 7537, YBC 4695, YBC 4697, YBC 4709, YBC 4710, YBC 4711 and YBC 4715 appear to treat rectangular areas, but these, in fact, deal with series of variants of certain basic systems of quadratic and linear equations; see Freudenthal 1977: 192–94.

Høyrup 1990: 38–39 suggested that mathematical problems which deal with lengths, widths and areas represent general paradigms for solving equations; their lengths, etc. represent real lengths as well as numbers.

15. Høyrup 1984: 59 suggested that the siege ramp problems "constitute a bridge between the normal algebra and 'refined' geometric problems."

life in Mesopotamia, are used to define the basic sub-categories of the practical problem texts, specifically as follows: (1) bricks and brickworks; (2) canals and irrigation; (3) defense works; (4) other constructions, excavations and earthworks; (5) *maḫiru* 'price equivalency'; (6) interest; (7) commerce, i.e., buying and selling; (8) inheritance or division of property; (9) clepsydra; (10) field plans; (11) herding and herd growth; (12) standard measuring vessels; (13) original quantity; (14) work other than work assignments in sub-categories (2), (3), and (4); (15) metalwork; (16) *kiššû* 'bundle of reeds' (and *nalbantu*); (17) wood placed on top of a wall and a pole or reed leaning against a wall; (18) logs; (19) forest and reed thicket; (20) doors; (21) bitumen or mud plaster; (22) foundation for a wall and pillar; and (23) Mari (tabular) texts.

The tablets may contain multiple problems involving more than one sub-category while others deal with only one subject. In the next chapter, the individual problems are extracted and classified according to each subject, and the individual sub-categories are examined to determine whether the mathematical values quoted in these tablets reflect actual values as noted in the administrative and economic texts, and whether they correspond to artifacts and objects from excavations and archaeological sites (when relevant). Finally, evaluation of the information presented by the texts in each sub-category will demonstrate that mathematics was applied in solving problems occurring in everyday life in the ancient Near East.[16]

SUMMARY

The purpose of the cuneiform mathematical texts, was most probably, the training of scribes. These texts, by virtue of their variety of topics and depth of knowledge, provided rigorous training for a scribe seeking one of the many scribal posts in the Mesopotamian administrative bureaucracy such as: *ṭupšar ṣābim* 'scribe for laborer groups', *šassuku* 'field scribe, land registrar, geometer', *ṭupšar ummāni* 'military scribe', *ṭupšar šarri* 'royal scribe', *ṭupšar nikkassi* 'scribe involved with accounts and balanced accounts' (my translation), *ṭupšar piqitti* 'administrator', **dub-sar a-ri/ru-a** 'district scribe', *zazakku* 'public

16. Cf. Neugebauer and Sachs *MCT*: 1 commented:
Problem texts sometimes exhibit only very remote connections, if any, with practical questions. They are school products intended to illustrate the rules for dealing with problems which are properly termed "algebraic."

secretary to high officials', *šandabakku* 'high administrative official', or even *ṭupšar immê* 'ordinary clerk' (Landsberger 1954; *AHw* s.v. *t/ṭupšarru*).[17]

As a bilingual proverb notes:

nam-dub-sar giš-šub-ba-sig$_5$-ga-tuku ^{d}lama-igi-zalág-ga nì-šà-ḫab é-gal-la-ke$_4$

ṭup-šar-ru-tú i-diš-šú da-ma-qi ra-še-e la-mas-si i-nu na-mir-tu ḫi-šiḫ-ti ekalli

'The scribal art, receiving a handsome fee, is a bright-eyed guardian, the need of the Palace.'

That is, the scribe at the height of his career provides his master with guidance on policy decisions and defends his interest with the vigilance of the magical *lamassû* at the palace gate; the scribe's reward is high-level civil service appointment within the bureaucracy (Gadd 1956: 37–38), because

nam-dub-sar-ra ama-gù-dé-ke$_4$-e-ne a-a um-me-a-ke$_4$-eš

'The scribal art is the mother of speakers, the father of scholars' (Laessøe 1953a: 218).[18]

17. We cannot determine whether advanced mathematical training included all or only some sub-categories of problem texts to prepare scribes for scribal positions. Michalowski 1983: 226 argued that we cannot ascertain how much training in practical matters was provided by the schools and how much was learned on-the-job, particularly since school curriculum was not standardized for all of Mesopotamia.

18. Note the scribe who wrote the mathematical text AO 6484 is also identified as the scribe of the astronomical series *Enūma Anu Enlil*. Is this scribe a scholar?!

CHAPTER 4

Analysis of the Practical Problem Texts

I. BRICKS AND BRICKWORKS

DISCUSSION (1) AO 8862 #5–8

Problems #5 to #8 belong to a group involving the transportation of 540 (9 sixties) bricks a distance of 30 **ninda** or 3 *ašlu*.
#5 and #6: In both problems the builder sets a rate of payment of 2 **bán** barley for each man who transports 540 bricks 30 **ninda**. The builder requests workers for the job. The problem is to find the number of bricks transported and the amount of barley paid. The answers to both problems are given in tabular form.
#7: The problem is to calculate the number of bricks, workers and work-days. The problem establishes certain relations among these quantities. The three items are added together to equal 2,20, and two-thirds of the number of workers is equal to the number of work-days. The answers are given in tabular form.
#8: The text is broken. Only the operation of adding earth and bricks (*e-pi-ri-ia* [*ù*] *li-bi-it-ti* [*ak*]*-mu-ur-ma*) is preserved on the tablet.

(2) AO 10822

All ten problems extant in this tablet concern bricks of types 1–5 (*MCT*: 93).
#1: The text is broken; references to bricks and layers of bricks are preserved.
#2: The text is broken. The number of layers of bricks, half-bricks and kiln-burnt bricks is given. The problem is to find how many of each of these bricks are needed.
#3: The problem is to find the number of bricks in a wall (**iz-zi** = *igāru*) of given dimensions.
#4: The problem is to find the thickness of a brick wall (**iz-zi** = *igāru*) of given dimensions.
#5: The problem is to find the number of bricks in a pile of bricks of given dimensions.
#6 and #7: Both problems concern a pile of bricks of given dimensions and a given number of bricks. The problems are unclear since the text is difficult to read.
#8: The problem is to find the number of half-bricks in a pile of bricks of given dimensions.
#9 and #10: Both problems concern a pile of half-bricks of given dimensions and a given number of bricks. The problems are unclear due to difficulty in reading the text.

	SUBCATEGORY		SOLUTION WORKED		ANSWER GIVEN		DRAWING			
TABLETS	ONE TYPE	MIXED	YES	NO	YES	NO	YES	NO	DATE	REFERENCE
(1) AO 8862										
#5–7:		x	x		x			x	OB	*MKT* I: 108–23
#8:		x	?	?	?	?		x		
(2) AO 10822	x			x		x		x	OB	*MKT* I: 123–26
(3) AOT 304	x			x	x			x	Ur III	Vaiman 1961: 246–47
(4) BM 85194										
#16,19,30:		x	x		x			x	Kassite or older†	*MKT* I: 142–93
(5) Haddad 104										
#9,10:		x	x		x			x	OB	Al-Rawi and Roaf 1984
(6) IM 54538	x		x		x			x	OB	Baqir 1951: 33–35: no. 3
(7) VAT 6598 #1–4:		x	x		x			x	OB	*MKT* I: 277–87
(8) YBC 4607	x			x	x			x	OB	*MCT*: 91–94
(9) YBC 4669										
#10–14:		x		x	x			x	Kassite or older†	*MKT* I: 514–16; *MKT* III: 26–29
(10) YBC 4673										
#1,3,6:		x		x		x		x	Kassite	*MKT* I: 506–08;
#2,4:		x		x	x			x	or older†	*MKT* III: 29–34
#5:		x		x	x(part)			x		
#7:		x		x	?	?		x		
#8:		x		x	x?			x		
(11) YBC 4708	x			x				x	Kassite	*MKT* I: 389–402
#1–5,10,16,21,34:	x			x	x			x	or older†	
#37,40,47,49–55:	x			x	x			x		
#6,11–15,17–20:	x			x		x		x		
#22–33,35,36,38,39:	x			x		x		x		
#41–46,48,56–60:	x			x		x		x		
#7–9:	x			x	?	?		x		
(12) YBC 7284	x		x		x			x	OB	*MCT*: 97–98

†In this and the subsequent charts of this chapter, tablets so indicated are dated to the Kassite period by Neugebauer, but to the OB period by Thureau-Dangin [*TMB*].

TABLETS	SUBCATEGORY ONE TYPE	MIXED	SOLUTION WORKED YES	NO	ANSWER GIVEN YES	NO	DRAWING YES	NO	DATE	REFERENCE
(13) YBC 7997	x		x			x		x	OB	*MCT*: 98–99
(14) YBC 10772	x		?	?	?	?		x	OB	*MCT*: 98

(3) AOT 304

The problem is to find the number of bricks in a pile of type 1 bricks. The coefficient 7;12 is used as in YBC 4708.

(4) BM 85194 #16, #19, #30

#16: The problem is to find the number of burnt (fired) bricks per layer in a cistern. The diameter and exterior circumference are calculated using $\pi = 3$. Thureau-Dangin (1932d: 78–80) explained this problem as dealing with a cross-section taken at the base of the cylinder.

#19: The problem is to find the interior and exterior circumference of a cistern made of burnt (fired) bricks. As in problem #16, the example deals with a cross-section taken at the base of the cylinder.

#30: The problem is to find the equivalent amount of barley for a ship carrying 1 **sar** of bricks. The solution to the problem is based on transforming the volume occupied by 1 **sar** = 12,0 bricks to units of capacity (**sìla** and **gur**). A coefficient of 5,0,0 is used to convert volume-**sar** to **sìla** (*MCT*: 96–97). The text demonstrates that the volume of one brick (of type 1) is 0;0,0,41,40 volume-**sar**.

(5) Haddad 104 #9, #10

Problems #9 and #10 concern brick-making and contain technical terminology for this task. Three operations are involved in brick-making, namely, *ḫabātum* 'digging', *balālum* 'mixing' and *labānum* 'molding'. For each task a certain volume of earth (called *allum* 'hoe'), is processed daily. *alli neqrebum* is used to calculate the total volume of earth (in **sar**) processed by one man in one day.

The following coefficients are used to solve problems #9 and #10 in Haddad 104:

0;20 *alli ḫabātim*
0;20 *alli labānim*
0;10 *alli balālim*
0;5 *alli neqrebum*

0;45 *muttalliktum*, which is used in calculating the distance for one man to transport a volume of earth in one day

2;13,20 *tupšikkum* which is used in calculating the volume of earth (in **sar**) which can be transported a given distance by one man in one day

The coefficients for *alli ḫabātim*, *alli labānim*, *alli balālim* and *alli neqrebum* occur only in Haddad 104.

muttalliktum, a type of pavement or service, is also found in the following coefficient lists:

45 **igi-gub** *mu-ut-ta-al-li-i[k]-tum* YBC 5022: 39

45 **igi-gub** *mu-ta-li-ik-tim* IM 49949: 16

tupšikkum 'basket' (particularly in connection with construction projects of corvée duty), is found in the following coefficient lists:

2,13,20 **igi-gub** **gišdusu** *šu-up-ši-ki-im* (*šupšikkum* is a gloss for **gišdusu** and a phonetic variant of *tupšikkum*) IM 49949: 3

2,13,20 *ù* 1,15 *ša* GIŠ-DUB-ÍL YBC 5022: 39

Also, in both problems the number of bricks which one man is expected to produce in a day (*iškarum*) is multiplied by 3 to give the total work load of the *allum*; 3 may indicate a team of three men, but the reason for this operation remains unclear.

#9: The problem is to find how many bricks one man makes in a day and how many bricks are in the *allum* of brick-making; the exact meaning of *allum* is unknown in this context. The size of the brick is given as ⅔ **kùs** for both length and width and 1 **kùs** for depth. The coefficients of *alli ḫabātim*, *alli labānim*, *alli balālim* and *alli neqrebum* are used in solving the problem.

#10: The problem is to find the work assignment or output of one man when material has to be transported a distance of 5 **ninda** or 1 *ṣuppam*. In addition to the coefficients used in problem #9, this example also applies the *muttalliktum* and *tupšikkum* coefficients in its calculations.

The *allum* values calculated for transporting material 5 **ninda** in Haddad 104 do not correspond with another list of *allum* values (Goetze 1951: 141–42: §E rev. 25–29):

a-na 40 **ninda** *a-za-bi-il* 2,13,20 *al-lu-um*

a-na 20 **ninda** *a-za-bi-il* 5 *al-lu-um*

a-na 15 **ninda** [*a-za-bi-il* x] *al-lu-um*

a-na 10 **ninda** *a-z*[*a-bi-i*]*l* 6,40 *al-lu-um*

a-na 5 **ninda** *a-za-*[*bi-i*]*l* 10 al-lu-um

According to the method of computation in Haddad 104,

for 40 **ninda** the *allum* is 2;30

for 20 **ninda** the *allum* is 5

for 15 **ninda** the *allum* is 6;40
for 10 **ninda** the *allum* is 10
for 5 **ninda** the *allum* is 20.

(6) IM 54538

This text belongs to a group involving the transportation of 540 (9 sixties) bricks a distance of 30 **ninda** or 3 *ašlu*. The problem is solved by using a quadratic equation.

(7) VAT 6598 #1–4

The problems deal with a wall of kiln-burnt bricks of types 4 or 5. 2;15, the coefficient of kiln-burnt bricks (*agurru*), is used to solve these problems (see *TMB*: 127–28).

In view of the recent join of BM 96957 to VAT 6598 made by Christopher Walker (Letter to author 1986), the problems are about **é-sig$_4$ sig$_4$-al-ur$_5$-ra** 'a wall of kiln-burnt bricks'. BM 96597 contains problems dealing with **é-sig$_4$** 'wall' and **é-sig$_4$ sig$_4$-al-ur$_5$-ra**, to which the term *kuburrû* 'thickness' and the coefficient of kiln-burnt bricks or 2;15 are applied, as in the continuation of the tablet on VAT 6598. Because Mr. Walker and Dr. Michael Roaf plan to publish this text, it will not be included here.

#1: The text is broken; references to kiln-burnt brick of 9 **sar** are preserved.

#2 and #3: Both problems concern a brick wall of kiln-burnt bricks and given volume and height. The problem each time is to find the length and thickness of the wall.

#4: The problem is to find the *šipkatu* (of which the exact meaning is unknown) and slope of a brick wall of kiln-burnt bricks of given height and thickness at the base.

(8) YBC 4607

The problems of this tablet deal with bricks of various dimensions. Five types are distinguished, all of which have the same height (**sukud**) of 5 fingers (**šu-si**) but differ as to the length (**uš**) and width (**sag**) (see *MCT*: 93).

Problems #1 to #5 deal with bricks of types 1 to 5, respectively, and with finding the area (**gagar**), the volume (**sahar**) and the capacity equivalent of the volume measurement. The transformation to capacity measure is obtained by multiplying the number of **sar** which constitute the volume (l·b·h) by the coefficient 5,0,0.

Problems #6 to #10 concern finding the number of bricks of types 6 to 10, respectively, in 1 **sar** area. The use of the unit **sar** for counting bricks in pre-Kassite periods is well attested outside of the mathematical texts but the specific relation 1 **sar** = 12,0 (720) bricks, regardless of brick type, is definitely proved in problems #6 to 10 (*MCT*: 94).

(9) YBC 4669 #10–14

The text of problems #10 to #14 is broken. The problems concern volume but the details are uncertain. References to **sig$_4$-gir$_3$-gub**, to be read **sig$_4$-anše gub** 'brick pile', or **gir$_3$-gub** = *kilzappu* 'step', are preserved.

(10) YBC 4673 #1–8

#1: The problem is to find the area and volume of type 1 brick.
#2: A worker receives a rate of payment of 1 **bán** barley for transporting 540 (9 sixties) bricks 30 **ninda** or 3 *ašlu*. The problem is to find the amount of barley paid if he transports 450 bricks of type 1.
#3: A builder receives a rate of payment of 1 **bán** barley for transporting 540 (9 sixties) bricks 30 **ninda** or 3 *ašlu*. The problem is to find the amount of barley paid if he transports 360 bricks of type 1.
#4: One worker can transport 540 bricks 30 **ninda** in one day. The problem is to find the number of workers needed to transport 1 **ubu** and 4 **sar** of bricks. The text proves that the brick-**sar** stands in the same relation to the brick-**ubu** as the volume- or area-**sar** to the volume- or area-**ubu**. That is, 1 **ubu** = 50 **sar** and 1 **sar** = 12,0 bricks; therefore, 1 **ubu** = 50 × 12,0 = 10,0,0 bricks.
#5: One worker can transport earth and bricks 30 **ninda**. The problem is to find what fraction of a day he spends in molding bricks and the number of bricks he molds. This problem may be a concisely worded variant of a main type which is not yet known. Lewy (1949b: 156–58) proposed solving this problem by using coefficients.
#6–8: The text is broken; references to bricks and brick piles are preserved.

(11) YBC 4708

All the sixty problems of this tablet concern brick piles of type 1 bricks. Problems #1 to #36 concern the number of bricks in a brick pile in the shape of a rectangular prism. The data are the same for this group. The problems are solved by using quadratic equations. The shape of the brick pile and the method of solution are the same in the ensuing problems; only the given data change in the following groupings: problems #37 to #39, problems #40 to #42, problems #43 to #48 and problems #49 to #52 (here the problems are to be solved by using third degree equations). Problems #53 to #56 concern the number of bricks in a brick pile in the shape of a truncated pyramid. The data are the same for this group. The problems are to be solved by using quadratic equations. In problems #57 to #60 the data change, though the shape of the brick pile remains a truncated pyramid. The problems, once again, are to be solved by using quadratic equations.

All these problems concern the finding of length (l), base (b), height (h) or the number of bricks in the pile (N), while the other parameters are given or assumed to be those in the previous problems. The coefficient 7;12 is used here in order to obtain N or the number of bricks in the pile (*MCT*: 94–95 and n. 245).

The term **sig$_4$ sig$_4$-anše** = *libitti amarim*, meaning 'bricks of a pile', is used in the text instead of the more usual form, **sig$_4$-anše sig$_4$** = *amar libittim* 'a pile of bricks'. Thureau-Dangin (1936b: 165) suggested that this construction may be influenced by Sumerian, where the rectus frequently precedes the regens for forming a composite noun.

(12) YBC 7284

This text is a small, round tablet which reveals the existence of the coefficient 12,0,0 **ma-na/sar**, used to transform volume of bricks to units of weight. Type 1 brick is implied in this problem.

(13) YBC 7997

The text does not specifically state the problem to be solved but the data and computations indicate the problem is to calculate the number of bricks of type 1 in the core of a cylindrical oven (*utūnu*). The example contains unrealistic elements, namely: the height of the cylindrical core seems excessive, and the bricks in the cylindrical core are rectangular prisms.

(14) YBC 10722

The text is broken and largely erased. The problem belongs to a group involving the transportation of 540 (9 sixties) bricks a distance of 30 **ninda** or 3 *ašlu*. Line 4 states that four men are hired. The numbers 7, 11, 13, and 14, in lines 5–8, may refer to the wages paid in line 3. The numbers 7, 11, 13, and 14 form a special series found in operations with reciprocals and *maḫīru* 'price equivalency' problems (see *MCT*: 17–18).

cf. CBM 12648

Friberg (1982: 57) suggested that this text probably deals with bricks of standard formats; however, we find the text too fragmentary to support his interpretation.

SUMMARY Ten types of bricks, classified according to size, are found in the mathematical texts (*MCT*: 93–94). The sizes are realistic (Dunham 1982: 41). The bricks are also referred to as **sig$_4$** = *libittu* 'brick', **sig$_4$-áb** = *arḫu* 'half-brick' and **sig$_4$-al-ùr-ra** = *agurru* 'kiln-burnt brick'. No drawings occur in these problems.

Dunham (1982) noted that YBC 9819 lists two additional types of bricks with different measurements not found in these mathematical problems dealing with bricks: (1) *ukurum*, (possibly a variant of *agurru*) and (2) *zarin*.

Oppenheim (1947b: 127) suggested that *agurru* can be traced back to the Sumerian(?) word **añur(a)** rather than to the Old Akkadian *ukkurum* (contra *MCT*: 96 n. 248a). He suggested that *ukkurum* was a special type of brick that entered Ur III Sumerian as a technical term. Goetze (1948: 34–35) agreed with Oppenheim. He even read 48 1,40 1,20 *ú-ku!-ur!-ru!-um* in YBC 7243: 43 (*MCT*: 136–37) and noted a different coefficient for ^{sig4}*a-gu-ru-um* in the next line, namely, 2(?),42 ^{sig4}*a-gu-ru-um* YBC 7243: 44. Goetze added that ^{sig4}*ù-ku-ru-um* in AO 7667 (Scheil 1915: 161–72) must be considered some kind of opposite of ^{sig4}*za-ri-in*. *za-ri-in* denotes a mediocre quality of bricks, copper or wool. *za-ri-in* also denotes a second-rate quality of dates and is considered the opposite of *damqum*. *zarinnum* (quality of) metal objects may indicate that they were to be used as scrap (see *CAD* s.v. *zarinnu* A). Goetze concluded that *ukurrum* is similar to *damqum* and signifies 'best' or 'choice quality'. Its form suggests a Semitic origin and may derive from Akkadian *waqrum*.

Regardless of their type, bricks are counted by the principle of 1 **sar** = 12,0 (= 720) bricks (see *MCT*: 94–95) in BM 85194 #30, YBC 4607 #6–10, YBC 4673 #4 and YBC 4708.

Powell (1982c: 122–23) suggested that this brick-counting system and associated coefficients presuppose flat bricks. The use of plano-convex bricks spans much of the Early Dynastic period; the shift to the use of flat bricks occurs at Girsu during the time of Entemena. Therefore, this portion of the OB mathematical texts can be dated to the period Entemena–Šar-kali-šarri.

Bricks are converted to units of weight in YBC 7284; the data in this problem are also used in calculations for bricks in the Ur III–OB periods (Powell 1982c: 119-123).

A group of texts is concerned with the transportation of 540 (9 sixties) bricks a distance of 30 **ninda** or 3 *ašlu*: AO 8862 #5–9, IM 54538, YBC 4673 #2–4 and YBC 10722.

The mathematical texts also deal with problems of carrying earth and making bricks in AO 8862 #8–9 and YBC 4673 #5 and in the coefficient lists, namely, 10(?),1,20 **igi-gub saḫar íl-íl sig$_4$ dù** '10(?),1,20 coefficient: carrying earth, making bricks' in A 3553: 35 (Kilmer 1960: 276–77).

In addition, Lewy (1949b: 158–61) noted coefficients for barges carrying different kinds of bricks and other items (see Kilmer 1960: 296: section VI D 1). She proposed that the coefficient list YBC 5022 (*MCT*: 132–36) may have been compiled by an architect in order to execute large scale construction projects that could not be satisfied by local production of bricks, hence the coefficients for barges carrying bricks.

The mathematical texts do not detail the brickmaking process. They refer only to the use of earth (mud) in making bricks. However, the Mesopotamians were aware of the advantages of adding organic substances to mud brick and mud mortar; for example, the building of the city wall of Assur is decribed in an account of Shalmaneser III as follows: "I added a layer whose height is 13 sun-dried bricks to the great construction of the city wall. I wetted its clay with honey, good ghee, oil of cedar(?), date-wine and grape wine. I made its sun-dried bricks in wagons and mouldings of cedar." (Safar 1951: 13, 20: col. iv 50–55; for studies on mud-brick construction see Bruno et al. 1968–69: 252–70, Dunham 1980: 1–28; Torraca et al. 1972: 259–86).

The mathematical problems offer numerous examples concerning brick piles in the following texts: AO 10822 #5–10, AOT 304, YBC 4669 #10–14, YBC 4673 #6–8 and YBC 4708. The piles are rectangular prisms or truncated square pyramids. Problems concerning brick walls are examined in AO 10822 #3 and #4 and VAT 6598 #1–4.

Bricks are important in the history of metrology because they are the only surviving artifacts for which textual evidence also exists (from the time between OAkk period to end of OB period), linking norms of length, area, volume, capacity and weight. Unfortunately, textual evidence for bricks cannot be corroborated by bricks from excavations and archaeological sites, because archaeological publications rarely show how many (datable/stamped) bricks are measured and to what degree of accuracy (Powell 1982c: 117).

II. CANALS AND IRRIGATION

DISCUSSION (1) AOT 305

All four problems of this tablet concern irrigation projects (**íd-a-dug**), which are triangular-based prisms. The word used for 'volume' in the text is the word for area (**a-šà**).

(2) BM 85196 #15, #16

#15: The problem is to find the original volume, the new volume, and the length that one man can dig when the sides (*taraḫḫu*) of a small branch canal (**pa$_5$-sig** = *atappu*) are enlarged. The problem is illustrated by a diagram in which a rectangle is used to depict a trapezoidal cross-section.

#16: The problem is to find the fraction of a day spent in digging each of the three levels of an irrigation canal ([íd]*namkarum*). The term **bal éš-kàr** 'fraction of an assignment' (BM 85196 rev. i 38) is used to denote the actual volume of earth

TABLETS	SUBCATEGORY ONE TYPE	SUBCATEGORY MIXED	SOLUTION WORKED YES	SOLUTION WORKED NO	ANSWER GIVEN YES	ANSWER GIVEN NO	DRAWING YES	DRAWING NO	DATE	REFERENCE
(1) AOT 305	x			x	x			x	Ur III	Allotte de la Füye 1907: 75–77
(2) BM 85196										
#15:		x	x		x		x		Kassite	*MKT* II: 43–59
#16:		x	x		x			x	or older†	
(3) Erm 15073										
#4:		x	x		x			x	OB	Vaiman 1961:
#5:		x	x		x		x			232–44
(4) *TMS*, Text 24	x		x		x			x	OB	*TMS*: 117–22
(5) *TMS*, Text 25 #1:		x	x		x			x	OB	*TMS*: 122–24
(6) VAT 7528	x			x	x			x	Kassite or older†	*MKT* I: 508–13
(7) VAT 8523	x		x		x			x	OB	*MKT* I: 373–80
(8) YBC 4186	x		x		x			x	OB	*MCT*: 91
(9) YBC 4666										
#1–22:	x			x	x			x	OB	*MCT*: 76–81
#23:	x			x		x		x		
(10) YBC 7164	x			x	x			x	OB	*MCT*: 81–90
(11) YBC 9874										
#1:	x			x		x	x		OB	*MCT*: 90–91
#2:	x		x		x			x		

excavated by one man in one day at each level. The first level of work is called *li*-LUM; the term for the second level is no longer extant, and the third level is described by *ḫad/ṭālum* and *iḫd/ṭil*. The details of this problem are corrupt as are many others in the same tablet.

(3) Erm 15073 #4, #5

#4: The problem is to find the number of workers necessary to excavate a canal (**íd**) of given dimensions with a water reservoir (**kun** = *zibbatu*). A coefficient of 10 (**gín**), the volume of earth to be excavated as the daily work assignment for

a laborer involved in construction and excavation of earthworks and canals, is used in solving the problem.
#5: The problem is to find the number of workers necessary to excavate a canal (**íd**) of given dimensions with a water reservoir(?) (**kun** = *zibbatu*). The same coefficient of 10 (**gín**) is used in solving the problem. The problem is illustrated by a drawing of this irrigation project as a partitioned trapezoid. Vaiman (1960) analyzed this illustration as a three-dimensional figure which appears to be two-dimensional when seen from above.

(4) *TMS*, Text 24

The text is broken; the nature of the problem to be solved is not entirely clear because the data are not extant. The text uses mathematical terms that are difficult to define: (a) *manâtu*, mathematical variable (see *CAD* s.v. *manâtu*; cf. Høyrup 1990: 300 for translation 'contribution'); (b) *takīltu* a mathematical term that changes meaning according to the context (*MCT*: 130; cf. *TMB*: 226 for translation 'factor'); (c) *talūku*, a previously given number (*AHw* s.v. *talūku*); and (d) *kajamānu*, the exact meaning of which is not known within a mathematical context. The problem is solved by using quadratic equations; however, the scribe chooses only the positive root and discards the negative root.

This text is the earliest documented use of the separation sign (**gam**) for medial zeroes (Friberg 1982: 90).
#1: The problem concerns the trapezoidal cross-section of a canal.
#2: The problem concerns the trapezoidal cross-section of a canal. References to both *atappu* 'small branch canal' and *kalakku* 'excavation' are preserved in this problem. Neugebauer and Sachs (*MCT*: 79) discussed similarities between certain *kalakku* and canal problems.

(5) *TMS*, Text 25 #1
#1: The problem is to find the depth and the width of a branch canal. The text is broken, and most of the data are no longer extant. Forty workers (or 40,0 = 2400) dig the canal and excavate a given volume of earth and water. 48, the coefficient of a branch canal, is used in the calculations. Powell (9188: 165–66) suggested that the problem concerns the allocation of water along a canal.

(6) VAT 7528

All nineteen problems extant in this tablet concern a branch canal (*atappu*). According to the colophon, the tablet contains twenty-four (or twenty-five) problems; it is the third tablet in a series.
#1: The problem is to find the length one man can dig when excavating a branch canal, which is a trapezoidal cross-section. The daily work assignment or output of one man is **⅓ sar**.

#2–4: The problem in each example is to find the amount of time or the number of workers needed to dig a branch canal, which is a trapezoidal cross-section. The daily work assignment or output of one man is ⅓ **sar**. The data are the same for these problems. Problems #3 and #4 are solved by using quadratic equations.
#5–15: The text is broken; the problems concern a branch canal, which is a trapezoidal cross-section. References to the number of workers and the time needed to complete the work are preserved in these problems.
#16 and #17: Both problems are broken; references to a branch canal and laborers are preserved on the tablet.
#18 and #19: The problem each time is to find the volume of a branch canal, which is a triangular (**sag-dù**) cross-section.

(7) VAT 8523

All six problems of this tablet concern damming (*kāsiru* 'dam') a canal by using a dike comprised of ⅔ brick and ⅓ soil. The text uses the term *šipru* 'work' to indicate the part of the dam which is constructed in bricks (*TMB*: 124 n. 4).

Problems #1 to #4 include the following data:

w = 0;20 **ninda** (width of the canal or **dagal**)
b = 5 **ninda** (width of the dam or **sag**)
h = 18 **kùš** (height of the dam which is the same as the depth of the canal or **sukud**)
v = 10 **ninda**^{2} × **kùš** (volume or **saḫar**)
k = 20 **ninda**^{2} × **kùš** (work or **kin** = *šipru*)

Each of these problems concerns finding one or more of the above parameters while the other parameters are variously given. The following relation holds: $wbh = K + V$. Powell (1988: 166) demonstrated that the problems are constructed with an eye to their solutions based on simple numerical ratios.

Problems #5 and #6 employ the same data as problems #1 to #4. New quantities are introduced for the length, width and height of a pile of bricks, which constitute the volume ($v = lbh$) of the daily work assignment for a laborer involved in damming up (*sekēru*) a canal.
#5: The problem is to find the number of man days needed to dam up the canal.
#6: The problem is to find the total volume of masonry work (K) on the dam; the problem checks the value found in problem #5 by performing the inverse operation to determine whether it satisfies the statement.

(8) YBC 4186

A cistern (**túl**), in the shape of a cube, is used to irrigate a field to a (uniform) depth of 1 **šu-si** (approximately 1⅔ cm). The length, width and depth of the cistern are 10 **ninda** each. The problem is to find the area of the field. Powell (1988: 162) indicated that the answer is "intuitively clear from the metrological relationships." The process of inundating the field is indicated by the term *nazālu* 'to pour, to distill liquid' or in technical terminology 'to pour liquid mass over or onto various materials' (*MCT*: 91 n. 230). The idealized irrigation as described in the problem is a light irrigation, sufficient to prepare the ground for weeding (Jacobsen 1981: 61).

(9) YBC 4666

All the problems concern excavation of a small branch canal (**pa$_5$-sig** = *atappu*). The problems are similar to examples that deal with excavations of volumes of earth (**ki-lá** = *kalakku*). The colophon indicates that the text includes 26 problems, but only 23 problems are found.

Problems #1 to #8 concern the rectangular cross-section of a small branch canal. The data are the same for this group. The volume of earth excavated by one man in one day (λ) is incorrectly given as ⅓ **gín**; the value of λ should be ⅓ **sar** as in VAT 7528 #1–4, #18 (see *MCT*: 79 and nn. 185, 186). Thureau-Dangin (1938a: 156–57) discussed similar inconsistencies in economic texts from the same time period.

Problems #9 to #11 concern the rectangular cross-section of a small branch canal. The data are different for each problem in this group. The volume of earth excavated by one man in one day (λ) is incorrectly given as ⅓ **gín**.

Problems #12 and #13 concern the rectangular cross-section of small branch canal. The base, height and volume of earth excavated by a laborer in a day (here incorrectly given as ⅓ **gín**) are the same in both problems. Additional magnitudes for the width, height, and daily wages per man are introduced in each example; the relationship assumed among all these quantities is unclear. Perhaps, problems #12 and #13 can be compared with *UET* 5: nos. 855 and 856, two Old Babylonian texts that describe the division of labor in constructing an irrigation canal. One text is based on preliminary measurements, while the other makes use of more accurate data (Friberg 1982: 86).

Problems #14 to #22 concern the trapezoidal cross-section of a small branch canal; the upper width is greater than the lower width. The data are the same for this group; in problems #21 and #22 information concerning the slope of the canal is added. The volume of earth excavated by one man in one day (λ) is 10 ***gín*** = 0;10 **sar**, half the work assignment or output of the preceding examples.

Problem #23 concerns the rectangular cross-section of a small branch canal for which only the data are given. This example is the "catch-line" for the first problem of YBC 7164, the tablet which follows.

All these problems concern finding one or more of the following parameters: length (l), base (b), height (h), area (A), volume (V), volume of earth excavated by one man in one day (λ), number of workers or **erim-ḫi-a** (m), wages of one man for one day's work (w, here 1 **bán** barley or 6 **še** silver), total wages expended (W), the length one or more workers digs (l_{1-30}), and time to complete the work (t), while the other parameters are variously given.

(10) YBC 7164

This tablet is a direct continuation of YBC 4666. All eighteen problems concern excavation of a rectangular cross-section of a small branch canal. The text contains numerous errors, omissions and abbreviations (*MCT*: 88). For example, *šu* is an abbreviation for *šuplum* 'depth' or *šupulšu* 'its depth'.

Problems #1 to #7 deal with the excavation of a rectangular cross-section of a small canal (**pa₅-sig**).

#1: The problem is to find the area (A), volume (V), number of workers or **erim-ḫi-a** (m), and total expenses in silver (E) for work on the rectangular cross-section of a small canal of given length, width and depth. The volume of earth excavated by one man in one day (λ) is 0;10 **sar**. The wages of one man for one day's work (w) is 6 **še** = 0;2 **gín** silver.

#2: This example is similar to #1; however, the volume of earth excavated by one man in one day (λ) is increased to 0;12 **sar**. The problem is to find the length of one man's work per day (l_λ).

#3: This example is similar to #2. The problem is to find the length of canal worked by 30 laborers in one day.

#4: This example is similar to #2. The problem is to find the time needed for 30 workers to finish work on the length of the canal. A 30 day calendar month is used to compute wages and time (*MCT*: 84).

#5: This example is the inverse of #4. The problem is to find the length of the canal when information concerning the length thirty workers dig and the time they need to complete the work is given.

#6: This example is similar to problem #1, with the depth being greater in problem #6. The volume of earth excavated by one man in one day (δ) varies from 0;20 **sar** at the upper level, called *šīlūtu*, to 0;10 **sar** at the lower level, called **dusu** meaning 'basket'. A third level of canal work at a still smaller rate of output, 0;7,30 **sar** (sic), also called **dusu**, is indicated. The problem is to find the daily assignment of the length to be dug by one man (l_λ) by computing the assignment at each level.

#7: This example is similar to problem #6. The problem is to find the volume and the parts of a man's work-day spent in two types of labor, **dusu** and *šīlūtu*.

Problems #8 to #16 deal with cleaning out the silted parts or, perhaps, even enlarging the original profile of the rectangular cross-section of an old, small canal (**pa₅-sig-libir-ra**). From the context of these problems, *taraḫḫu* (abbreviated to *ta* in problems #9 to #16) seems to mean the sides of the canal and *terdītu* 'increase' (*MCT*: 88). The data are the same for problems #8 to #12.

#8: The problem is to find the volume (V) to be dug on the rectangular cross-section of an old, small canal when the width is increased on each side. The length, width, depth and amount of increase are given. The answer is given incorrectly (in units of volume) because the scribe misunderstood the order of magnitude.
#9: This example is similar to #8. The problem is to find the length one worker digs in a day. The assigned rate of output for the volume of earth to be excavated by one man in one day (λ) is ⅓ **sar** = 0;20 **sar** (incorrectly given as ⅓ **gín** here).
#10: This example is similar to #9. The problem is to find the length ten workers dig in a day.
#11: This example is similar to #9. The problem is to find the time needed to complete the whole work by ten men.
#12: This example is the inverse of #11. The problem is to find the length of the canal when information is given concerning the length ten workers dig and the time they need to complete the work.
#13: The problem is to find the volume (V) to be dug on the rectangular cross-section of an old, small canal when both the width and depth are increased. The length, width, depth and amount of increase are given. The assigned rate of output for the volume of earth to be excavated by one man in one day (λ) is ⅓ **sar** = 0;20 **sar** (incorrectly given as ⅓ **gín** here).
#14: This example is similar to #13. The problem is to find the length one man digs in a day.
#15: This example is similar to #13, but new parameters are given. The problem is to find total volume (V) excavated when an old, small canal is enlarged.
#16: This example is similar to #13. The problem is to find the length one man digs in a day. The volume of earth excavated by one man in one day (λ) varies from 0;20 **sar** at the upper level to 0;10 **sar** at the lower level. The solution involves computing the complementary fractions of a day spent working at each level.
#17 and #18: Both problems concern the rectangular cross-section of a small canal. The data are different for each problem. The width and depth are given but not the length. References to waters of watercourses (*me-e pa-ta-a-tim*) are preserved. The problems to be solved are unclear due to the fragmentary nature of the text. Powell (1988: 165) suggested that the coefficient 48 may be applicable in #17 as indicated in the following:

48 **a-meš** *pat-ti* CBS 10996 obv. iii 24'
48 *me-e* **a-šà**: YBC 7243: 26
48 *šà* **pa$_5$-sig**: *TMS*, Text 3: 33
igi 48 **igi-gub pa$_5$** *pu-ṭú-ú*[*r*]: *TMS*, Text 25 obv. i 7.

(11) YBC 9874

The obverse and reverse of the tablet may each contain independent, but similar problems.

obv: The figure given is a schematic drawing of the rectangular cross-section of a canal, which is enlarged. The numbers inscribed in the diagram indicate the dimensions of the canal; the numbers at side fulfill certain numerical relations, though 20 could be interpreted as 0;20 **sar**, the usual assigned rate of output for the volume of earth to be excavated by one man in one day at the upper level.

rev.: The problem is to find the actual output of work in the upper level (*šīlūtu*). The problem then checks this value by performing the inverse operation to find the usual work assignment for the volume of earth excavated by one man in one day at the upper level, which is 0;20 **sar**.

SUMMARY Irrigation was necessary for agriculture in Mesopotamia, a land where salinity was a problem from early times (Jacobsen 1981: 5–8; Adams 1965: 7–12). Water was channelled to the fields from the major watercourse (**íd**, literally 'river, canal'), through branch canals and feeders (**pa$_5$**), often running along the tops of artificial dikes (**e**) to preserve the preferred elevation (Steinkeller 1988: 74). The width of the primary watercourses could be 120 m. or more; the branch canals were as narrow as 1.00 to 1.50 m. in width and 0.50 to 2.25 m. in depth, with a length slightly under 2 km. (1,800 m.). These data are also found in the mathematical problems. Weirs (^{giš}**kešda**, **kun-zi-da**) were built in order to raise the water level in the main stream. Bunds (**eg**?) were constructed to protect against floods (Pemberton et al. 1988: 212–17; Steinkeller 1988: 73–74); from earliest times on breaches in them were repaired with earth and bundles of reeds. Protective quay walls (**kar**) of baked bricks set in bitumen were built to guard against erosion on the canal banks at critical points. To prevent scour, pebbles or stones were used to cover a canal bed (a practice recorded as early as 2400 B.C.; Jacobsen 1981: 62–63).

The work assignment or rate of work for the volume of earth excavated by one man in one day varies from 0;20 **sar** at the upper level, called *šīlūtu*, to 0;10 **sar** at the lower level, called **dusu** meaning 'basket', indicating, perhaps, that the digging at lower levels required shovelling the excavated dirt into baskets that had to be raised to the top and emptied while the workers at the upper level merely had to dig out the earth and cast it from the shovel to the ground (*MCT*: 84–86, 88). Lewy (1949: 51) proposed that workers were not confined to work an entire day at either upper or lower stratum (see BM 85196 #16 and YBC 7164 #7 and #16).

The mathematical problems indicate payment of 6 še silver or 1 **bán** (= 10 **sìla**) barley as the wages of a laborer for one day's work in excavating canals. Such payments are actually attested for Ur III (*MCT*: 74 n. 179; Schwenzer 1915:

102; Landsberger 1937: tablet 6 iii 11–12). A calendar month of 30 days is used in computing wages and time (*MCT*: 84).

The mathematical problems offer numerous examples concerning excavating or enlarging a canal in the following texts: BM 85196 #15, Erm 15073 #4 and #5, VAT 7528 #1–15, YBC 4666 #1–23, YBC 7164 and YBC 9874 obv. The profile of a canal is represented by a rectangular or trapezoidal cross-section. Drawings of canals are rare. Schemata with inscribed numbers and calculations are used to illustrate enlarged canals in BM 85196 #15 and YBC 9874. The water reservoir of a canal is accompanied by a drawing of a partitioned trapezoid in Erm 15073 #5.

Other irrigation projects appear in the mathematical problems. One text concerns damming up a canal by using a dike of ⅔ brick and ⅓ soil (VAT 8523), and an idealized problem depicts irrigating a field from a cistern (YBC 4186).

The maintenance of major canals was the responsibility of the kings and rulers, but cleaning and dredging accumulated silt in other canals was under local authority (Renger 1990: 38–39; Hunt 1988: 196–202). In fact, accounts of dredging work on canals near Girsu with dredging assignments minutely detailed to meter and half meter of canal course are found from as early as 2400 B.C. (Jacobsen 1981: 62; see Maekawa 1987: 51–60). The parameters of the canals and branch canals in the mathematical texts are realistic (Powell 1988: 163) and often parallel the data of the Girsu accounts.

III. DEFENSE WORKS

DISCUSSION (1) BM 85194 #4, #25, #26

#4: The problem is to find the lower and upper widths and height of a dike, which is in the shape of a trapezoidal cross-section; the problem is also to find the circumference of this dike, which encircles the moat of a city. The slope of the dike is 1 **kùš** per 1 **kùš** or 45°. The daily work assignment or output of one man is 0;10 **sar**, which is the usual amount found in problems dealing with construction and excavation of earthworks and canals.

The problem is illustrated by a drawing of concentric circles to represent the dike, the moat and the city. It is the only defense works problem illustrated by a drawing. Vaiman (1960) analyzed this drawing as a three-dimensional figure in a simple perspective as seen from above.

The text uses the terms *dakāšu*, 'to inscribe a figure within a figure', and *nesû*, 'to circumscribe', in order to explain the accompanying illustration. Contra Friberg (1981b: 58), Kilmer (1964: 140–46; 1990b: 85) proposed that *dikšum /dakištum/dikištum* is used to indicate the regions between two circles or two squares in problems dealing with circular or square borders; Friberg (1981b: 58)

TABLETS	SUBCATEGORY ONE TYPE	SUBCATEGORY MIXED	SOLUTION WORKED YES	SOLUTION WORKED NO	ANSWER GIVEN YES	ANSWER GIVEN NO	DRAWING YES	DRAWING NO	DATE	REFERENCE
(1) BM 85194										
#4:		x	x		x		x		Kassite	*MKT* I: 142–93
#25,26:		x	x		x			x	or older†	
(2) BM 85210 #1–7:		x	x		x			x	Kassite or older†	*MKT* I: 219–33

argued that the size of the border in most mathematical problems is assumed to be known once the side of the square is known. However, such assumed dimensions are not applicable to mathematical problems of cities.

The phrase "I seized a city hostile to Marduk..." usually introduces the defense works of a city, which is described as protected by a wall with a sloping, rising dam. Though BM 85194 #4 is not delineated by this characteristic phrase, it clearly parallels descriptions of defense works. Powell (1982b: 60 n. 7) discussed inconsistencies in the text, noting that Thureau-Dangin (1932a: 65–66) suggested that this problem is the work of two different editors. Powell concluded that both Thureau-Dangin and Neugebauer (*MKT* I: 142–93; 1933a: 307–310) ignored the fact that the dike would be floating out over the empty ditch if constructed according to the specifications of the text.

Gandz (1948: 18) pointed out that this problem is preserved in Diophantus II 10.

#25: The city is protected by a defensive dam which is not the same height as the city wall. The problem is to find how much must be 'paced off' (*kabāsu*) from the foot of the dam in order to reach the city wall and also to find the height of the wall. The second part of the problem checks the answer found in the first part to determine whether it satisfies the statement by performing the inverse operation.

The problem is solved by using a quadratic equation; Thureau-Dangin (1932e: 139–142) demonstrated that the scribe chose the the negative square root since the positive root gives an absurd result. This example confirms that Babylonian mathematicians were aware of the double value (positive and negative square roots) as well as the double solution that comprises a second degree equation.

#26: This example is similar to #25. The problem is to find the height of the city wall and the distance between the dam and the city wall.

(2) BM 85210 #1–7

Neugebauer (*MKT* I: 229) and Thureau-Dangin (1932e: 142) regarded BM 85210 #1–4 as variants of BM 85194 #25 and #26. The data are the same for problems #1 to #4. However, Waschow (1936–37: 246) suggested that the problem concerns a siege staircase.

#1: A walled enemy city with a staircase (**giš-kun$_4$** = *simmiltu*) is seized. The staircase is against the (fortified) city wall; each step (*nakbasum*, also shortened to *na*, or *kilzappu*) is of given dimensions. The problem is to find how much must be 'paced off' (*kabāsu*) from the foot of the staircase in order to reach the city wall so that the city hostile to Marduk can be seized.

#2: This example is similar to #1. The problem is to find the height of the city wall.

#3: This example is similar to #1. The problem is to find the height of the staircase.

#4: This example is similar to #1. The problem is to find the length of the staircase.

#5: The text is broken; calculations concerning a walled city are preserved.

#6: The problem is to find the length of the moat (*ḫurḫurru*) so that the city hostile to Marduk can be seized.

#7: The problem is to find how much must be 'paced off' (*kabāsu*) from the foot of an earth structure in order to reach the city wall and also to find the height of the wall.

SUMMARY The descriptions of municipal defense works in the mathematical texts involve constructions of considerable proportion for which the measurements are plausible. The height of a city wall is given variously as 3 **ninda** (c. 18 m.; BM 85210 #1-5) or 45 **kùš** (c. 22.5 m.; BM 85194 #25, #26 and BM 85210 #6, #7), and the length of the moat as 8 **ninda** (c. 48 m.; BM 85194 #25, #26 and BM 85210 #6 and #7). The problems often deal with a distance of 40 **ninda** (c. 240 m) from the foot of the dam to the city wall (BM 85194 #25, #26 and BM 85210 #6, #7). In BM 85210 #1-4 the length from the foot of the (siege) staircase to the city wall is 4;30 **ninda** (c. 27 m.), and the height of each step is 1;40 **kùš** (c. 80 cm.); the resulting staircase would be approximately 32.5 m. in length.

Powell (1982b: 59–61) suggested that BM 85194 #4 is similar to a metrological passage relating to the defensive measures taken by the Chaldean king Merodach-Baladan at Dur-Jakin in 709 B.C. A moat was built 200 cubits wide (c. 100 m.), 18 cubits deep (c. 9 m.) and a distance of 1 *ašlu* (c. 60 m.) from the city wall; bridges were constructed across the moat, which was flooded by making a cut to the Euphrates, in order to surround the city with rivers. When the Assyrians advanced, the bridges were cut and Merodach-Baladan brought his forces to the space between the city wall and the moat. The Assyrians crossed the moat using a ramp, and a battle was waged outside the city wall. The Assyrians prevailed and then seized the city. However, in BM 85194 #4 the city is only 120 m. in diameter or c. 10,800 m.^{2} (1 hectare =10,000 m.^{2}) yet the work requires 24,300 man-days of labor (Powell 1982b: 60; Thureau-Dangin 1933c: 187). Powell's discussion of the extensive efforts involved in defense constructions and his use of parallel passages in other texts underscores the accurate portrayal of municipal fortifications in the mathematical problems.

In Mari the district governors constructed defense works to protect main urban centers, protecting them by an inner (*dūrum*) and outer (*šalḫum*/*šulḫum*) city wall with a gate (*abullum*) allowing entry. The walls were constructed of piled earth and fortified with bricks, and a moat (*ḫirītum*) filled with water was added for further protection (Sasson 1969: 3–7).

Oates (1985: 586–87) showed that space existed within the city walls at various sites for the accommodation of considerable numbers of extra people in an emergency, namely the inhabitants of undefended agricultural villages. Sites from the Mari period vary in size from 7 to 90 hectares, with sites up to 30 hectares having polygonal, sometimes almost precisely hexagonal, outlines with a central mound.

IV. OTHER CONSTRUCTIONS, EXCAVATIONS AND EARTHWORKS

DISCUSSION (1) BM 85194 #1, #2, #3, #5, #17, #18, #28

#1: The problem is to find the volume of a ramp (*arammu*) that gives access to a city. The ramp apparently crosses an enclosed or dammed canal; measurements are given from the beginning of the dam underground to the end of the dam in front of the city gate. The volume is calculated as follows: volume = average of widths × average of heights × length. The daily work assignment or output of one man is 0;10 **sar**.

#2: The problem concerns a chapel or cult socle (*kikurru*), which is a prismatic volume. The *kikurru* may be shaped as follows: (a) ╬ or (b) ╦ with (a) being more likely. The problem contains no drawings. The daily work assignment or output of one man is 0;10 **sar**. The problem is to find the length one man can dig in one day. The volume is calculated and divided by the work assignment of one man.

#3: The problem concerns a chapel or cult socle (*kikurru*); it is similar to problem #2. To the data of problem #2, problem #3 adds that 0;40 **sar** is the length that one man digs in a day and 15 workers are involved in the construction. The problem is to find the length one man digs in half of the cult socle.

#5: The problem concerns a city wall (**bàd** = *dūru*), of trapezoidal cross-section according to the data and computations. The problem contains no drawings. The daily work assignment or output is 0;10 **sar**. Volume is calculated by the formula:

$$V=\frac{b_1+b_2}{2}\times h\times l$$

The problem is to find the length one man digs in one day.
#17: The text is broken. The problem concerns a city wall (**bàd** = *dūru*); it is similar to problem #2 but the data are different. The slope is given. The problem is to find the length of the lower base and the volume and the length of a dividing line parallel to the base.

The term **bal** is used in the computations to express a mathematical ratio, between the corresponding sides of similar polygons or polyhedrons (Thureau-Dangin 1935b: 188; *TMB*: 232 s.v. **bal**).
#18: The problem concerns a city wall (**bàd** = *dūru*); it is similar to problem #17. The problem is to find the length of the upper and lower bases.
#28: The problem concerns a moat (*ḫirītu*) in the shape of an inverted truncated pyramid whose bases are square. The problem is to find the volume of moat and the length of the lower base (ZA-ZUM). Von Soden (1939: 146–47) suggested the reading *sà-súm* = *sassum* 'base' for ZA-ZUM. Contra Thureau-Dangin (1932d: 87–88), Neugebauer (*MKT* I: 187) argued that the scribe made no errors and used the correct formula to find the volume.

(2) BM 85196 #1, #3, #4, #6, #7, #8, #10, #17
#1: The problem concerns an abutment or platform (**kí-sa** = *kisû*), which is protected by a spur of land (*appum*) in order to create a breakwater. The resulting structure is in the shape of a triangular (underwater) prism (see Powell 1988: 167–68; Dunham 1982: 37–39). The volume of the underwater structure, denoted by the term **kí-sa kin**, is ⅔ the volume of the entire structure. The problem is to find the volume of the *appum*. Powell (1988: 168) suggested that term should be read **gi-sa** = *kiššû*, and that part of the *appu* was made of 'reed-bundles'. Thureau-Dangin (1935a: 4) noted that in "volume" problems, length and width are expressed in **ninda** (= 12 **kùš**) and height in **kùš**, which when multiplied together yield a volume expressed in **sar**.

The problem is illustrated by a drawing in which a triangular prism is depicted as a triangle. Vaiman (1960) analyzed this illustration as a three-dimensional figure which appears to be two-dimensional when seen from above.

TABLETS	SUBCATEGORY ONE TYPE	MIXED	SOLUTION WORKED YES	NO	ANSWER GIVEN YES	NO	DRAWING YES	NO	DATE	REFERENCE
(1) BM 85194										
#1–3,5:		x	x		x			x	Kassite	*MKT* I: 142–93
#17,18,28:		x	x		x			x	or older†	
(2) BM 85196										
#1,17:		x	x		x		x		Kassite	*MKT* II: 43–59
#3,4,6,7,8,10:		x	x		x			x	or older†	
(3) BM 85210 #8:		x	x		x			x	Kassite or older†	*MKT* I: 219–33
(4) Böhl Coll. 1821	x		x		x		x		OB	Kilmer 1964: 142–43
(5) Const. 236	x			x	x			x	OAkk	Allotte de la Füye 1907: 77–78
(6) Erm 15073										
#6,7:		x	x		x		x		OB	Vaiman 1961:
#8:		x	x		x			x		232–44
(7) IM 53961	x		x		x			x	OB	Baqir 1951: 35–36: no 4
(8) IM 54011	x		x		x			x	OB	Baqir 1951: 45: nos. 10,11
(9) *LBAT*: no. 1647 #1:	?	?	x		x?			x	LB	Sachs 1955: no. 1647
(10) Strssbg 362 #6:		x		x		x		x	OB	*MKT* I: 239–43
(11) *TMS*, Text 14	x		x		x			x	OB	*TMS*: 84–88
?(12) *TMS*, Text 24	x		x		x			x	OB	*TMS*: 117–22
(13) *UET* 3: no. 1386	x			x	x			x	Ur III	Vaiman 1961: 244–46
(14) YBC 4657										
#1–6,8–16,20–31:	x			x	x			x	OB	*MCT*: 66–76
#7,17–19:	x			x		x		x		
(15) YBC 4662 #19–28	x		x		x			x	OB	*MCT*: 71–76
(16) YBC 4663 #1–8:	x		x		x			x	OB	*MCT*: 69–76

TABLETS	SUBCATEGORY ONE TYPE	MIXED	SOLUTION WORKED YES	NO	ANSWER GIVEN YES	NO	DRAWING YES	NO	DATE	REFERENCE
cf. YBC 4669 #B1:		x	x		x			x	Kassite or older[†]	*MKT* I: 514–16; *MKT* III: 26–29
(17) YBC 4673										
#9–11:		x		x	x			x	Kassite	*MKT* I: 506–08;
#12:		x		x	?	?		x	or older[†]	*MKT* III: 29–34
#13:		x		x		x		x		
#14,15:		x		x	x			x		
(17) YBC 5037	x			x				x	OB	*MCT*: 59–66
#2,3,5–9,11,12:	x			x	x			x		
#25,27,31,34–44:	x			x	x			x		
#1,4,10,13–24:	x			x		x		x		
#26,28–30,32,33:	x			x		x		x		
(19) YBC 8588	x		x		x			x	OB	*MCT*: 75–76

#3: The problem concerns a dam (*tamlâtu*), which is a triangular prism. It is in the shape of a ramp with a gradual slope (Thureau-Dangin 1935a: 7). The problem is to find the volume of a dam of given dimensions. The scribes compute by dividing by 2 rather than multiplying by 0;30, which is clearly an easier operation.

#4: The problem concerns a dam (*tamlâtu*); it is similar to problem #3 but the lateral planes of the triangular prism are not parallel. The problem is to find the volume of the dam. An incorrect formula is used to find the volume of the dam. The term ZA-ZUM or *sà-súm* is used to indicate lower width.

#6: The problem is to find the volume of a cylinder containing grain (Thureau-Dangin 1935a: 10). The coefficient 0;5, an approximation of $1/4\pi$, is applied in the calculations.

#7: The text is broken; the problem is to find the volume and grain capacity of a ^{giš}*sikkatu* 'peg, foundation or wall cone' of given dimensions. Thureau-Dangin (1935a: 11) suggested that ^{giš}*sikkatu* is a funnel used to form a pile of grain of given volume (cf. *CAD* s.v. *sikkatu*). The shape is not certain but may be a triangular prism; however, the measurements of the 'peg' in BM 85196 #7 are much greater than the pegs found in building walls (see Powell 1984a: 36 n. 17)

#8 (?): The text is broken; the problem is to find the volume of a trapezoidal cross-section according to the data and computations.

#10: The problem is to find the volume and grain capacity of a storage box (^{giš}*sussulu*), a quadrangular prism with trapezoidal bases. The calculations provide information of the relation between capacity (**sìla**) and volume (**kùš^{3}**, **šu-si^{3}**).

#17: The problem is to find the volume of a ramp (*arammu*) before a city gate. It is in the shape of a triangular prism whose lateral planes are not parallel. The volume is calculated by the formula:

$$V=\frac{1}{2}\left(\frac{a+b}{2}+\frac{a'+b'}{2}\right)\times h\times l$$

The problem is illustrated by a drawing of a twice partitioned trapezoid, which Vaiman (1960) described as a "symbolic" drawing with no clear perspective.

(3) BM 85210 #8
#8: The text is broken; the problem is to find the volume of **ki-gub** = *manzazu* "an emplacement, stand," etc. and also a mathematical term (*CAD* s.v. *manzazu*).

Thureau-Dangin (1932e: 138) suggested that the problem concerns a square from which a half circle has been cut; he did not translate the term **ki-gub** (*TMB*: 51). Neugebauer (*MKT* I: 230) viewed the figure as a truncated pyramid and treated **ki-gub** as a mathematical term. Bruins (1950: 374–75) proposed that **ki-gub** designates a change in a variable in order to simplify calculation. The text is replete with mathematical terms that are difficult to explain (see Thureau-Dangin 1932e: 138 and Neugebauer *MKT* I: 230–31).

(4) Böhl Coll. 1821

A new city encircles an old city thereby increasing the residential area. The problem is to find the circumferences of the new and old cities. The problem is illustrated by a drawing of concentric circles with inscribed numbers. Kilmer (1964: 143 n. 17) recognized that the scribe had made a mistake in the calculations by taking a short cut; however, the correct answer appears at the top of the tablet. The data are the same in BM 85194 #4, a problem dealing with the defense works of a city; but this time the computations are without error.

(5) Const. 236

The problem is to find the volume of a wall (**iz-zi** = *igāru*) of given dimensions.

(6) Erm 15073 #6, #7, #8
#6: The problem is to find the volume of a construction or excavation, which is a trapezoidal cross-section, and the problem is illustrated by a drawing of a trapezoid (Vaiman 1960: 379–80).
#7: The text is broken; the problem concerns the volume of a dike (*iku*), which is a trapezoidal cross-section. The problem is illustrated by a drawing of a trapezoid.

#8: The problem concerns the division of labor (in building an earth wall?). The nature of the problem to be solved is not entirely clear. The problem applies the coefficient of 0;3,45 **sar** in its calculations (rev. iii 6, 8, 9, 11, 14, 17, 18, 20, 21, 23). This coefficient, which is the volume of the daily work assignment for a laborer constructing a wall of rammed earth (*pitiqtu*), is found in other mathematical problems concerning *pitiqtu* and in the following coefficient lists: A 3553: 8; IM 49949: 13; IM 52301 rev. 21; *TMS*, Text 3: 43; YBC 5022: 17; and YBC 7243: 13.

(7) IM 53961

The problem is to find the work assignment (*iškaru*) of one man in building a wall of rammed earth (*pitiqtu*) of given dimensions. The text uses the term *rupšu* for width and *mēlû* for height. The coefficient of 0;3,45 **sar** is used in solving the problem.

(8) IM 54011

#1: The text is broken. The problem is to find the volume of a wall of rammed earth (*pitiqtu*) of given dimensions and the number of days needed to complete the job. The coefficient of 0;3,45 **sar** is used to solve the problem.

#2: The text is broken. The problem is to find the work assignment or output (*iškaru*) of each man in building a wall of rammed earth (*pitiqtu*) of given dimensions. The coefficient of 0;3,45 **sar** is used to solve the problem. The problem contains no drawing.

(9) *LBAT*: no. 1647 #1

#1: The text is broken; references to **gur**$_7$ = *karû* 'pile of barley (prepared for storage)' are preserved (*CAD* s.v. *karû*).

(10) Strssbg 362 #6

#6: The problem is to find the volume and height of a ramp before a city gate (*arammu*). 3 *šakkanakkū* 'governors' supervise a given length of the ramp and a given number of men. The number of workers assigned to each section forms an arithmetic progression.

(11) *TMS*, Text 14

The problem is to find the length, width and height (*qaqqadu*, literally 'head') of a pile of barley, which is prepared for storage (**gur**$_7$ = *karû*). Analysis of the text is difficult because different quantities have the same numerical value and all units are not indicated. The value 8 is introduced as grain capacity coefficient (*našpak karê*); however, the coefficient lists indicate 6,0 as the grain capacity coefficient, an empirical constant used to convert the volume in **sar** to units of capacity in **bariga** (see IM 49949: 9; IM 52301 rev. 21 and *TMS*, Text 3: 49). References to the value 6,0 occur in problems #1–5 in Haddad 104 and in AO 6484 obv. 23.

Bruins and Rutten (*TMS*: 85–88) suggested that a silo is constructed by digging a structure in the shape of a rectangular solid below the ground and by building a triangular base prism on top above the ground. The shape of the silo is determined according to the data and computations. However, the text does not indicate that the structure is built both below and above ground. The authors note that seals from Susa depict silos as cylindrical structures above the ground to which access is gained by ladder.

(12) *TMS*, Text 24. See II. Canals and Irrigation.

(13) *UET* 3: no. 1386

The text is broken; the problem concerns volumes of earth used in building a wall (**iz-zi** = *igāru*). The problem applies the coefficient of 0;3,45 **sar** in its calculations.

(14) YBC 4657

(15) YBC 4662 and

(16) YBC 4663. These three texts form a group of 31 problems concerning excavations of volumes of earth (**ki-lá** = *kalakku*). YBC 4657 contains the data and most of the answers to 31 problems, which are worked out in three other tablets of which two are preserved, YBC 4662 and YBC 4663.

Problems #1 to #30 concern an excavation in the shape of a rectangular base prism. The data is the same for this group. Work on the excavation includes the following parameters:

λ = 0;10 **sar** (the daily work assignment of one man or **éš-kàr**)
w = 6 **še** = 0;2 **gín** wages (**á**) per man per day paid in silver
t = 9 (duration of the work in number of days)
m = 30 (number of workers or **erim-ḫi-a**)
M = 4,30 number of man days or **erim-ḫi-a**)
E = 9 **gín** total expenses in silver

All these problems concern finding one or more of the following parameters: length (l), base (b), height (h), area of the base (g), volume (V), the daily work assignment of one man (λ), wages (w), time (t), number of workers (m), number of man days (M), or total expenses (E). The parameters are given are in various combinations. Linear equations are used to solve problems #9 to

#12, #15 and #16, and quadratic equations are use to solve problems #7, #8, #13, #14 and #17 to #20.

Problem #31 concerns an excavation in the shape of a square base prism. The problem is to find the area, volume, number of workers and total expenses in barley. The length of the side of the base, the height, the daily work assignment and the wages per man per day in barley (10 **sìla** = 1 **bán**) are given. Problem #31 may be the first problem of another tablet which follows.

cf. YBC 4669 #B1
#B1: The problem is to find the daily work assignment or output for a worker (**éš-kàr**) and the number of workers necessary to complete an unspecified task. The computations involve a daily work assignment per man of 10 **gín**, which is usually found in problems dealing with constructions and excavations of earthworks and canals. See XIV. Work.

(17) YBC 4673 #9–15
#9–12: All the problems concern a wall of rammed earth (*pitiqtu*) of given dimensions. The coefficient of 0;3,45 **sar** is used in solving the problems. Each time the problem is to find the work which one laborer completes in a day expressed in terms of the length of wall he constructs.
#13: The problem concerns a dike (**e** = *iku*) of given dimensions. Thureau-Dangin (1937d: 89; *TMB*: 212 n. 4) suggested the problem deals with a vertical cross-section of a dike in the shape of a right-angled isosceles triangle. The slope of the dike is 1 **kùš** per 1 **kùš** or 45°. The problem is to find the volume of earth heaped up (*tabāku*) for a length of ½ **ninda** to construct the dike.
#14 and #15: Both problems concern an old dike (**e libir-ra**) of given dimensions which is increased in both the width and height. The problem is to find the volume of earth heaped up for a length of ½ **ninda**, the volume of earth added to the old dike and the original volume of the old dike.

(18) YBC 5037

All the forty-four problems of this tablet concern excavated volumes of earth (**ki-lá** = *kalakku*).
Problems #1 to #3 concern an excavation in the shape of a square prism. The data are the same for this group.
Problems #4 to #9 concern an excavation in the shape of a rectangular prism. The data are the same for this group.
Problems #10 to #34 concern an excavation in the shape of a rectangular prism. The data are the same for this group.

Problems #35 to #44 concern an excavation in the shape of a square truncated pyramid. The volume is calculated by an approximative formula, $V=\frac{1}{2}(x_l^2+x_u^2)z$ =1;5 **sar**, where x is the length (**uš**), y the width (**sag**), z the depth (**bùr**), x_l the lower square side (**ib-si$_8$ ki-ta**) and x_u the upper square side (**ib-si$_8$ an-ta**) of the truncated square pyramid.

In all the examples the wages (w) are 6 **še** or 0;2 **gín** silver; the daily work assignment or output for a laborer (λ) is 7;30 **gín** in problems #1 to #3 and 10 **gín** in problems #4 to #34. Wages (w) and expenses (E) are not given in problems #35 to #44.

All these problems concern finding one or more of the following parameters: length (x), width (y), height (z), area (x^2 or xy), volume (V), the daily work assignment (λ), number of workers (m), wages (w), or total expenses (E). The parameters are given in various combinations. Problems #35 to #44 deal with the lower square side (x_l) and the upper square side (x_u) instead of length (x) and width (y). Problems are often solved by using quadratic equations (problems #8, #9, #11–34, and #39–44).

(19) YBC 8588

The problem is to find the depth of an excavation of a given volume of earth (**ki-lá** = *kalakku*). The text indicates 'in one excavation there are nine excavations (of equal size?)' (*i-na iš-te-en ka-la-ak-ki-im* 9 *ka-la-ak-ku*) (*MCT*: 65–66, 76 and n. 179a; *CAD* s.v. *kalakku*). Despite these enigmatic lines, the calculations are clear and the results are checked in the text. The daily work assignment or output of one man (**éš-kàr**) is 0;15 **sar**.

SUMMARY The mathematical texts include a variety of topics concerning constructions, excavations and earthworks as follows:

(1) *ālu* (= **uru**[ki]) 'city' (Böhl Coll. 1821)
(2) *appum* 'spur of land, causeway' (BM 85196 #1)
(3) *arammu* 'ramp which gives access to a city' (BM 85194 #1 and BM 85196 #17)
(4) *dūru* (= **bàd**) 'city wall' (BM 85194 #5, #17 and #18)
(5) *ḫirītu* 'moat, ditch' (BM 85194 #28)
(6) *igāru* (= **iz-zi**) 'wall' (Const. 236 and *UET* 3: no. 1386)
(7) *iku* (= **e**) 'dike' or **e-libir-ra** 'old dike' (Erm 15073 #6, #7 and YBC 4673 #13, #14)
(8) *kalakku* (= **ki-lá**) 'excavation of a volume of earth' (YBC 4657, YBC 4662, YBC 4663, YBC 5037 and YBC 8588)
(9) *karû* (= **gur$_7$**) 'pile of barley (prepared for storage)' (*LBAT* no. 1647 #1 and *TMS*, Text 14)

(10) *kikurru* 'chapel or cult socle' (BM 85194 #2, #3)
(11) *kisû* (= **kí-sa**) 'abutment or platform' (BM 85196 #1)
(12) *manzazu* (= **ki-gub**) 'emplacement or stand' (BM 85210 #8)
(13) *pitiqtu* (= **im-dù-a**) 'wall of rammed earth' (IM 53961, IM 54011 and YBC 4673 #9–12)
(14) *sikkatu* 'peg, foundation or wall cone' (BM 85196 #7)
(15) *sussulu* 'storage box or trough' (BM 85196 #10)
(16) *tamlâtu* 'dam' (BM 85196 #3, #4)
(17) miscellaneous structures such as a cylinder containing grain (BM 85196 #6) and a trapezoidal cross-section (BM 58196 #8 and Erm 15073 #6).

The measurements in the mathematical problems may describe a part or a whole of each construction or excavation. All are solid geometric shapes according to the data and computations. The problems are sometimes accompanied by drawings of planar figures. The daily work assignment for a laborer is usually 0;10 **sar** (see Foxvog 1986: 59–75); however, the daily work assignment for a laborer constructing an earth wall is 0;3,45 **sar**. The wages are 6 **še** silver or 10 **sìla** of barley. The problems concerning construction and excavation of earthworks are usually of simple mathematical character, similar to the mathematical problems concerning construction and excavation of canals.

V. *maḫīru* 'PRICE EQUIVALENCY'

DISCUSSION These texts deal with the quantity of goods that can be bought for a unit weight of silver or the weight of silver necessary to purchase a unit quantity of goods. *maḫīrum* meaning 'the equal, the equivalent of' is used in commercial transactions to indicate the concept of price.

(1) Haddad 104 #6
#6: Friberg suggested to the author that this is a price equivalency problem though the word *maḫīrum* 'price equivalency' does not occur. He interpreted the problem as follows: 2 **gur** grain is obtained when equal numbers of containers (*parsikātum*) are sold at different prices, costing 1 *parsiktum*, 5 **bán** and 4 **bán**, respectively; the problem is to find the price equivalent of each container.

Al-Rawi and Roaf were less successful in understanding the problem, which they interpreted to mean 2 **gur** of grain is obtained by filling three containers of different sizes (1 *parsiktum*, 5 **bán** and 4 **bán**). The problem is to find the amount of grain each container holds.

The verb *pašāru* used in Haddad 104 #6 (rev. iii 33, 36–38), MLC 1842: 7 and *TMS*, Text 13: 4, is also found in VAT 6469 obv. i 5 and VAT 6546 obv. ii 7 and may link the latter two texts with *maḫīru* texts; however, the

TABLETS	SUBCATEGORY ONE TYPE	SUBCATEGORY MIXED	SOLUTION WORKED YES	SOLUTION WORKED NO	ANSWER GIVEN YES	ANSWER GIVEN NO	DRAWING YES	DRAWING NO	DATE	REFERENCE
(1) Haddad 104 #6:		x	x		x			x	OB	Al-Rawi and Roaf 1984
(2) IM 54464	x		x		x			x	OB	Baqir 1951: 43–44: no. 9
(3) MLC 1842	x		x		x			x	OB	*MCT*: 106–07
(4) Strssbg 362 #2:		x	?	?	x			x	OB	*MKT* I: 239–43
(5) *TMS*, Text 13	x		x		x			x	OB	*TMS*: 82–83
(6) VAT 7530	x			x		x		x	OB	*MKT* I: 287–89
(7) YBC 4698 #14–16:		x		x		x		x	Kassite or older†	*MKT* I: 513; *MKT* III: 42–45

fragmentary state of VAT 6469 and VAT 6546 has led us to classify them as "commerce" (see below). The meaning of *pašāru* as 'to sell' is clarified by the contrasting parallel term *šâmu* 'to buy' in MLC 1842: obv. 6–7 **ganba** *a-ša-am ù ki-ia ap-šu-ur* '(At what) price equivalent did I buy and how much did I sell for?'; and in *TMS*, Text 13: 3–4 *ki ma-ṣí a-ša-am u ki ma-ṣí ap-šu-ur* 'How much did I buy and how much did I sell?'

(2) IM 54464

The problem is to find the price equivalency between 1 shekel of silver and given amounts of lard (*naḫû*) and oil (*šamnu*).

(3) MLC 1842

The cost of 30 **gur** of barley is calculated when the price rises and falls. The problem is to find the price equivalency and also something having to do with selling ($\sqrt{}$*pašāru*).

(4) Strssbg 362 #2

#2: The text is broken; references to 4 **bán**, which is described as the price equivalent, is preserved.

(5) *TMS*, Text 13

2 **gur** 2 **bariga** 50 **sìla** of oil is purchased for 1 **gín** (shekel) of silver. When selling, 4 **sìla** of oil is deducted, resulting in a profit of ⅔ **ma-na**. The

problem is to find the price at which the merchant buys and the price at which he sells. The problem is solved by using a quadratic equation.

(6) VAT 7530

All six problems of this tablet deal with four or more kinds of goods. The problem each time is to find the respective prices of the same number of **ma-na** of each of four or more different goods. Neugebauer and Sachs (*MCT*: 18–19) found a constant difference of 0;0,1,40 **gín** or 1/12 **še** between the price equivalency given in the text and the price equivalency calculated from the data. They suggested that this small difference in magnitude is below the lowest fraction of a **še** considered a reasonable metrological unit and consequently is disregarded in the problems of this text.

Thureau-Dangin (1936b: 162–63) in analyzing problem #3 noted that the four kinds of goods (A, B, C and D) involve the numbers 7, 11, 13 and 14 **ma-na** so that a number exactly divisible by 7, 11 and 13 is required in order to get a finite sexagesimal number as the answer. YBC 11125 and YBC 7353 (MCT: 17–19), reciprocal tables involving operations with the numbers 7, 11, 13 and 14 would have been useful in solving the price equivalency problems of this text.

However, Powell (1978: 220–22) suggested that the solution to this problem is similar to the silver coil problem in BM 85196 #18.

(7) YBC 4698 #14–16

#14: What the problem asks to be solved is not clear. The problem deals with the prices of nine commodities. The problem contains an enumeration of 1, 2, 3, ...9 similar to the counting of 1, 2, 3...10 **ma-na** in VAT 7530 problem #5. Friberg has suggested to the author that the numbers 1 to 9 really represent sixties.

#15: The text is broken. References to the price of (baskets of?) fish are preserved. Salonen (1966: 278, 290; 1970: 80–82) showed fish were measured with standard measuring containers.

#16: The problem concerns the prices of lead and silver. What the problem asks to be solved is not clear.

SUMMARY The mathematical problems dealing with price equivalency include commodities such as barley (MLC 1842 and Haddad 104 #6), lard and oil (IM 54464 and *TMS*, Text 13), fish (YBC 4698 #15), silver and lead (YBC 4698 #16) and unspecified goods (YBC 4698 #14 and VAT 7530). In the OB period the following ratios between units apply: 1 **gur** = 5 **pi**, 1 **pi** = 6 **bán**, 1 **bán** = 10 **sìla** and 1 **ma-na** = 60 **gín** and 1 **gín** = 180 še.

In MLC 1842 two quantities of barley are bought and sold at different market prices, namely, 1 **gur** of barley at 0;7,30 **ma-na** of silver and 1 **gur** of barley at 1 **ma-na** of silver. However, OB economic texts show prices of 1 **gur** of barley for either 1 **gín** or 2 **gín** of silver (Schwenzer 1915: 102; for a price range of 150–180 **sìla** of barley for 1 **gín** of silver, see Meissner 1937: 5). Haddad 104 #6 involves barley containers which are sold at different prices but the price equivalent is also in barley. The problem may refer to non-standard norms which do not rely on standard ratios linking systems of length, area, volume and weight. Non-standard norms may reflect either measuring practices or accounting practices which are difficult to explain (Powell 1990: 494).

In *TMS*, Text 13 the purchase price of 11 **sìla** of oil is 1 **gín** of silver, and the selling price of 7 **sìla** of oil is 1 **gín** of silver. In IM 54464 the price of 15 **sìla** of lard corresponds to that of 10 **sìla** of oil or 1 **gín** of silver. The price of oil in the OB economic texts is 20 or 30 **sìla** of oil for 1 **gín** of silver (Schwenzer 1915: 103; Meissner 1937: 8). The price of lard can only be corroborated by Ur III data at a median price of c. 17 **sìla** corresponding to 1 **gín** of silver (Snell 1982: 153-54); the median price of oil for the Ur III period is c. 12 **sìla** corresponding to 1 **gín** of silver (Snell 1982: 151-52). Thus, the correlation between prices quoted in the mathematical tablets and those in administrative and economic texts of the OB period is difficult to verify. Studies have shown that price data from all periods are lacking and rarely show trends (Snell 1982: 6-10). The mathematical problems teach the concept of price equivalency, useful in the accounting procedures of the ancient Near East (see Snell 1982), but without incorporating corresponding prices from the OB period.

Some mathematical problems provide exercises in price equivalency without specifying the commodities and prices. In YBC 4698 #14–#16 commodities are sold but their corresponding prices are unclear, even when the items are identified (in #15 fish for which the price and unit of measurmement are unclear and in #16 unspecified quantities of lead and silver equal to 7 **ma-na**). In VAT 7530 unidentified quantities are sold at unit equivalents of silver.

VI. INTEREST

DISCUSSION (1) AO 6770 #2

#2: The problem is to find the number of years for 1 **gur** of grain to double when loaned at the assumed rate of 20% interest for a 360 day calendar year. The answer is not completely worked out, and a linear interpolation with an interest table seems indicated. The answer is an interval between three and four years, with 2;33,20 months subtracted from four years.

(2) *TMS*, Text 22

The problem is to find the principal when a three year period is used to amortize a debt and payments are made with interest annually. The medium of

	SUBCATEGORY		SOLUTION		ANSWER					
	ONE		WORKED		GIVEN		DRAWING			
TABLETS	TYPE	MIXED	YES	NO	YES	NO	YES	NO	DATE	REFERENCE
(1) AO 6770 #2:		x	x		x			x	OB	*MKT* II: 37–42; *MKT* III: 62–65
(2) *TMS*, Text 22	x		x		x			x	OB	*TMS*: 111–14
(3) VAT 8521	x		x		x			x	OB	*MKT* I: 351–67
(4) VAT 8528	x		x		x			x	OB	*MKT* I: 353–67; *MKT* III: 59–61
(5) YBC 4669 #B11:		x		x	x			x	Kassite or older†	*MKT* I: 514–16; *MKT* III: 26–29
(6) YBC 4698 #1,2:		x		x		x		x	Kassite or older†	*MKT* I: 513; *MKT* III: 42–45

exchange is assumed to be silver. Bruins (1962: 289–91) suggested that the calculations in this problem yield a more accurate result than the linear interpolation in AO 6770 #2.

(3) VAT 8521

All four problems in this text concern silver which is loaned at the usual rate of 20% interest and repaid at the end of the year (Thureau-Dangin 1936d: 70). In each problem the calculated value is checked by performing the inverse operation(s).

#1: The problem is to find the principal when the rate of interest at the end of the year is a square or n^2.

#2: The problem is to find the principal when the rate of interest at the end of the year is a cube or n^3.

#3: The problem is to find the principal when the rate of interest at the end of the year is a square or n^2.

#4: The problem is to find the principal when the rate of interest at the end of the year is equal to $n \times n(n-1)$. Coincidentally, this formula is used for solids which are equal on two sides and slightly smaller on one side, called "diamonds" by Greek arithmeticians (see Thureau-Dangin 1936d: 74).

(4) VAT 8528

#1: 1 **gú** 4 **ma-na** silver represents principal plus interest for money loaned at a rate of 20% interest. The problem is to find the amount of time for which the money is loaned. In the second part of the problem the answer is checked by performing the inverse operation; that is, the problem is to find the amount accrued (principal plus interest) for 30 years at the rate of 20% interest.

The scribe calculates the interest for one year and notes that the capital doubles every 5 years, that is, *i-na* **mu 5-kam kù <-babbar>** *ù* **máš-bi** *it-ta-am-ḫa-ru* 'in 5 years the money and its interest are equal' (obv. i 12–13). He then uses a logarithmic formula for five years, then 25 years (or 5^2) and finally adds the results from the formulas for 5 years + 25 years = 30 years. In the proof logarithms are used to calculate the rate of interest for 1, 5, 10, 15, 20, 25, and 30 years; this long-winded method of computation implies that the scribe did not have a table of powers at his disposal (*MKT* III: 60). Thureau-Dangin showed that the term **ba-si** refers to a system of logarithms with base 2 in this text. Logarithmic tables would, of course, be useful in calculating interest (see Lewy 1947: 305–310).

#2: The problem is to find the amount of grain the creditor receives when 10 **gur** of barley is loaned at a rate of 12% interest. The creditor is repaid 1 **bán** barley each day for a year (of 360 days); this rate does not include interest.

Thureau-Dangin (1936d: 70–71) described the method of solution in this problem as a summary of a decreasing arithmetic progression comprised of six terms, the first being the interest in *sūtu* per year and the last, interest in *sūtu* per day. The solution is found by multiplying the number of terms by the average of the sum of the two extreme terms.

(5) YBC 4669 #B11

#B11: The problem is to find the principal when money is loaned at a rate of 20% interest and 1 **gín** is received at the end of 3 years.

(6) YBC 4698

#1: The problem is to find the rate of interest when 1 **bariga** is received as the annual interest on the principal of 1 **gur** of grain. The rate of interest is 20%.

#2: The problem is to find the rate of interest when 1 **bariga** 4 **bán** is received as the annual interest on the principal of 1 **gur**. The rate of interest is 33⅓%.

SUMMARY The following tablets illustrate problems dealing with the rate of interest or *ṣibtu*: AO 6770 #2, *TMS*, Text 22, VAT 8521, VAT 8528, YBC 4669 #B11 and YBC 4698 #1, #2. The term *ḫubullu*, also denoting 'the rate of interest', is found only once in the problem texts dealing with interest in VAT 8528 rev. i 10.

The rate of interest is generally 12 shekels per mina per year or 20%, as shown in AO 6770 #2, VAT 8521, VAT 8528 #1, YBC 4698 #1 and YBC 4669 #B11. Landsberger (1937) noted also a rate of 20% interest in Ai tablet ii 41–43.

The rate on borrowing grain is the customary 12% in VAT 8528 #2. However, in YBC 4698 #1 and #2 this rate of interest is 20% and 33⅓%, respectively.

A 360 day calendar year is used in calculating interest as in AO 6770 #2 and VAT 8528 #2 (*MKT* I: 360). Interest is paid at the end of the year.

The problems reflect both the use and non-use of "logarithmic" tables tables in calculating interest (see *MCT*: 36 and n. 96d). The interest problems in VAT 8521 and VAT 8528 form a bridge between the table texts and the problem texts in their method of solution.

VII. COMMERCE

DISCUSSION (1) VAT 6469

The text is broken; references to *nēmelu* 'commercial profit' are preserved.

(2) VAT 6546

The text is broken; references to *nēmelu* 'commercial profit' are preserved.

	SUBCATEGORY		SOLUTION		ANSWER					
	ONE		WORKED		GIVEN		DRAWING			
TABLETS	TYPE	MIXED	YES	NO	YES	NO	YES	NO	DATE	REFERENCE
(1) VAT 6469	x		x		?	?		x	OB	*MKT* I: 268–69
(2) VAT 6546	x		x		?	?		x	OB	*MKT* I: 268–69
(3) YBC 4698										
#3–5:		x		x		x		x	Kassite	*MKT* I: 513;
#6,8–11:		x		x	x			x	or older[†]	*MKT* III: 42–45
#7:		x		x		x		x		
#12,13:		x		x		x		x		

(3) YBC 4698 #3–13

Problems #3 to #13 contain a series of equations, which use the ideogram ANŠE of which the exact meaning is unknown. The term ANŠE cannot be identified with the measure *imeru* (which contains 100 **sìla**) inasmuch as this meaning does not elucidate the text. Neugebauer (*MKT* III: 43) argued that ANŠE may mean any weight similar to the general term for stone weight, **na**$_4$ = *abnu*. Powell suggested in private communication that ANŠE = **anše(-r)** = *amaru* 'pile'; in Ur

III texts the meaning is 'reed pile' and later 'brick pile' (that is, **sig$_4$-anše** = *amaru*). Both 'reed pile' and 'brick pile' are measured in linear and volume units, and transformation into capacity units still does not clarify the text. No satisfactory explanation has been found as yet to explain this complete series of equations dealing with the economics of commerce.
#3: One stone weight (**na$_4$**) of given volume is four times the given volume of another (NA$_4$-KA), resulting in 1 **gín** of silver. The problem is to find the **na$_4$** and NA$_4$-KA when the ANŠE is squared.

Thureau-Dangin (1937d: 89–90) read ANŠE as **ì-giš** = *šamnum* 'oil', thus seeing in problem 3 the ratio of first quality oil, *šamnum rēštûm*, to ordinary oil, *šamnum*, as 3:12, as in the Old Babylonian period. In problems #6 and #8 the ratio of grain to oil is 30:1. Furthermore, Thureau-Dangin's reading does not satisfy problems #4 to #13.
#4: 1,30 **dug** of iron and 9 **dug** of gold result in 1 **ma-na** silver. The problem is to find iron and gold when the ANŠE is 1 **gín**. **dug** is a dry measure of 20 or 30 **sìla**. The ratio of gold to iron is 10:1.

According to Schwenzer (1915: 31) gold was not a medium of exchange during the time of Hammurapi and was used only for ornamental ware. By working through the correspondences, Neugebauer (*MKT* III: 44) determined the ratio of gold to iron as 48:1 or 24:1. Bjorkman 1989 indicated that the problem does not give an early price for iron; in fact, there are no prices for iron before the first millennium B.C.
#5: The same data as problem #4 result in ½ **ma-na** silver when the ANŠE is 1 **gur**.

Problems #6 to #11 form a homogenous group containing the technical terminology of commerce. The problems concern a total amount of barley which is thirty times the estimated weight loss (ANŠE). The actual weight loss (**bal**) varies. The surplus silver (**kù-babbar dirig**) is determined by a complex set of operations among the total amount of barley, the estimated weight loss and the actual weight loss. The profit (**sum**) is the result of subtracting the actual weight loss from the estimated weight loss. Neugebauer (*MKT* III: 44) explained the terms and charts the values in these problems.
#6: The problem is to find the amount of surplus silver.
#7: The problem is to find the initial amount of silver.
#8: The problem is to find the amount of surplus silver.
#9: The problem is to find the estimated weight loss (ANŠE) and the actual weight loss (**bal**)
#10: The problem is to find the amount of surplus silver.
#11: The problem is to find the estimated weight loss (ANŠE) and the actual weight loss (**bal**).
#12: The problem is to find the ANŠE, which is squared.

#13: One stone weight (**na$_4$**) of given volume is half the given volume of another (NA$_4$-KA), resulting in 1 **gín** of silver. The problem is to find the **na$_4$** and NA$_4$-KA when the ANŠE is squared.

SUMMARY The problems dealing with commerce, that is, buying and selling, are either broken or fraught with textual difficulties, because they contain as "givens" terms and conventions of economics not understood. There are no drawings in these problems.

VIII. INHERITANCE OR DIVISION OF PROPERTY

DISCUSSION (1) AO 17264

Six brothers inherit a field in the shape of a trapezoid. The two oldest get equal shares, the third and fourth get equal shares as do the fifth and sixth. The problem is to find the length of each side of the trapezoid, which is partitioned into six proportional parts.

Gandz (1948: 23–30) analyzed this problem as a "composite trapezoid partition problem" based on chains of Babylonian triples or solutions to the indeterminate equations $n^2-q^2=q^2-m^2$. Thureau-Dangin (1934c: 66–69) suggested further that the problem supposes knowledge of Thales' theorem of similar triangles, which states that a line parallel to one of the sides of a triangle divides the other sides into proportional parts.

Caveing (1985: 21–23) proposed that the ancient scribe was able to invent the formula for finding the length of a transversal because he already knew the numerical solution.

(2) IM 31210

The tablet is a badly damaged text of six columns of which the first and last columns are almost completely destroyed. All eight problems deal with division of money, and some contain arithmetic progressions. Bruins (1971: 236–37) suggested that *meḫrum* 'equivalent' means 'false value' in the text and is used to solve the problems by the method of false position.

#1: The text is broken. The problem is to find the amount of each share when money is divided among six brothers; the shares form an arithmetic progression.

#2: The problem is to find the amount of each share when 1,10 **ma-na** is divided among seven brothers.

#3: The text is broken. The problem is to find the amount of each share when money is divided among three brothers.

TABLETS	SUBCATEGORY ONE TYPE	MIXED	SOLUTION WORKED YES	NO	ANSWER GIVEN YES	NO	DRAWING YES	NO	DATE	REFERENCE
(1) AO 17264	x		x		x			x	MB, Kassite	*MKT* I: 126–34
(2) IM 31210	x		x		x			x	OB	Bruins 1954: 57–61
(3) K 8705	x		x		?	?		x	Late Assyrian	*MCT*: 140
(4) K 11097	x		x		?	?		x	Late Assyrian	*MCT*: 139–40
(5) *LBAT*: no. 1648										
#1:		x	x		x			x	LB	Sachs 1955:
#2:		x	x		x			x		no. 1648
(6) Strssbg 362 #1:		x	x		x			x	OB	*MKT* I: 239–42
(7) *TMS*, Text 10	x		x		x			x	OB	*TMS*: 69–74
(8) *TMS*, Text 26	x		x		x			x	OB	*TMS*: 124–29
(9) *UET* 5: no. 121 #1:		x		x	x			x	OB	Vaiman 1961: 248–50
(10) VAT 6597	x		x		x			x	OB	*MKT* I: 274–77
(11) VAT 7531	x			x		x		x	OB	*MKT* I: 289–94
(12) VAT 7621										
#1:	x			x		x	x		OB	*MKT* I: 290–94
#2:	x			x		x		x		
(13) VAT 8522 #2:		x	x		x			x	OB	*MKT* I: 367–73
(14) YBC 4608 #5:		x	x		x			x	OB	*MCT*: 49–53
(15) YBC 9856 #2:		x		x	x			x	OB	*MCT*: 99–100

#4: The problem is to find the amount of each share when 1,31;40 **ma-na** silver is divided among 10 brothers.

#5: The problem is to find the amount of each share when a unit-quantity of silver is divided among eight brothers. The oldest gets twice as much as the second and third, and the remaining five brothers get equal shares.

#6: The problem is to find the amount of each share when 1⅔ **ma-na** is divided among ten brothers
#7: The problem is to find the amount of each share when 1½ **ma-na** is divided among four brothers; the solution is not possible in finite sexagesimal fractions.
#8: The text is broken, and the problem can not be reconstructed.

(3) K 8705

The text is part of a collection of tablets with instructions and calculations concerning the lore of the *bārû*-priests, who examine sheep livers. The texts are often characterized by the phrase **dub ḫa-la** 'tablet of inheritance' (or, 'tablet concerning a dividing line'), found in the introduction, catchline or subscript. The term *ba-ru-ti* 'act of divination, craft of the diviner' appears in line 2. **udu-nitá** (= *immeru*) 'rams' appears in lines 8, 9, 10 and 15.

The text is broken, but the terminology of the sheep's liver omens can be restored from the duplicate text, K 8865 (Borger 1957: 190–91, 193–95). Enigmatic mathematical calculations occur in both the text and its duplicate; however, the computations may involve a kind of numerical determinism for formulating the division of property, as indicated by the word *zittu* and the terminology of inheritance in complementary texts such as K 11097.

(4) K 11097

The text is part of a collection of tablets with instructions and calculations concerning the lore of the *bārû*-priests, who examine sheep livers.

The text is broken; references to an inheritance share (**ḫa-la** = *zittu*) are preserved in lines 1 and 4. The phrases, *abu* (= **ad**) *ana māri-šú ša i-ram-mu* 'the father to the son whom he loves...' in line 3 and *aḫu* (= **šeš**) *ana aḫi* (= **šeš**) '(one) brother to the (other) brother...' in line 6, confirm that the problem deals with division of property. The terminology of liver divination can be restored from similar texts (see Borger 1957: 192).

(5) *LBAT*: no. 1648 #1, #2
#1: The text is broken; references to the division of money are preserved.
#2: The text is broken; references to the division of money among five brothers are preserved; the shares may form an arithmetic progression.

(6) Strssbg 362 #1
#1: The problem is to find the difference between each share when 1⅔ **ma-na** silver is divided among ten brothers so that the shares form an arithmetic progression.

(7) *TMS*, Text 10

The problem concerns dividing 1 **ma-na** silver between two brothers; the text contains different methods of solution to the same problem. To simplify computation, irregular numbers are removed from the denominators.

(8) *TMS*, Text 26 #1–3

The text of problems #1 to #3 is broken; in each case, the problem concerns a trapezoidal field which is divided equally between two brothers. In problem #1 the Pythagorean theorem is used in the solution.

(9) *UET* 5: no. 121 #1

#1: The problem is to find the amount of each share when 26 **ma-na** 15⅔ **gín** 15 **še** silver is divided among five brothers; the shares form an arithmetic progression.

(10) VAT 6597

#1: The text is broken; references to division of money or property are preserved.
#2: The text is broken. The problem is to find the amount of each share when money is divided among three brothers. The shares form an arithmetic progression.
#3: The problem is to find the amount of each share when money is divided among four brothers. The shares form an arithmetic progression.
#4: The problem is to find the amount of each share when 1 **ma-na** is divided among three brothers. The purpose of the diagram at the beginning of the problem is unclear
#5: The text is broken; references to the division of money among three brothers are preserved.

(11) VAT 7531

All four problems concern the allocation of fiefs to soldiers (**uku-ús** = *rēdû*). Each field is in the shape of a trapezoid (see Vaiman 1961: 258–63). The measurement of each side of the trapezoid is given. The property, 'as far as his field extends' (**a-šà-bi** *ma-la ma-ṣú*), is to be divided among a given number of people. Each problem concludes with the phrase **uku-ús** *sí-ik-ka-as-su ku-ul-li-im-šu* 'and show each soldier his stake' (see Thureau-Dangin 1924a: 34–35: no. 41). *kullumu* (with *sikkatu*) 'to show someone the peg driven in a field' denotes the legal act accompanying the transfer of a field (*CAD* s.v. *kullumu*).
#1: The problem concerns a field in the shape of a trapezoid, which is divided equally among three brothers; each soldier is shown his stake.

#2: The problem concerns a field in the shape of a trapezoid, which is divided among 300 people (**erim-ḫi-a**), of whom 120 people get equal shares; each soldier is shown his stake.
#3: The problem concerns a field in the shape of a trapezoid, which is divided equally among 1000+ people (**erim-ḫi-a**); each soldier is shown his stake.
#4: The problem concerns a field in the shape of a trapezoid, which is divided equally among five brothers; each soldier is shown his stake.

(12) VAT 7621

Both problems #1 and #2 concern the allocation of fiefs to soldiers (**uku-ús** = *rēdû*).
#1: The problem concerns a field in the shape of a trapezoid, which is illustrated by a drawing of a partitioned trapezoid. The two parts of the field are divided among nine male children (*māru*); each soldier is shown his stake.
#2: The text is broken. The problem concerns the division of property equally among three people (**erim-ḫi-a**), later qualified by the term male children (*māru*). Each soldier is shown his stake.

(13) VAT 8522 #2
#2: The problem concerns the division of 59 **ma-na** silver among five brothers (**šeš-meš**).

(14) YBC 4608 #5
#5: Six brothers inherit a field in the shape of a triangle of given height and area. The area is divided by equidistant lines parallel to the base of the triangle. The problem is to find the difference between the shares.

(15) YBC 9856 #2
#2: The problem is to find the amount of each share when 1 **ma-na** silver is divided among five brothers. The shares form an arithmetic progression.

SUMMARY There is some third millennium evidence for patrilinear descent, but by the OB period the family was clearly patrilinear and the line of descent male. Accordingly, a son and heir was of great significance, both to support his father in old age and to appease the spirits after the father's death. For childless couples, adoption was an option, or the husband could recognize the children of one of his slave-girls (CH §170). The wife could also avoid divorce by providing her own slave-girl to be her husband's concubine, with whom he could father children, who would then be regarded as the wife's own (Postgate 1992: 92).

In the patrilinear system, property was divided between the sons or each surviving male line, because the male offspring of a dead brother would also inherit (Postgate 1992: 96–97). The size of shares varied according to time and places in the ancient Near East. In northern Old Babylonian practice (particularly Sippar) all sons shared equally. In central and southern Babylonia and Mari, the eldest received a larger share, sometimes even a double share (for example, in Larsa, Kutalla and Mari). In the Middle Babylonian period in Nippur and in the southern Babylonian cities of Ur and Kutalla, the eldest received an additional share of 10% of the total estate. The eldest might even be given the preference of choosing his share, while the other brothers drew lots (Kraus 1969: 10; Klíma 1940: 21–33; Postgate 1992: 98).

In Old Babylonian-period documents, brothers divided the estate after the death of the *pater familias*. The estate could include land, house, furniture, animals and slaves, temple offices (prebends), debts and *ilkum*-duties (Postgate 1992: 96). The eldest son inherited temple offices and usually the family house, most probably because the house was connected with his duties to the ancestral tombs (often buried beneath the house) and the family shrine. If the house were large enough, it was sometimes divided—with appropriate alterations to room use and means of access—in order to accommodate the brothers separately. Many OB house-sale documents are actually "paper transactions" to compensate heirs for differences in size in the part(s) of the house inherited (Postgate 1992: 98–99).

The practice of all the sons inheriting *per stirpes* (CH §165) could be modified by the head of the household during his lifetime by giving separate gifts, which were not included in the division of estate. According to this arrangement, daughters were allotted an appropriate dowry by the father or by the sons if the father did not fulfill this obligation prior to his death (CH §166). If the girl died or left her husband's family without sons, this property reverted to the patrimonial estate (CH §163–164). The daughter's sons belonged to another patrimonial group (CH §167). In the case of a priestess, her brothers were obligated to support her; upon her death they recovered the property, unless she had given the property to adopted daughters (Postgate 1992: 97–98).

Wills from Emar and Nuzi reflect the practice of investing daughter(s) with the status of sons, whereby the testator could will his entire property and his gods to them. Even a son-in-law could be adopted in order to attach the children to the lineage; this practice was an attractive proposition for a son-in-law who might receive a smaller share of his own father's inheritance. Perhaps men cited with matronymics rather than patronymics were the children of women viewed legally as men (Grosz 1986: 81–84).

In the mathematical problems dealing with inheritance and division of property, money and fields are apportioned. The size of the shares varies, reflecting equal division of property as in *TMS*, Text 26 #1–3, or unequal division based on various mathematical principles. For example, the shares may form an arithmetic progression as in IM 31210 #1, Strssbg 362 #1, *UET* 5: no. 121 #1,

VAT 6597 #2 and YBC 9856. In IM 31210 #5 the oldest gets twice as much as the next two brothers, etc., and in AO 17264 the oldest two get larger shares with other divisions for shares of the next four brothers. Gandz (1938: 322) maintained that an absolute, independent algebra, which is "at liberty to disregard and ignore the rules of inheritance established by law and religion" cannot exist. Unfortunately, since most of the mathematical tablets were purchased and their provenance cannot be established, the mathematical problems neither complement nor corroborate the provisions of the inheritance law based on geographic comparisons.

When money is divided, the mathematical problems deal with amounts adding up to 1 **ma-na** silver in the following texts: IM 31210 #5, *TMS*, Text 10, VAT 6597 #4 and YBC 9856. Amounts greater than 1 **ma-na** but less than 2 **ma-na** are divided in IM 31210 #2, #4, #6, #7 and in Strssbg 362 #1.

Sometimes the property to be divided is described as a partitioned trapezoid, namely in the mathematical texts, AO 17264, *TMS*, Text 26, VAT 7621 #1 (which also includes a drawing) and YBC 4608 #5 (on the shapes of fields, see X. FIELDS AND FIELD PLANS).

No documents recording land sales are extant from the Ur III period, and later South Babylonian field sales are rare. Field leases from Old Babylonian Ur confirm the joint ownership of the extended family, with the average group of holders being 8 men with full property rights. In one inheritance text from Nippur, the division of a city plot between brothers was recorded but actually carried out a generation later (Postgate 1992: 95–96).

Generally, in the mathematical problems, two to ten people receive shares. The smallness of the inheritance often makes property division seem theoretical, but administrative texts of the period confirm this information (see Kraus 1969: 7).

Another form of property division, allocation of state land, is represented in VAT 7531 and VAT 7621; in both problems each soldier is shown the boundaries of his stake (see Dunham 1986: 57 n. 115; 59; Thureau-Dangin 1924a: 3–5). These texts refer to the well-known institution of *ilkum*, whereby military service was rewarded with state land (see CH §26–40). This duty could be re-assigned to the sons of the original holder, who inherited the rights to the land at the same time (CH §28). As with other prebend land, *ilkum* lands could be rented out and even sold (see CH §36–37).

Finally, K 8705 and K 11097 belong to a group of tablets called *niṣirti bārûti* 'the secret knowledge of *bārû*-divination'. The group begins with the phrase **be dub ḫa-la** *niṣirti* [lú]*bārûti* 'if a tablet of inheritance, (representing) the secret lore of the *bārû*-priest' (Borger 1957: 190). *niṣirtu*, a synonym for *awatum* 'word', is a term designating an oral omen result; it rarely occurs in OB compendia but is frequently used in NA (Jeyes 1989: 16–18). In the terminology of the liver, the interpretative system includes symbolism in which people, institutions (here, inheritance or division of property), aspects of human life, events and even abstract ideas were represented (Jeyes 1989: 96). The combination of hepatoscopic terminology with numerical equations evokes Landsberger's expression, "liver mathematics" (Reiner: 1985: 592). Like the mathematical texts from Mari, the *niṣirti bārûti* tablets form a bridge between pure mathematical problem texts and economic/administrative tablets.

IX. CLEPSYDRA

DISCUSSION (1) BM 85194 #6, #7, #8, #34, #35

Problems #6 to #8 concern the volume of water discharged from a clepsydra ([giš]**dib-dib** = *dibdibbu*) which is filled to a marked line. Each of these problems concerns the finding of one of the following parameters:

A = the volume of water measured in **sìla** which is discharged from a marked line on a clepsydra

B = 1 **sìla** of water which corresponds to a given height on the clepsydra

i = the amount (measured in **šu-si**) which the water level sinks when 1 **sìla** of water is discharged

h = the complete height of the water clock

α = the difference in height between the old water level level and the new level as the water is discharged

In these problems the following relations hold:

$$\alpha = \frac{1}{h} \times A \times i$$

$$\alpha = \frac{Ai}{Bh}$$

#6: The problem is to find the difference in height between the old water level and the new level as water runs out (α).
#7: The problem is find the amount the water level sinks when 1 **sìla** is discharged (i). The data are the same as problem #6.
#8: The problem is to find the difference in height between the old water level and the new level as water runs out (α).
#34 and #35: The text of problems #34 and #35 is broken. Both problems concern a *pisannu*-container. The calculations demonstrate that these examples are related to the clepsydra problems. The problems are solved by using the

circumference coefficient 0;5, which is an approximation of $1/4\pi$ (see Kilmer 1960: 284; 1985). A cylinder appears to be the assumed shape.
Problem #34 concerns finding the difference in height between the old water level and the new level for a container with a circumference of 3 **šu-si**.
Problem #35 concerns finding the height of a 1 **sìla** container with a circumference of 10 **šu-si**.

TABLETS	SUBCATEGORY ONE TYPE	MIXED	SOLUTION WORKED YES	NO	ANSWER GIVEN YES	NO	DRAWING YES	NO	DATE	REFERENCE
(1) BM 85194										
#6–8:		x	x		x			x	Kassite	*MKT* I: 142–93
#34,35:		x	x		x			x	or older†	
(2) BM 85210 #13:		x	x		x			x	Kassite or older†	*MKT* I: 219–33

(2) BM 85210 #13
#13: The problem concerns finding the difference in height between the old water level and the new level as water runs out (α). See BM 85194.

SUMMARY All information on water clocks is indirect. There are no drawings of a clepsydra in the mathematical problems, and the shape of the water clock remains in dispute. Thureau-Dangin (1933a: 51) suggested that in both Babylonia and in Egypt the water clock was cylindrical. Neugebauer (1947: 39–40) demonstrated that in the mathematical texts the water clock was either of a cylindrical or prismatic shape, and time was computed by measuring the flow of water through a hole in the bottom of a water clock of a given height. Professor A. Aaboe noted in our discussion of this issue that Neugebauer later rejected these shapes as impractical without giving an alternative. Smith (1969: 77) argued that, based on the mathematical problems, the Babylonian water clock was a rectangular or cylindrical tank. Water clocks could not have been conical vessels, because calculations assume that the volume of the vessel is proportional to the height (Neugebauer 1947: 39). The problems in the mathematical texts do not present a clear picture because the calculations deal with the proportion between a given volume and height and not the shape of the vessels.

The clepsydra problems involve filling a vessel with water to a marked line or reference point and then letting a certain quantity of liquid escape. The Babylonians did not measure time by the difference in water levels within the clepsydra but rather by the volume (or weight) of the water discharged and received in a graduated vase as in BM 85194 #34, #35 (Thureau-Dangin 1932e:

135; 1933a; 1937a; Neugebauer 1947: 39). In the mathematical texts the volume of the water flow is measured in **sìla**, and in the astronomical compendium [mul]APIN, the water flow is measured by its weight in **ma-na**.

To date, only one extant artifact from the ancient Near East, a bowl, has been identified as connected with the measurement of time using water: in the Nimrud collection, Smith (1969: 77) identified a bowl with a central hole in the base, similar to European clock bowls, which offer a more practical method for measuring small time intervals than a tank type water clock. When placed on the surface of water in another receptacle, these bowls sink in a very short time, due to the weight of the bowl and the size of the hole; variations of atmospheric pressure, etc., are negligible for practical purposes. A set of such bowls can measure time intervals from under five minutes to a minute or slightly less. Smith suggested that these bowls were used in conjunction with the water clock to measure the short and long portions, respectively, of a watch (*maṣṣartu*).

X. FIELDS AND FIELD PLANS

DISCUSSION (1) AO 6484 #5, #7

The area of the fields (A) is computed in units of square cubits (**kùš**^{2}). The final value obtained for the area is multiplied by a factor 0;0,0,21,36 (c), which is used to convert areas measured in **kùš**^{2} into areas measured in units of (100 **kùš**)^2. The transformation to **še** units of barley or a secondary measurement is obtained by multiplying the value cA by a factor 1,48,0,0 (*MCT*: 143–44).

The convention of expressing the area of a field in (corresponding) quantities of seed points to agricultural practices for its origin (see Bruins 1962: 298; Nemet-Nejat 1982: 3–4). Powell (1984a: 40–41, 64–66; 1990: 483) demonstrated that the system of seed measure is based on the assumption of 120 furrows laid out in 20 NB-LB finger (**šu-si**) intervals in a square measuring 100 cubits (**kùš**) on each side, such that equal amounts of seed are used for planting the area and feeding the oxen.

#5: The problem is to find the barley measurement for a square area. The term for field (**a-šá** = *eqlu*) is not mentioned.

#7: The problem is to find the barley measurement for a triangular area. The term for field (**a-šá** = *eqlu*) is not mentioned. The Pythagorean theorem is used to calculate the altitude and the area of the triangle.

Bruins (1962: 299) showed that the field in this example is in the shape of an isosceles triangle; the altitude bisects the base of this isosceles triangle to form two 3-4-5 rational right angle triangles, which, he argued, can hardly be considered a "common occurrence" in agriculture.

TABLETS	SUBCATEGORY ONE TYPE	SUBCATEGORY MIXED	SOLUTION WORKED YES	SOLUTION WORKED NO	ANSWER GIVEN YES	ANSWER GIVEN NO	DRAWING YES	DRAWING NO	DATE	REFERENCE
(1) AO 6484 #5,7:		x	x		x			x	Seleucid	*MKT* I: 96–107
(2) AO 6770 #5:		x	x			x		x	OB	*MKT* II: 37–42; *MKT* III: 62–65
(3) BM 85194 #9,27:		x	x		x			x	Kassite or older[†]	*MKT* I: 142–93
(4) CUA 34	x			x	x			x	OB	Goetze 1957: 40: no. 33
(5) IM 53965	x		x		x			x	OB	Baqir 1951: 39–40: no. 7
(6) IM 54010	x		x		x			x	OB	Baqir 1951: 38–39: no. 6
(7) Strssbg 362 #5:		x		x	x			x	OB	*MKT* I: 239–42
(8) Strssbg 366 #4:		x	x		x			x	OB	*MKT* I: 257–59
(9) Strssbg 368	x		x		x			x	OB	*MKT* I: 311–14
(10) *UET* 5: no. 864	x		x		x			x	OB	Kilmer 1964: 140–42
(11) VAT 7532	x		x		x		x		OB	*MKT* I: 294–303
(12) VAT 7535										
#1,2:	x		x		x		x		OB	*MKT* I: 303–10
#3:	x			x		x	x			
(13) VAT 7848										
#1,3,4,6,7:	x		x		x			x	Seleucid	*MCT*: 141–45
(14) VAT 8389	x		x		x			x	OB	*MKT* I: 317–35
(15) VAT 8391	x		x		x			x	OB	*MKT* I: 319–35

Triangular fields are common in Mesopotamia and are depicted in field and estate plans from the Neo-Sumerian period to the Late-Babylonian period (Nemet-Nejat 1982: 14). In Late Babylonian field plans, triangular fields are even mentioned by name, that is, **sag-dù** = *santakku* 'triangle' (Nemet-Nejat 1982:

285). According to the calculations in the field plans, the triangular fields are treated as right triangles (for example, Liverani 1990: 150–54: nos. 7, 9, 10, 11, 14, 22, 28, 25, 26, 27, 16, 20; Owen 1972: no. 1; Stephens 1953: 3; Scheil 1895: 33; Nemet-Nejat 1982: Text 69 obv. plan #4); triangular fields that are not right triangles give the altitude (for example, Liverani 1990: 150–54: nos. 7, 26; Owen 1972: no. 1; Scheil 1902: no. 718; Nemet-Nejat 1982: Text 59 plan #5, Text 69 rev. plan #4, Text 70 plan #6).

(2) AO 6770 #5
#5: The problem is to find the original length of a (broken) reed (**gi** = *qanû*) which is used to measure an area. The problem is solved by using a quadratic equation. The term for field (**a-šá** = *eqlu*) is not mentioned.

(3) BM 85194 #9, #27
#9: The problem is to find the area of a building plot (*ašbatu*) which is in the shape of a square. The steps used to calculate the solution are not clear.
#27: The problem is to find the amount of grain needed to seed an area of 1 **bùr** when the ratio of seed to land is defined as 4 furrows per ½ **ninda**. Eight to twelve furrows per **ninda** represents the range of sowing rates in traditional Ur III practices (Powell 1984a: 46–54, 63–64; 1985: 30–33). The text is a harvest problem, which may concern the sale of a field or crop. The calculations result in an unrealistic 120-fold harvest; however, the firm position of the eight furrow ratio is demonstrated in applied mathematics (Powell 1972: 180–84).

(4) CUA 34
obv.: An area measurement is given, which bears no relationship to the areas listed on the reverse.
rev.: Four area computations, based on the formula length × width = area, are arranged in columns with the headings **uš** 'length', **sag** 'width' and **a-šà** 'area'.

The length and width are measured in **ninda**, and the areas of the fields are expressed in the basic unit of area, **sar**. The areas of #1 and #2 are rounded off. The area of #3 is qualified by the term *ši-ka-tum* 'harrowed', and the area of #4, by the term *ma-ia-ru* 'plowed'; both terms are found in CH §43 which requires the tenant farmer (*errēšum*) to plough and harrow the field which he neglected (*eqlam ša iddû mayyarī imaḫḫaṣ išakkak*).

(5) IM 53965

The problem is to find the length and width of a rectangular field which is measured by a broken reed. The area of the field is given, but the dimensions of the reed are not known. The problem is solved by using a quadratic equation.

(6) IM 54010

The text is broken; references to harvest and reed are preserved.

(7) Strssbg 362 #5

#5: The problem is to find the length measured by a reed (**gi** = *qanû*) using data concerning the reed which form a geometric progression.

(8) Strssbg 366 #4

#4: The text is broken; the problem is to find the original length of a (broken) reed (**gi** = *qanû*).

(9) Strssbg 368

The problem is to find the original length of a broken reed which is used to measure a rectangular field. The problem is worked out by using the method of false position (also called false value) to solve a quadratic equation.

(10) *UET* 5: no. 864

The problem is to find the size of the indentation (*dikištum*) of a task field (**a-šá giš-kar**) of Gula, which is 4 **gán** in area; the figure described in this text is a square within a square. However, Friberg (1979b: 7 and 1981b: 58) suggested *dikištum* as an outward extension of the original figure. Kilmer (1964: 140–42) showed that the size of the border is assumed to be known once the side of the square is calculated.

(11) VAT 7532

The problem is to find the original length of a broken reed which is used to measure a trapezoidal field. The problem is worked out by using the method of false position to solve a quadratic equation (Thureau-Dangin 1937c: 17; 1936b: 161; *MKT* III: 58). The problem is illustrated by the drawing of a trapezoid.

(12) VAT 7535 #1–3:

All three problems are similar to VAT 7532. The problem each time is to find the original length of a broken reed which is used to measure a trapezoidal field. Each problem is worked out by using the method of false position to solve a quadratic equation. Each problem is illustrated by the drawing of a trapezoid.

(13) VAT 7848

#1: The text is broken; the problem is to find the area (A) of a field. Multiplication by a factor 0;0,0,21,36 (c) and the barley measurement for the area

are preserved. The value cA seems to be multiplied by 36 in order to convert the area to **sìla** units of barley.
#2: The problem is to find the hypotenuse of a field in the shape of a right triangle. The answer is found by using the Pythagorean theorem.
#3: The problem is to find the area of a trapezoidal field. The altitude is computed; then the area is calculated and multiplied by a factor 0;0,0,21,36. The barley measurement for the area, also given, may be multiplied by 36.
#4: The problem is to find the area of a circular field. The area of the circle is calculated by multiplying the circumference squared by the coefficient 0;5 which is an approximation of $1/4\pi$ (*MCT*: 9, 44). The value obtained for the area is multiplied by a factor 0;0,0,21,36.
cf. #5: The text is broken; the example seems to treat a cube with each side measuring 10 **kùš**. However, the final value of 1,40, which is then designated as 100 **kùš**, indicates the problem concerns area.
#6: The problem seems to treat volume, but only two dimensions of 10 **kùš** each are involved in the calculations. The barley measurement for the area is given with either **ninda** or **bán** as the unit.
#7: The text is broken; reference to *zaq-pi* 'planted', perhaps referring to barley, is preserved (Nemet-Nejat 1982: 424–25).

(14) VAT 8389 #1–4 and (15) VAT 8391 #1–6

All ten examples concern two fields and their respective yields. The fields (**a-šà** = *eqlu*) are described in each problem by the term **líl**, which has two Akkadian equivalents, *tamirtu* 'land around a city' and *ṣēru* 'hinterland' (Hallo 1970: 59). Jacobsen (1935: 33 n. 7) translated **líl** as 'commons...the field and arable lands held in common by the citizens of the city'.

The data are the same for all problems. The problems are solved by using the method of false position (Thureau-Dangin 1937c: 20–23; cf. Bruins 1971: 230–36; 1962: 287) or, more specifically, double false position (Vogel 1960: 90–91).

Problems concern the finding of the area of the fields (F_1, F_2), the amount of barley harvested (γ_1, γ_2) in each field from an area measuring 1 **bur** or 30,0 **ninda**^{2} (ϕ_1, ϕ_2), the yield of barley per **ninda**^{2} of each field, which represents the share of rent collected (g_1, g_2), and/or the actual yield in **sìla** of each field (G_1, G_2) while the other parameters are given or assumed to be given in the previous problems.

In the Old Babylonian period rental payments for land were made by paying a share of the crop or an amount agreed in advance; some contracts only specify that the payment will be 'like neighboring fields' (Postgate 1992: 185).

SUMMARY The land registrar, *šassuku*, was primarily concerned with assessing taxes on real estate but was also engaged in survey work (Wiseman 1972: 146–47). In his role as geometer, the *šassuku* worked with rods of different (standardized) lengths or a measuring rope (Wiseman 1972: 146–47; Powell 1984a: 36). In the mathematical texts a (broken) reed (**gi** = *qanû*) is used in measuring fields (AO 6770 #5, IM 53965, Strssbg 362 #5, Strssbg 366 #4, Strssbg 368, VAT 7532, VAT 7535 #1–3 and possibly IM 54010).

Other evidence for the mechanics of surveying is sparse. In Šulgi B: 199, **sag-gag** 'triangle' and **gi-ninda** 'measuring rod', both essential tools for field surveying, are mentioned in connection with building cities. In one of the compositions depicting school life, "The Disputation between Enkimansi and Girnishag," the measuring rope (**éš-gána**), the nindan (cubit) rod (**gi-**DIŠ**-ninda**) and the peg (**giš-gag**) are described as necessary in surveying a field (see Chapter 2). The term **gi-ninda** denotes two metrological units, **gi** and **ninda**, as a concrete object for field surveying (Klein 1986: 5 n. 27).

The *terminus technicus* **kak...dù** in third millennium real estate sales may have reflected the original practice of driving in a peg (to hold the measuring line) and later the ceremonial aspect of embedding a clay peg with the recorded sale in the house wall (Powell 1984a: 36 n. 17).

temen was used to describe a method of field surveying in which a fairly regular polygonal area was marked off and measured in order to approximate an irregular area (Dunham 1986: 33). In discussing **temen** in Ur III field plans, Dunham (1986: 37) suggested:

> **temen**'s drawn in the field plans were actually laid out on the fields with pegs and ropes. Certainly there are clear references to using the measuring cord on a field from Early Dynastic times on. Hence, the **temen** would be a figure made of ropes stretched between pegs.

Like Ellis (1968: 78–93), Dunham rejected the idea that the peg deposits could have been surveying pegs since archaeological evidence shows inconsistencies wherever peg deposits have been found *in situ*. Dunham (1986: 39) cautioned "that very little is known about how ancient Mesopotamians oriented and measured off the ground plans and precincts of their temples, although we do know such measuring was important enough to be mentioned in their royal inscriptions and religious texts."

In the mathematical texts, the fields (surveyed) are triangles, trapezoids, rectangles, squares and circles. Field plans are drawn only in VAT 7532 and VAT 7535; in both cases the fields are depicted as trapezoids. With the exception of squares and circles, the shapes of the fields are also confirmed by drawings and/or descriptions of fields in economic and administrative documents (Nemet-Nejat 1982: 14, 278; Joannes 1990).

Gandz (1947: 107–113) demonstrated that the problem of the relation of perimeter to area was systematically investigated in Babylonian schools. The standard form for quadratic equations in Babylonian mathematics was as follows: for a given perimeter $x+y=a$ and area $xy=b$, find the length, x, and width, y (see Chapter 2, Algebra and Number Theory). He argued that the standard type of quadratic equations originated so that land surveyors could compute area and protect the public. An examination of field plans supports Gandz's theory. In field plans dating from the same period as the mathematical texts, the Old-Babylonian to the Late-Babylonian periods, as many as four measurements are given so that actual area could be calculated; whereas, in Neo-Sumerian field plans only sufficient measurements are given to compute the aproximate area. The OB–LB surveys offer more comprehensive measurements, reflecting historical changes in the ownership of land, so that private owners could clearly determine the size of properties (Nemet-Nejat 1982: 18–20).

In the NB–LB periods the areas of fields could be measured in barley or reed (Nemet-Nejat 1982: 285, 318–319; *MCT*: 143–45; Lewy 1949a: 1–11). **še-numun** 'barley' and **gi-meš** 'square **gi** or reed' are listed together as areal measurements in summary paragraphs of a metrological table in Hilprecht 1906: pl. 20 [CBS 8539] obv. iii 15–18; cf. Goetze 1951: 145–46). Powell (1984a: 36) summarized the difference between the two units as follows: areas over a *sūt* (= 27,46;40 or 1666.666 square cubits) were measured in barley and those under a *sūt* in reeds.

XI. HERDING AND HERD GROWTH

DISCUSSION (1) BM 85194 #31

#31: The text is broken; references to *pa-ru-um*/*pa-ri-im* 'cow' are preserved in rev. iii 13–14.

(2) *UET* 5: no. 121 #2–4

Problems #2 to #4, written entirely in Sumerian, concern finding the number of sheep entrusted to each shepherd.

#2: 1,01,01,01 (= 219,661) sheep are consigned to 13,13 (= 793) shepherds, resulting in each shepherd's herd having 4,37 (= 277) sheep.

#3: 1,01,01,01 sheep are consigned to 13 shepherds, resulting in each shepherd's herd having 4,41,37 (= 16,897) sheep.

#4: 1,01,01 (= 3661) are consigned to 7 shepherds, resulting in each shepherd's herd having 8,43 (= 523) sheep.

TABLETS	SUBCATEGORY ONE TYPE	SUBCATEGORY MIXED	SOLUTION WORKED YES	SOLUTION WORKED NO	ANSWER GIVEN YES	ANSWER GIVEN NO	DRAWING YES	DRAWING NO	DATE	REFERENCE
(1) BM 85194 #31:		x	x		x			x	Kassite or older†	*MKT* I: 142–93
(2) *UET* 5: no. 121 #2–4:		x		x	x			x	OB	Vaiman 1961: 248–50
(3) VAT 8522 #3:		x		x	x			x	OB	*MKT* I: 367–73
(4) YBC 4669 #B7,B8:		x		x		x		x	Kassite or older†	*MKT* I: 514–16; *MKT* III: 26–29
(5) YBC 7273	x			x	x			x	OB	*MCT*: 131
(6) YBC 7326	x			x	x			x	OB	*MCT*: 130–31

(3) VAT 8522 #3

#3: The tablet is broken but calculations indicating herd growth in a sheep-fold are preserved on the tablet. The lambs number 80% of the original sheep in the fold.

(4) YBC 4669 #B7, #B8

#B7: The problem is to find the number of sheep and the number of lambs when the number of sheep (**u$_8$-udu-ḫi-a** = *ṣēnu*) added to the number of lambs (**sila$_4$-gub** = *lillidu*) is 540 (9 × 60).

#B8: The problem is to find the number of sheep and the number of lambs when a number of sheep substracted from a number of sheep is 60 and and a number of lambs substracted from a number of lambs is a number no longer preserved on the tablet. Other numbers in the problem are not completely clear. The problem assumes a known relation exists between the number of sheep and and the number of lambs.

(5) YBC 7273

The obverse of the tablet contains only a diagram with calculations indicating herd growth in two sheep-folds. The lambs number 80% and 70% of the original sheep in each of the two folds. The reverse has been erased except for two personal names.

(6) YBC 7326

The problem is to find the original number of sheep and the number of lambs bred in two folds. The lambs number 80% and 70% of the original sheep in each of the two folds.

The obverse does not state the problem fully, but the numbers given on the diagram on the reverse complete and clarify the problem, although these numbers make sense only when the text is compared to similar problems in VAT 8522 #3 and YBC 7273. From these problems, *tarbaṣu* (= **tùr**) means both sheep-fold and the number of sheep in the fold.

SUMMARY In third millennium Mesopotamia, cattle (and equids) were considered draft animals; sheep were raised for their wool, goats for their hair, and cattle, goats and sheep for their skins. Because of spoilage, dairy products (milk, cheese, ghee) played a limited role in the ancient economy. Slaughtered animals were eaten mainly by the wealthy; the meat of animals that died due to old age, sickness or accident, was distributed to workers and dogs (Gelb 1967: 68–69).

Herds of cattle were maintained by the temple or palace, most probably because they needed to be moved to winter pastures (Adams 1966: 49; Oppenheim 1962: 45). Sheep and goats, also kept in large flocks belonging either to the state, temple or private owners, were supervised by shepherds *(rē'û)*, who sub-contracted the actual herding to "shepherd boys" (*kaparru*). Contracts between owner and shepherd from the Old Babylonian period detail the responsibilities of the shepherd and his compensation, which often included a fixed share of the proceeds (Finkelstein 1968; Postgate 1992: 159–60; CH §261–67).

In *TCL* II 5499 the number of cattle and the amount of butter and cheese they produced are charted for ten successive years in the Ur III period. The low birth-rate of calves (50%) and the production of butter and cheese as given in this text correspond to the low production of wool in ancient Mesopotamia (Gelb 1967). In modern times, by contrast, one cow can produce one calf per year (see Grizmek 1972: 370–71; 380–81).

In a group of texts from Old Babylonian Larsa, the lambing percentages are 75% or 80% of the herd and the production of wool is 2 **ma-na** (c. 1 kg.) per sheep (Kraus 1966: 121–33, 139-40). Sheep breed as early as nineteen months old, with a gestation period of approximately five months, depending on the species. Generally, a ewe gives birth to one lamb, but twins sometimes occur. Shortly before giving birth the ewes chase off their previous year's offspring because they have only two working teats. In modern times lambing percentages can exceed 100% in farm flocks and produce from 2 kg.–9 kg. of wool, depending on the species (Grizmek 1972: 501–04).

One problem referring to cows is too fragmentary to give any useful information (BM 85194 #31). The herding problems (*UET* 5: no. 121 #2–#4) are fanciful plays on the numbers of 1,01,01,01 and 1,01,01; any correlation with the actual size of a herd, which varied from 4 to over 200 animals, is coincidental (see Postgate 1992: 159). However, the problems dealing with a growth rate of 70% to 80% for sheep (VAT 8522 #3; YBC 4669 #B7, #B8; YBC 7273; YBC 7326) correspond to the actual lambing percentages and the low production of wool in Mesopotamia discussed above.

XII. STANDARD MEASURING VESSELS

DISCUSSION (1) BM 85194 #22, #24

The standard measuring vessels in problems #22 and #24 are square-based prisms. Thureau-Dangin (1932b: 189–90) correctly analyzed these problems as dealing with the measure of the **sìla**; however, he mistakenly believed the vessels mentioned in both problems are cylinders, even though π is not used in the calculations (*MKT* I: 181–82).

#22: The problem is to find the depth of a **^{giš}ba-rí-ga**(sic) and the depth of 1 **sìla** in a **^{giš}ba-rí-ga**.

#24: The problem is to find the depth of 1 **sìla** in a **^{giš}bán** (which holds 10 **sìla**).

(2) BM 85196 #14

#14: What the problem asks is unclear due to the difficulty in reading the text. The problem concerns a basket (**^{gi}gur** = *pānu*) which is used to collect 10 **gur** 1 **bán** barley for taxes or interest. The basket is checked (√*latāku*) and its measure is found to be inexact (*lā kīnu*)

(3) Haddad 104 #4, #5

The solutions to problems #4 and #5 are based on applying 6,0, the grain capacity coefficient, and $\pi = 3$ to calculate the circumference of the cylindrical measuring vessels.

#4: The problem is to find the depth of a *qû* measuring vessel of given diameter, which holds 1 **sìla** of grain. The second part of the problem checks the answer found in the first part by performing the inverse operation; that is, the problem is to find how much grain a *qû* measuring vessel of given diameter and depth holds.

#5: The problem is to find how much to reduce the depth of a *sūtu* measuring vessel of given diameter, which contains 1 **bán ⅔ sìla** of grain, so that the vessel holds exactly 1 **bán** of grain.

TABLETS	SUBCATEGORY ONE TYPE	MIXED	SOLUTION WORKED YES	NO	ANSWER GIVEN YES	NO	DRAWING YES	NO	DATE	REFERENCE
(1) BM 85194 #22,24:		x	x		x			x	Kassite or older[†]	*MKT* I: 142–93
(2) BM 85196 #14:		x	x		x			x	Kassite or older[†]	*MKT* II: 43–59
(3) Haddad 104 #4,5:		x	x		x			x	OB	Al-Rawi and Roaf 1984
(4) *LBAT*: no. 1649 #4:		x	x		x			x	LB	Sachs 1955: no. 1648
(5) YBC 4669 #1–9,B9:		x		x	x			x	Kassite or older[†]	*MKT* I: 514–16; *MKT* III: 26–29

(4) *LBAT*: no. 1648 #4
#4: The text is broken; references to ^{giš}**bán-meš** are preserved.

(5) YBC 4669 #1–9, #B9

Problems #1 to #9 all contain the following: (a) the name of a standard measuring container, (b) the size of the transversal (**dal** = *pirkum*), (c) the request to find the height and (d) the answer.

Both Neugebauer (*MKT* I: 514–15) and Thureau-Dangin (1936e: 60) showed that the volume in problems #1 to 9 is measured according to the formula V = **dal**^{2} × h and the relation 1 **sìla** = 3,36 **šu-ši**^{3} = (6 **šu-ši**)^3. Neugebauer (*MKT* III: 28) suggested that these vessels are square-based prisms (cf. **dal** ^{giš}**<ba>-rí-ga** in CBS 10996 obv. ii 22'–27'). The problems concern the following standard measuring vessels: problem #1 ^{giš}**ba-rí-ga**, problems #2 and #3 ^{giš}**ba-an** and problems #4 to #9 ^{giš}**ninda**.

#B9: The problem is to find the volumes of two measuring containers, ^{giš}**ba-rí-ga**, when the difference between them is 6⅓ **sìla** (*MKT* III: 28; 9⅓ **sìla** in *MKT* I: 515).

SUMMARY Capacity measures evolved from the need for containers of special shapes and sizes, according to the commodities being gathered, transported, processed, stored, etc. The earliest capacity measures and the ratios between units must have developed from the standard sizes of these containers (Powell 1990: 492–93). All kinds of liquids and solids, particularly barley, were measured with measuring vessels (Sachs 1945: 26–27, contra Lewy 1945: 25–26). Even fish were measured with standard measuring containers (Salonen 1970: 80–82).

The vessels ranged in size from small containers of 1/15 **sìla** to large tubs, used for measuring hundreds of **sìla**. The customary measuring vessels held 1 to 10 **sìla** (Salonen 1966: 278). However, the mass production of 1 **sìla** vessels far exceeded other sizes (Gelb 1982: 589).

The measuring vessels mentioned in the mathematical problems are **gišba-an** = *sūtu*, **gišba-rí-ga** = *parsiktu*, **gišsìla** = *qû*, **gišninda** = *akalu* and **gigur** = *pānu*. There are no drawings of measuring vessels in these problems.

gišba-rí-ga in BM 85194 #22 and in YBC 4669 #1 and #B9 is both a capacity measure and a standard measuring vessel (Salonen 1966: 292). In the lexical lists **ba-rí-ga** is equated with both *parsiktu* and *sūtu*, thereby, suggesting the basic meaning is 'measuring vessel' (Powell 1990: 505).

gišba-an/bán in BM 85194 #24, Haddad 104 #5, YBC 4669 #2 and #3 and the plural, **gišbán-meš** in *LBAT* no. 1648 #4, refer to the capacity measure *sūtu*. The vessel may be made from stone, silver, gold, clay or copper and is used mostly to measure barley, oil, etc.; it is listed among household objects to be divided for inheritance purposes (*CAD* s.v. *sūtu*; Powell 1984b: 57–61).

gišsìla in Haddad 104 #4 is a vessel used to measure barley, dates, etc. and liquids such as oil, wine, beer, etc. **gišsìla** is portrayed as funnel-shaped in its pictogram; however, Salonen (1966: 292–93) suggested that originally **gišsìla** was a bowl. It has been described as a golden bowl and even a silver bucket (*CAD* s.v. *qû* B 2). The few surviving containers that have been scientifically studied confirm that the **sìla** was approximately 1 liter (Powell 190: 503–04). Thureau-Dangin argued (1936e: 60) that the measure of the **sìla** remains uncertain as long as the shape of the receptacle is unclear (see Thureau-Dangin 1937b: 80–86; Postgate 1978: 71–75; for the early periods, see Damerow and Englund 1989). Powell (1984a: 42) demonstrated that the assumption of cylindrical rather than rectangular vessels incorrectly supposes a **sìla** equaling 0.762 liter.

Capacity norms depend on the measurement of the **sìla**; texts that establish a standard of measure refer to the **bán** or **ba-rí-ga** as containing a specific number of **sìla** (Powell 1990: 494). The **sìla** is never qualified, thereby suggesting fluctuation within small margins as acceptable (Powell 1990: 503).

gišninda in YBC 4669 #4–9 is a vessel that can be made from copper (Salonen 1966: 348). The **ninda** sign resembles a bevelled-rim bowl, the most common of Near Eastern pots. Bevelled-rim bowls may have been used to distribute rations of corn or flour for daily rations, from which the extended meaning 'bread' or 'food' developed; or, these bowls may have been used as bread moulds (Millard 1988).

gigur 'basket', may have been used to measure barley or to collect taxes in BM 85196 #14. **gigur** may also refer to a **gur** measuring vessel made from bronze (Salonen 1966: 285–86). In the history of capacity measures, the introduction of the Akkad-OB **gur** was accompanied by metrological reform, as suggested by the application of the qualifiers 'royal' (**lugal**), 'true' (**si-sá**) and 'greater' (**maḫ**) to the standards (Powell 1990: 497–98).

In the mathematical problems the shapes of the **ᵍⁱˢba-rí-ga**, **ᵍⁱˢba-an/bán** and **ᵍⁱˢninda** are square based prisms (BM 85194 #22, #24; YBC 4669 #1–9), but the shapes of these vessels are still a matter of controversy among scholars. Salonen (1966) showed that standard measuring vessels were made from a variety of materials such as gold, silver, stone, clay, copper, bronze and reed (basket), but the shapes of these containers were rarely mentioned outside the mathematical tradition.

XIII. ORIGINAL QUANTITY

DISCUSSION (1) AO 6770 #3

#3: The problem is to find the original weight of a stone (**na**$_4$ = *abnu*); four parameters are given, which lead to a linear equation with irregular coefficients. The calculations are mathematically faulty in this problem.

In Sumerian and Akkadian the term for weight may refer to the (original?) material of manufacture, namely 'stone'; bronze weights from later periods are in the shape of ducks and lions. Whether bronze weights are referrred to by the term 'stone' or another term is not known (Powell 1971: 242). For a list of weight specimens, see Thureau-Dangin (1921: 138–42) and Powell (1971: 249–73); for a discussion of weighing techniques see Powell (1971: 237–43) and Petruso (1985).

(2) BM 85194 #10–13

The problems concern finding the width (i) or depth (k) of a container (**im-lá**) in problems #10, #12, and #13 and the depth (i) of a mass for weighing (**ki-lá**) in problem #11. In these problems **im-lá** 'container' is also used to mean width, and **gam** is used to mean depth (as in NBC 7934 obv. 3, 8; rev. [3]). Computations do not consider the three dimensional shapes of the objects but, instead, lead to an equation of the first degree in one variable, involving the following relations:

$$i=\frac{18}{4\times 3}\times k=1;30k$$

$$k=\frac{4\times 3}{18}\times i=0;40i$$

#10: The problem is to find k when i = 1,0.
#11: The problem is to find i when k = 40.
#12: The problem is to find k when i = 30.
#13: The problem is to find i when k = 20.

	SUBCATEGORY		SOLUTION WORKED		ANSWER GIVEN		DRAWING			
TABLETS	ONE TYPE	MIXED	YES	NO	YES	NO	YES	NO	DATE	REFERENCE
(1) AO 6770 #3:		x	x			x		x	OB	*MKT* II: 37–42; *MKT* III: 62–65
(2) BM 85194 #10–13 :		x	x		x			x	Kassite or older†	*MKT* I: 142–93
(3) BM 85210 #11,12:		x	x		x			x	Kassite or older†	*MKT* I: 219–33
(4) IM 53957	x		x		x			x	OB	Baqir 1951: 37–38: no. 5
(5) MM.86.11.404 #2:	?	?	x		x			x	Seleucid	*MCT*: 140–41
(6) YBC 4652	x			x	x			x	OB	*MCT*: 100–03
(7) YBC 4669 #B3–B5:		x		x	x			x	Kassite or older†	*MKT* I: 514–16; *MKT* III: 26–29
(8) YBC 4698 #17:		x		x	x			x	Kassite or older†	*MKT* I: 513; *MKT* III: 42–45

(3) BM 85210 #11, #12

Both #11 and #12 are simple arithmetical problems concerning **im-lá** 'container'. The parameters and computations are similar to BM 85194 problems #10 to #13.

#11: The problem is to find k when i = 40. In this case 40 is divided by 10 to yield 4 **šu-si**.

#12: The problem is to find i when k = 4 **šu-si**.

(4) IM 53957

The problem is to find the original quantity of barley (**še**).

(5) MM.86.11.404 #2

The tablet contains at least four fragmentary examples but only in problem #2 (obv. 5–12) can some numerical relations be established. References to hoes (**al** = *allu*) and a pitchfork(?) (*rāzu* perhaps to be identified with *ra'izu*) are preserved. The problem is to find: 'What is the original number or weight of my hoe(s)?' (**sag al**-*ia*).

(6) YBC 4652 #3–10, #19–22

The colophon indicates the tablet contains 22 problems; only 12 problems are, at least, partly preserved, of which 6 can be fully restored. All the problems in the text are clearly of the same type. Each problem begins with **na$_4$ ì-pà ki-lá nu-na-tag(-ga)** 'I found a stone but did not weigh it' and concludes with **sag na$_4$ en-nam sag na$_4$**... 'What was the original weight of the stone? The original weight of the stone was...' The tablet deals with a stone weight to and from which various quantities are added and subtracted to realize a weight of 1 **ma-na**; the problems are solved by using linear equations for one unknown quantity.

The terms **ki-lá-tag(-ga)** 'weighed' and **ki-lá-nu-tag(-ga)** 'unweighed', are used in Ur III economic texts dealing with bolts of cloth (*MCT*: 103).

(7) YBC 4669 #B3–B5

#B3: The problem is to find the original quantity of a food ration (**šuku**).

#B4 and #B5: The problem each time is to find the original quantity of barley (**še**).

(8) YBC 4698 #17

#17: The problem is to determine the original weight of a stone. The statements referring to gold, silver and the commercial term ANŠE are unclear.

SUMMARY These problems contain no drawings and deal with objects and materials from everyday life such as stone weights, reeds, hoes, containers, barley, and food rations. Although the quantity or measurement of the items may arise from daily life (for example, a 1 **ma-na** weight in YBC 4652), the intent of the problem is to teach linear equations for one unknown, to which complex parenthetical expressions are sometimes added and subtracted as in AO 6770 #3 and YBC 4652.

XIV. WORK

DISCUSSION (1) BM 85194 #23

#23: A weaver can weave 0;20 or ⅓ (cubit) of cloth a day. The problem is to find the length of time needed to weave a garment 48 (cubits) long. Powell noted in private communication that the cubits are assumed to be the intended measure since cubits are used to refer to cloth in economic texts.

TABLETS	SUBCATEGORY ONE TYPE	SUBCATEGORY MIXED	SOLUTION WORKED YES	SOLUTION WORKED NO	ANSWER GIVEN YES	ANSWER GIVEN NO	DRAWING YES	DRAWING NO	DATE	REFERENCE
(1) BM 85194 #23:		x	x		x			x	Kassite or older†	*MKT* I: 142–93
(2) BM 85196 #12,13:		x	x		x			x	Kassite or older†	*MKT* II: 43–59
(3) YBC 4669										
#B1:		x	x		x			x	Kassite	*MKT* I: 514–16;
#B10:		x		x	x			x	or older†	*MKT* III: 26–29
(4) YBC 4673										
#16:		x		x		x		x	Kassite	*MKT* I: 506–08;
#17–21:		x	x		x			x	or older†	*MKT* III: 29–34
(5) YBC 9856 #1:		x		x	x			x	OB	*MCT*: 99–100

(2) BM 85196 #12, #13

#12: The problem is to find the amount of time needed to manufacture new rope ½ **danna** long. The "daily work assignment per man" is given as 10 **ninda** or 120 **kùš**.

#13: The text is badly broken. The problem is to find the amount of time (or man-days) needed to transport sheaves (*kurullu*) a given distance at a given rate. The data in this problem cannot be verified by economic and administrative texts.

(3) YBC 4669 #B1, #B10

#B1: The problem is to find the daily work assignment per man or **éš-kàr** and the number of workers necessary to complete an unspecified task. The computations involve a daily work assignment per man of 10 **gín**, which is usually found in problems dealing with construction and excavation of canals and other earthworks. Problem #B1 is the only problem in YBC 4669 in which the calculations are explained (see Thureau-Dangin 1937d: 88).

#B10: The problem is to find what fraction of a day one man spends in demolition and transportation when a distance of 20 **ninda** is involved. The daily work assignment per man is 5 **gín** for demolition and 5 **gín** for transportation.

(4) YBC 4673 #16–21

#16: The nature of the problem to be solved is not entirely clear. Calculations concern three kinds of wool for which numbers (of workers?) would be inserted.

For a discussion of different kinds of woolen textiles, see Hallo (1979: 4–10). Thureau-Dangin (1895: 118–139) indicated that quality is differentiated in wool and beer. However, the absence of defect is noted only for grain, dates, animals and beverages.

Problems #17 to #21 involve a series of equations containing the term **ú-kin**, which refers to the volume assignment per man or **éš-kàr**. **éš-kàr** is used instead of **ú-kin** in problem #21.

The computations presented in the text are as follows:

#17: ⅔ **ú-kin** × 7½ **sar** (volume) = 5 **sar ú-kin**

⅓ **sar** (volume) × 7½ **gin**(?) = 2½ **sar** (volume)

#18: 1/5 volume **ú-kin** × 5,0 volume **ú-kin** = 1,0 workers

#19: 1/6 volume **ú-kin** × 5,0 **sar** (volume) = 50 workers

#20: 3/2 **ú-kin** × 5,0 **sar** (volume) = 7,30 workers

#21: 1/10 **éš-kàr** × 2½ **sar** (volume) = 15 workers

ú-kin is, perhaps, to be identified with **kin** = *šipru* 'work'. In the mathematical texts **kin** is used in conjunction with volume. The expression **kin** (*ù* **saḫar**) 'work and volume' is found in VAT 8523 obv. i 5, 11, 13, 14, 17, 20, 24, 25; obv. ii 5, 9, 10, 18, 22, 23, 31; rev. 5, 18, 21, 26 (*MKT* I: 373–80). The term **ki-sá kin** 'volume of an (underwater) platform' appears in BM 85196 obv. i 7 (*MKT* II: 43–59).

(5) YBC 9856 #1

#1: The problem is to find the parts of a work day spent in performing three kinds of work which are executed at different speeds.

Friberg, in discussing this text with me, suggested the possibility that this problem is a 'price equivalency' or *maḫīru* problem. He proposed PI-*ya* should be read **pi**(-*ya*) = **ba-rí-ga**, which is a standard measuring vessel. Friberg's explanation does not take into account the total 1,0 which refers both to *iš-ka-ra-am* 'work assignment' and to the total number of PI-*ya*. Furthermore, his suggestion does not eliminate the difficulties in translating *ku*-ZI-*im* in this context.

SUMMARY The category "work" refers to a miscellaneous group of tasks, as the following:

(1) weaving cloth (BM 85194 #23)
(2) manufacture of new rope (BM 85196 #12)
(3) transportation of sheaves a given distance (BM 85196 #13)
(4) demolition and transportation (YBC 4669 #B10)
(5) production(?) of three kinds of wool (YBC 4673 #16)
(6) work of an unspecificed nature (YBC 4669 #B1, YBC 4673 #17–21 and YBC 9856 #1).
It does not include problems that deal with work on canals, excavations and other constructions.

Certain information contained in the mathematical problems cannot be verified by administrative and economic texts from the second millennium, namely, the time (that is, man-days) required for weaving, for manufacturing new rope, for transporting sheaves, and for demolition and transportation.

XV. METALWORK

DISCUSSION (1) BM 34568 #16
#16: A container weighing 1 **ma-na** is made of an alloy of gold and bronze mixed at a ratio of 1:9, respectively. The problem is to find the weight of each metal.

(2) BM 85196 #18
#18: The problem is to find the weight of two coils of silver such that

$$\frac{x}{7}+\frac{y}{11}=1 \text{ } gín$$

$$x-\frac{x}{7}=y-\frac{y}{11}$$

Powell (1978: 220–22) suggested alternate methods for solving this problem. He showed that the solution is in multiples of ⅛, and, if the scribe knew this, the answer could have been calculated more easily than multiplying by 7 and 11.

Friberg (1982: 121–22) proposed that the solution to this problem is similar to the *maḫīru* problem in VAT 7530. Both problems are linked by similar phrases, that is, [*lu*]-*ú* 1 **gín** *ša ta-aḫ-ru-*[*ṣ*]*ú li-li ù li-r*[*i-i*]*d-ma* **ḫar kù-babbar** *li-im-ta-aḫ-ru* 'May the silver coil you have cut off go up or down (Powell: 'let it vary') so that the silver coils become equal' in BM 85196 rev. ii 19–20, and **kù-babbar** *li-li ù li-ri-id*<*-ma*> *ma-ḫi-ru* (or **ganba**) *li-im-ta-ḫar* 'May the silver go up or down so that the price becomes equal' in VAT 7530 obv. 10, 20–21 (*li-im-ta-ḫi-ra*); rev. 6–7.

TABLETS	SUBCATEGORY ONE TYPE	MIXED	SOLUTION WORKED YES	NO	ANSWER GIVEN YES	NO	DRAWING YES	NO	DATE	REFERENCE
(1) BM 34568 #16:		x	x		x			x	Seleucid	*MKT* III: 14–22
(2) BM 85196 #18:		x	x		x			x	Kassite or older†	*MKT* II: 43–59
(3) M 15192 + M 15236	x			x	x			x	OB	Soubeyran 1984: 41–45: no. 8
(4) YBC 4669 #B6:		x		x	x			x	Kassite or older†	*MKT* I: 514–16; *MKT* III: 26–29

(3) M 15192 + M 15236

The text, similar to *ARM* 24: no. 146, records the amount of gold and silver lost (*maṭû*) from masses of precious metal weighing more than 1 **ma-na** but less than 2 **ma-na**. The data are tabulated by repeated use of the phrase *ina* x **ma-na** y **gín kù-babbar** (or **guškin**) z **gín** *imṭi* 'for x mina y shekels silver (or gold), z shekels are deducted'. In lines 1–10 the metal used is silver, and in lines 11–19, gold. The text may be a record of precious metal lost during manufacture of articles because the formula is followed by **ur-maḫ** 'lion' in line 6. Alternatively, the text may be a record of metal deducted as taxes for washing and heating.

(4) YBC 4669 #B6

#B6: The problem is to find the area of a *ruqqu*, perhaps a standardized sheet of metal, of given length and width, and to find the amount of silver needed in order to beat (*maḫāṣu*) a *ruqqu* of 3 **šu-si** square.

A relation can be established between 1 **gín** silver and an area of 3 **šu-si** square for a *ruqqu*. 3 **šu-si** square can be explained in units of the area-**sar** as (0;0,30 **ninda**)^2 = 0;0,0,15 **sar**. The reciprocal of 0;0,0,15 namely 4,0,0, may be the 4 listed as a coefficient of a *ruqqu* of silver (YBC 5022: 26), though 26,40 is also listed as a coefficient for a *ruqqu* of silver (YBC 7243: 51).

SUMMARY The mathematical texts concerning metalwork include a variety of topics, representing various stages of manufacture and use.

A problem concerning an alloy of gold and bronze is found in BM 34568 #16. A few grains of bronze could be added to gold for a red color (Limet 1960: 44–45; cf. Moorey 1985: 69–70).

A silver coil problem in BM 85196 #18 reflects the problem of early coinage. Before coinage, silver was cast as standard ingots or cut from ingots and took distinctive shapes, most commonly coils (*ḫullu*, *šemeru*), often weighing 5 **gín**, or miniature axeheads (*ḫaṣṣinu*); spiral bands were even found in grave deposits (Moorey 1985: 116–18; Powell 1978: 213–215; Limet 1960: 212–13; Hallo 1963: 138; Thureau-Dangin 1935a: 27). The extant earliest coins of Babylonian origin (c. 331–28 B.C.), staters (*istatirru*), were not standard weights and, therefore, payments were made by weighing the staters rather than by their face value (Moorey 1985: 116; Powell 1978: 217).

A Mari tabular text, M 15192 + M 15236, with 13 or 14 entries, records the loss of precious metal either during manufacture or through taxation. The methods for working (and recycling) gold and silver are similar, and a deficit or loss may occur in manufacturing or refining (*CAD* s.v. *maṭû*); administrative texts provide an account of the metal during the process (Moorey 1985: 73, 119).

Silver used for payment was not of standard quality and required controls as the use of such qualifiers as **gin** 'normal, reliable, true' and other terms for assaying and refining demonstrates. In Mari, silver in the form of scrap and manufactured items was the conventional form of payment; the silver was weighed (Moorey 1985: 116–17). In the Neo-Babylonian period payments made to the temple or palace consisted of uncoined silver dropped into a box. The contents were smelted in an official foundry and then cast into ingots, which were deposited in the treasury (Oppenheim 1947a).

A problem concerning a *ruqqu*, perhaps a standardized sheet of metal, is introduced in YBC 4669 #B6 (see Lewy 1949b: 161–62). The mention of area, not volume, seems to indicate a flat object, and the relation of silver to area seems to indicate standardized dimensions (Neugebauer and Sachs MCT: 138–39). Sheet-metal work was used for vessels, fittings (for example, to exhibit repoussé work) and as overlays on statues, furniture, and architectural elements such as palace and temple doors (Moorey 1985: 46–50). The coefficient lists, YBC 5022 and YBC 7243, may have provided useful formulas for estimating the amount of material needed for a *ruqqu* (see Lewy 1949b: 161–70).

XVI. *kiššû* 'BUNDLE OF REEDS' and *nalbantu* '...'

DISCUSSION: *kiššû* Sachs (1944: 33–36) showed that *kiššû* 'bundle of reeds' is a truncated cone of standardized dimensions; it is 6 **kùš** long, with a circular base having a circumference of 24 **šu-si** and a diameter of about 8 **šu-si**, and a circular top with a circumference of 12 **šu-si** and a diameter of about 4 **šu-si**.

TABLETS	SUBCATEGORY ONE TYPE	MIXED	SOLUTION WORKED YES	NO	ANSWER GIVEN YES	NO	DRAWING YES	NO	DATE	REFERENCE
(1) BM 85194 #14,15:		x	x		x			x	Kassite or older†	*MKT* I: 142–93
(2) BM 85196 #2,5:		x	x		x			x	Kassite or older†	*MKT* II: 43–59

(1) BM 85194 #14, #15
#14: The problem is to find the volume of a reed bundle.
#15: The problem is to find the circumference (sic) of a bundle of reeds at a height of 3 **kùš** or half-way up.

(2) BM 85196 #2, #5
#2: The problem is to find the volume of a reed bundle. The text is a duplicate of BM 85194 #14 with insignificant variations of phraseology.
#5: The problem is to find the number of reed bundles a cargo boat (*malallû*) can hold. The boat is a rectangular prism of given dimensions; the length of a bundle of reeds, 6 **kùš**, is the same as the height of the cargo boat.

Values for cargo boats carrying reed bundles appear in the coefficient lists, but do not elucidate the problem, namely:

30 **gišma-lá gi-níg-sa-ḫi-a**
75 **gišma-lá gi-níg-sa-ḫi-a e** CBS 10996 ii 8', 9'

DISCUSSION: *nalbantu* The exact meaning of *nalbantu* is not known but must have something do with brickmaking based on the root $\sqrt{}$*labānu*. Thureau-Dangin (1935a: 16) suggested that *nalbantu* may be a platform for molding or drying bricks.

nalbantu is not to be confused with *nalbanu* 'brick mold', which has coefficients relating to the size of bricks (see *MCT*: 137–38; Powell 1982: 116–123; YBC 5022: 3, 4, 6, 9, 11, 13, 14; *TMS*, Text 3: 45; Haddad 104 #9, #10). The dimensions of *nalbantu* have nothing to do with the size of bricks.

(1) BM 85196 #11
#11: The problem is to determine the volume and base of *nalbantu*. The text applies an incorrect formula, namely, $V=\frac{h}{2}(s+s')$, instead of $V=\frac{h}{2}(s^2+s'^2)$, in solving the problem.

TABLETS	SUBCATEGORY ONE TYPE	MIXED	SOLUTION WORKED YES	NO	ANSWER GIVEN YES	NO	DRAWING YES	NO	DATE	REFERENCE
(1) BM 85196 #11:		x	x		x			x	Kassite or older†	*MKT* II: 43–59
(2) BM 85210 #9:		x	x		x			x	Kassite or older†	*MKT* I: 219–33

(2) BM 85210 #9
#9: The problem is to find the volume and height of a *nalbantu*. The correct formula is applied in this example.

SUMMARY *kiššû* and *nalbantu* should be translated according to their primary meanings, *kiššû* 'bundle of reeds' and *nalbantu* '...,' a brickmaking term; they should not be treated as geometrical terms, 'truncated cone' and 'truncated pyramid', respectively.

The Babylonians used the things around them to conceptualize solid geometric forms. In the mathematical problems an object may have more than one shape, as *iṣu* 'log', which is a cylinder in Haddad 104 #1 and a truncated cone in Haddad 104 #2 and #3. Conversely, a given shape may be associated with more than one object; that is, a truncated pyramid is recorded as *nalbantum* in BM 85196 #11 and BM 85210 #9, as *ḫirītu* 'moat' in BM 85194 #28, and as *manzazu* 'socle' in BM 85210 #8.

XVII. WOOD PLACED ON TOP OF A WALL AND A POLE OR REED LEANING AGAINST A WALL

DISCUSSION (1) AO 6484 #3, #4
#3: The problem is to find how far to step back from a wall (*igāru*) of a given height in order to see the wood on top.
#4: The problem is to find the height of the wall on which a piece of wood is placed. The data are the same as in problem #3.

TABLETS	SUBCATEGORY ONE TYPE	MIXED	SOLUTION WORKED YES	NO	ANSWER GIVEN YES	NO	DRAWING YES	NO	DATE	REFERENCE
(1) AO 6484 #3,4:		x	x		x			x	Seleucid	*MKT* I: 96–107
(2) BM 34568 #12:		x	x		x			x	Seleucid	*MKT* III: 14–22
(3) BM 85196 #9:		x	x		x			x	Kassite or older†	*MKT* II: 43–59

(2) BM 34568 #12
#12: A reed (**gi** = *qanû*), leaning against a wall (**i-zi** = *igāru*), is moved a given distance down the height of the wall so that it is a given distance from the base of the wall. The problem is to find the length of the reed and the height of the wall. The Pythagorean theorem is used to solve the problem.

(3) BM 85196 #9
#9: A support beam (^{giš}*palû*), leaning against a wall, is moved a given distance down the height of the wall. The problem is to find how far the bottom end of the beam is from the wall. The solution is worked out and then checked. The Pythagorean theorem is used to solve the problem (Thureau-Dangin 1935a: 13).

SUMMARY These problems illustrate a rare instance in which Babylonians used the objects around them to visualize a plane geometric form, here the triangle, in order to teach a geometric technique, the Pythagorean theorem (BM 34568 #12 and BM 85196 #9). They contain no drawings.

In all cases the measurements given are plausible; that is, the height of the wall is given variously as 10 **kùš** (c. 5 m.; AO 6484 #3, #4), 12 **kùš** (c. 6 m.; BM 34568 #12) and 0;24 **ninda** (c. 2.4 m.; BM 85196 #9); the height of the wood as 1 **kùš** (c. 0.5 m.; AO 6484 #3, #4); and the height of the support beam as 0;30 **ninda** (c. 3 m.; BM 85196 #9). Only the length of the reed, 15 **kùš** (c. 7.5 m.; BM 34568 #12), is unrealistic.

XVIII. LOGS

DISCUSSION (1) Haddad 104 #1–3

Problems #1 to #3 may provide evidence for standard measuring vessels. Two coefficients are used to solve the problems, namely: 0;5, the circumference coefficient used to calculate the area of a circle when the square of the circumference is given; and 6,0, a grain capacity coefficient used to transform the volume in **sar** to the capacity in **bariga**.

TABLETS	SUBCATEGORY ONE TYPE	SUBCATEGORY MIXED	SOLUTION WORKED YES	SOLUTION WORKED NO	ANSWER GIVEN YES	ANSWER GIVEN NO	DRAWING YES	DRAWING NO	DATE	REFERENCE
(1) Haddad 104 #1–3:		x	x		x			x	OB	Al-Rawi and Roaf 1984
(2) VAT 8522 #1:		x	x		x		x		OB	*MKT* I: 367–73
(3) YBC 8600	x		x		x			x	OB	*MCT*: 57–59

#1: The problem is to find the amount of grain a log in the shape of a cylinder holds. The diameter and height are each equal to 0;5 **ninda**. The log holds 1 **bariga** 5 **bán** 2½ **sìla** of grain.
#2: The problem is to find the amount of grain a log in the shape of a truncated cone holds. The log holds 1 **gur** of grain. The text uses an incorrect formula to solve the problem so that the calculated volume is smaller than the actual volume. For someone involved in buying and selling grain, this difference affects profit margins and suggests the theoretical nature of the problem (Al-Rawi and Roaf 1984: 188).
#3: The value of (the grain contained in) a log in the shape of a truncated cone is 1 **gú**. The log holds 10 **gur** of grain. The problem is to find the amount of grain worth 1 **ma-na** of silver and to determine how much should be cut from the top or bottom of the log. The method of solution is similar to problem #2.

(2) VAT 8522 #1
#1: The value of a cedar tree (*erinnu*) of given dimensions is 3⅓ **ma-na**. The problem is to find how much cedar must be cut to be worth ⅓ **ma-na** of silver. The price of cedar wood in terms of silver cannot be confirmed by outside sources. Two coefficients are used to solve the problem, namely, 0;5 and 6,40, another grain capacity coefficient (BM 47860 obv. 7; IM 52301 rev. 20; *TMS*, Text 3: 50; VAT 8522 obv. i 6a; see Friberg 1982: 113–14).

The truncated cone described in the problem is illustrated by a drawing of a partitioned trapezoid. *kabāru* 'to be thick, fat (said of logs and trees)' is used to describe the thickness in **sìla** of a log.

(3) YBC 8600

The problem is to find the thickness (*kubru*) of a log. The text checks the answer by finding the circumference of the log. The coefficient 4,48 (rather than the coefficient 5,0) is used to convert the units in **ninda**² into **sìla** when measuring the thickness of a log. The coefficient 5,0, is not applied because it results in inconvenient fractions of **sìla**-thickness when the circumference of a log is an integer number of **kùš** (*MCT*: 58). The coefficient lists, YBC 5022: 58 and YBC 7243: 35, indicate 4,48 as the coefficient value for the thickness of a log.

SUMMARY The category of logs (*iṣu*, literally 'wood' or 'tree') involves two kinds of problems: (1) the thickness of a log measured in **sìla** in VAT 8522 #1 and YBC 8600 and (2) possibly standardized measuring vessels in Haddad 104 #1-3 (see Lewy 1947: 310; Neugebauer 1949: 426).

"Log" metrology began as the measurement of cylindrical solids based on the following definition: a log with a circumference of 1 **kùš** contains 2 **sìla.** "Log" metrology unites length, volume and capacity to form a cluster of metrological relationships such that 2 **sìla** = 1 **kùš** circumference by 0;12 **kùš** (or 0;1 **ninda**) height, 2 **bariga** = 1 **kùš** circumference by 0;12 **kùš** (or 1 **ninda**) height, 2 **gur** = 1 **kùš** circumference by 1,0 **kùš** (or 5 **ninda**) height. These calculations suggest practical experiments with measures of length and capacity. The advantage of defining thickness in **sìla** is that it eliminates the necessity of transforming volume units into capacity units (Powell 1984a: 42-43; cf. *MCT*: 58).

XIX. FOREST AND REED THICKET

DISCUSSION AND SUMMARY (1) *TMS*, Text 25 #2

The text is broken, and no drawing accompanies the problem. References to *apu* 'canebrake', *qištu* 'forest', and the problematic passage [...]-*ru šu-up-lu* 24 **saḫar** *ak/q-ba-a*[*r*..], are preserved on the tablet; however, neither *kabāru* 'to be thick, fat (said of logs and trees)' nor *qebēru* 'to bury (objects)' has an "a" thematic vowel. Though Bruins and Rutten (*TMS*: 124) suggest a similarity between this problem and and the log problem in VAT 8522 #1, none of the coefficients applied in solving log problems is used in the calculations of the text.

TABLETS	SUBCATEGORY ONE TYPE	MIXED	SOLUTION WORKED YES	NO	ANSWER GIVEN YES	NO	DRAWING YES	NO	DATE	REFERENCE
(1) *TMS*, Text 25 #2:		x	x		x			x	OB	*TMS*: 122–24

XX. DOORS

DISCUSSION (1) *TMS*, Text 15

The problem is to find the diagonal of a door which is a rectangle. Two possible solutions are calculated involving Pythagorean triples 3a-4a-5a; this problem is not accompanied by a drawing. Friberg (1982: 89) suggested that the problem actually concerns a double gate and restored **[min-]ká**.

TABLETS	SUBCATEGORY ONE TYPE	SUBCATEGORY MIXED	SOLUTION WORKED YES	SOLUTION WORKED NO	ANSWER GIVEN YES	ANSWER GIVEN NO	DRAWING YES	DRAWING NO	DATE	REFERENCE
(1) *TMS*, Text 15	x		x		x			x	OB	*TMS*: 88–91
(2) VAT 6598 #5,6:		x	x		x		x		OB	*MKT* I: 277–87

(2) VAT 6598 #5, #6

#5 and #6: Both problems concern finding the diagonal of a rectangular door. The data and accompanying drawings are the same for both problems. Each problem gives an approximation for the diagonal of a door.

Walker (Letter to author 1986) enclosed a copy of his join, BM 96957 + VAT 6598, which contains two problems concerning doors. One problem is to find the width of a door of given length and diagonal. The other problem is to find the length of a door of given width and diagonal. Both problems are illustrated by drawings, and the data are the same as in VAT 6598 #5 and #6.

SUMMARY The problems involve finding the diagonal of a door (**ká** = *bābu*) or using the diagonal in the computations. In VAT 6598 the measurements given for the door (of a building, palace, temple, etc.) are plausible, namely 0;40 **ninda** (c. 4 m.) for the height and 0;20 **ninda** (c. 1 m.) for the width. No units of measurement are given in *TMS*, Text 15.

Calculations involving a diagonal would mitigate against the practical application of these problems. Rather these problems illustrate the rare cases in which objects from everyday life are used to teach a plane geometry concept.

XXI. BITUMEN OR MUD PLASTER

DISCUSSION (1) AO 6770 #4

#4: The problem is to find the amount of bitumen (*kupru*) needed to cover a given area (to a thickness of 1 mm). The coefficient 0;15 **sìla** is applied; this value is also found in coefficient lists (A 3553:13, YBC 5022: 18, YBC 7243: 31; for additional values see *TMS*, Text 3: 67 and YBC 7243: 24).

TABLETS	SUBCATEGORY ONE TYPE	MIXED	SOLUTION WORKED YES	NO	ANSWER GIVEN YES	NO	DRAWING YES	NO	DATE	REFERENCE
(1) AO 6770 #4:		x	x		x			x	OB	*MKT* II: 37–42; *MKT* III: 62–65
(2) Haddad 104 #7,8:		x	x		x			x	OB	Al-Rawi and Roaf 1984

(2) Haddad 104 #7, #8

#7 and #8: Both problems concern finding the amount of mud (*ṭiddum*) needed to plaster (see *CAD* s.v. *ṣêru*) a given area to a specified thickness. The thickness of the mud plaster (1.7 cm.) is the same in both problems but the shape of the area is a square in problem #7 and a rectangle in problem #8. The coefficient 5 *middum* is used to convert the volume (of mud plaster) in **sar** to units of capacity in **bariga**.

SUMMARY Despite disadvantages, mud brick was used for building in Mesopotamia because the raw materials were readily abundant and the resulting structures could easily be repaired. A coat of mud plaster protected the mud bricks of a wall from rain because the surface plaster would be washed away before the rain could strike the bricks themselves. Summer was usually the season for building. Modern experiments have shown that the heat of the sun increases the compressive strength of clays, as well as their plasticity and their water resistance, thereby rendering the buildings more weather resistant (see Dunham 1980: 24 n. 1–2). In addition, the Mesopotamians were aware of the benefits of adding organic substances to mud brick and mud mortar. They also knew how to safeguard the bottoms of their walls from moisture by using damp-proof materials or by raising the buildings above the ground (see Dunham 1980: 15–24). Bitumen, used for waterproofing and paving, was not employed in a pure state but rather in combination with other minerals and vegetation.

The amounts calculated in the mathematical problems for plastering and paving are realistic (see Thureau-Dangin 1936d: 80 n. 2; Al-Rawi and Roaf 1984: 203). Knowing how to calculate these amounts would be useful for scribes dealing with deliveries of plaster and bitumen for corvée work.

XXII. FOUNDATION FOR A WALL AND PILLAR

DISCUSSION (1) NBC 7934

The problem is to find the volumes of three rectangular prisms when the length (l), base (b) and height (h) are given. The tablet also lists each volume, the total of two volumes and the grand total of all three volumes.

TABLETS	SUBCATEGORY ONE TYPE	MIXED	SOLUTION WORKED YES	NO	ANSWER GIVEN YES	NO	DRAWING YES	NO	DATE	REFERENCE
(1) NBC 7934	x			x	x			x	OB	*MCT*: 55–56

SUMMARY Rectangular prisms form a basic building shape in building plans and in excavations (Heinrich and Seidl 1967; Dunham 1986: 31, 63–64). In our discussion of this text, Sally Dunham noted that the basic building shapes between walls and in house plans are ╡╞ and ╤╤, representing two possible configurations for the three rectangular prisms in NBC 7934. The tablet contains no drawing.

In order to explain the terminology the three depths (**gam**-2-**kam** in obv. 3 and **gam** in obv. 8, rev. [3]), each measuring ½ **kùš**, Neugebauer and Sachs (*MCT*: 56) suggested that the three volumes may represent the foundation for a wall and pillar. The purpose of the figure may have been stated on the edge, which contains a broken line of text.

XXIII. MARI TEXTS

DISCUSSION (1) M 6829

The text presents a religious context for a mathematical text; it tabulates the number of sheep consecrated for an *ikribu* offering and the number of sheep set free. The proportion of living sheep to consecrated sheep does not follow either a geometric or an arithmetic pattern. The text presents a progressive tariff which may have been useful in the temple accounting procedures.

M 6829, similar to *ARM* 24: no. 260, is broken and does not contain the introductory and concluding remarks present in *ARM* 24: no. 260, namely, "when a man consecrates a great number of sheep to Iltum in his *ikribum* offering," and "the one who sacrifices shall stand before Iltum and mark both sides(?) of the sheep with the sign of Iltum," respectively.

TABLETS	SUBCATEGORY ONE TYPE	SUBCATEGORY MIXED	SOLUTION WORKED YES	SOLUTION WORKED NO	ANSWER GIVEN YES	ANSWER GIVEN NO	DRAWING YES	DRAWING NO	DATE	REFERENCE
(1) M 6829	x			x	x			x	OB	Soubeyran 1984: 35–41: no. 7
(2) M 6861	x			x	x			x	OB	Soubeyran 1984: 46–47: no. 9
(3) M 8613	x			x	x			x	OB	Soubeyran 1984: 30–35: no. 6
(4) M 15192 + M 15236	x			x	x			x	OB	Soubeyran 1984: 41–45: no. 8

(2) M 6861

The text illustrates the accumulation of a quantity of grain according to an official fixed tariff but at an unusual rate. This text documents the arrival of two shipments of grain to the palace and the subsequent delivery by the royal offices to Dadum, as indicated by the formula **šu-ti-a** PN 'goods received by PN'. Two measurements, **giš** 60 (**sìla**) **še** (line 2) and **giš** 20 (**sìla**) **še-ba** (line 11), are used to record the shipments. **še-ba** 'ration' (line 11) may indicate food rations redistributed by Dadum using a standard measuring vessel of 20 **sìla**.

(3) M 8613

The text contains an exponential progression of the number of grains accumulated each day of a month of 30 days. The tabulation is based on the formula 2^n...n, or 2 grains on the first day, 4 grains on the second day, etc., with n being the number of days and a grain (**še** = *uṭṭetum*) being 1/180 *šiqlu* (Powell 1990: 499–500, 511; Sachs 1946). This metrological unit appears in our text either with or without a phonetic complement as **še**, **še**-*tum*, **še**-*ṭe*$_6$-*tam*.

(4) M 15192 + M 15236

The text shows the amount of precious metal lost (*maṭûm*) from the masses of gold and silver weighing more than 1 **ma-na** but less than 2 **ma-na**. The data tabulated may be a record of metal lost during manufacture or a record of metal deducted as taxes for washing and heating. It is similar to *ARM* 24: no. 146. See XV. METALWORK.

SUMMARY So far the Mari archive shows no evidence of pure mathematical problems. Three tablets are tabular texts (M 6829, M 8613 and M 15192 + M 15236). Computations and drawings do not occur on any of the tablets.

Mari mathematical texts (Soubeyran 1984: 19–48: nos. 6–9) deal with administrative topics and form a bridge between economic/administrative texts and mathematical texts. The connection between the two types of texts may be explained by a reference to the *zazakku*, a high administrative official from the scribal hierarchy connected with the assessment of taxes: PN **dub-sar-zag-ga** *ša* ana PN_2 *[išš]aknu ana ṭupšarrūtim utīrma ana* **é-dub-ba** *ī[rub]* 'He sent the *zazakku* scribe who had been assigned to PN (to assist the tax officials [**enku**: written **zag-ḫi-a**] in their assessment of taxes) back to study, and he attended the *edubba*' (*PBS* 7: 89: 36–37).

CHARTS

All bibliographical references are listed in alphabetical order by museum number with the rare exceptions of *LBAT*, *TMS*, *UMB* and *UET*. A standard set of pages for each tablet can be found in the list of tablets in the charts.

Tablets containing mixed problems are indicated by * in the chart of tablets containing algebraic/geometric problems, in the chart of tablets containing practical problems and in the chart of tablets containing mixed problems. Mixed problems are included in the charts of algebraic/geometric and practical problems for completeness. In the date column † notes the date suggested by Neugebauer; however, Thureau-Dangin [*TMB*] suggested that the tablet be dated to the Old Babylonian period.

LIST OF TABLETS: MUSEUM NUMBERS AND BIBLIOGRAPHIC REFERENCES

A 3553	Kilmer 1960: 275–78
A 24194	*MCT*: 107–19
A 24195	*MCT*: 119–29
AO 6484	*MKT* I: 96–107
AO 6770	*MKT* II: 37–42; *MKT* III: 62-65
AO 8862	*MKT* I: 108–23
AO 10642	*MKT* I: 123
AO 10822	*MKT* I: 123–26
AO 17264	*MKT* I: 126–34
AOT 304	Vaiman 1961: 246–47
AOT 305	Allotte de la Füye 1907: 75–77
BM 13901	*MKT* III: 1–14
BM 15285	*MKT* I: 137–42 Saggs 1960: 131–45
BM 34568	*MKT* III: 14–22
BM 47860	Livingstone 1986: 18–23
BM 80209	Friberg 1981b
BM 85194	*MKT* I: 142–93
BM 85196	*MKT* II: 43–59
BM 85200 + VAT 6599	*MKT* I: 193–219
BM 85210	*MKT* I: 219–33
Böhl Coll. 1821	Kilmer 1964: 142–43
CBM 12648	*MKT* I: 234–35
CBS 11318	Neugebauer and Sachs 1984: 247
CBS 10996	Kilmer 1960: 278–83
Const. 236	Allotte de la Füye 1907: 77–78
CUA 34	Goetze 1957: 40: no. 33
Db_2-146	Baqir 1962
Erm 15073	Vaiman 1961: 232–44

Erm 15188	Vaiman 1958
Erm 015189	Vaiman 1955
Haddad 104	Al-Rawi and Roaf 1984
IM 31210	Bruins 1954: 57–61
IM 31247	Bruins 1953c: 249–53
IM 31248	Bruins 1953d: 256–57
IM 43996	Bruins 1953d: 255–56
IM 49949	Edzard 1970: no. 236
IM 52301	Baqir 1950b
IM 52304	Goetze 1951
IM 52672	Bruins 1954: 61
IM 52685	Goetze 1951
IM 52916 (IM 52685, IM 52304)	Goetze 1951
IM 53953	Baqir 1951: 31–32: no. 2
IM 53957	Baqir 1951: 37–38: no. 5
IM 53961	Baqir 1951: 35–36: no. 4
IM 53963	Bruins 1953c: 249
IM 53965	Baqir 1951: 39–40: no. 7
IM 54010	Baqir 1951: 38–39: no. 6
IM 54011	Baqir 1951: 45: nos. 10, 11
IM 54464	Baqir 1951: 43–44: no. 9
IM 54472	Bruins 1954: 56
IM 54478	Baqir 1951: 30–31: no. 1
IM 54538	Baqir 1951: 33–35: no. 3
IM 54559	Baqir 1951: 41–42: no. 8
IM 55357	Baqir 1950a
Ist. 04360	*MKT* I: 235–36
Ist. 04552	*MKT* I: 236
K 8705	*MCT*: 140
K 11097	*MCT*: 139–40
LBAT: no. 1647	Sachs 1955: no. 1647
LBAT: no. 1648	Sachs 1955: no. 1648
M 10	Sachs 1952: 152–53
M 6829	Soubeyran 1984: 35–41: no. 7 // *ARM* 24: no. 260
M 6861	Soubeyran 1984: 46–47: no. 9
M 8613	Soubeyran 1984: 30–35: no. 6
M 15192 + M 15236	Soubeyran 1984: 41–45: no. 8 // *ARM* 24: no. 146
MAH	Bruins 1955a: 47–48
MLC 1354	*MCT*: 56–57
MLC 1731	Sachs 1946
MLC 1842	*MCT*: 106–07
MLC 1950	*MCT*: 48–49

MM 86.11.404	*MCT*: 140–41
NBC 7934	*MCT*: 55–56
NBC 8082	*MCT*: 10
NCBT 1913	*MCT*: 10
Plimpton 322	*MCT*: 38–41
Strssbg. 362	*MKT* I: 239–43
Strssbg. 363	*MKT* I: 243–48
Strssbg. 364	*MKT* I: 248–56
Strssbg. 366	*MKT* I: 257–59
Strssbg. 367	*MKT* I: 259–63
Strssbg. 368	*MKT* I: 311–14
TMS, Text 1	*TMS*: 22–23
TMS, Text 2	*TMS*: 23–24
TMS, Text 3	*TMS*: 25–34
TMS, Text 5	*TMS*: 35–49
TMS, Text 6	*TMS*: 49–51
TMS, Text 7	*TMS*: 52–57
TMS, Text 8	*TMS*: 58–62
TMS, Text 9	*TMS*: 63–69
TMS, Text 10	*TMS*: 69–74
TMS, Text 11	*TMS*: 74–78
TMS, Text 12	*TMS*: 78–81
TMS, Text 13	*TMS*: 82–83
TMS, Text 14	*TMS*: 84–88
TMS, Text 15	*TMS*: 88–91
TMS, Text 16	*TMS*: 91–95
TMS, Text 17	*TMS*: 95–97
TMS, Text 18	*TMS*: 97–101
TMS, Text 19	*TMS*: 101–5
TMS, Text 20	*TMS*: 105–08
TMS, Text 21	*TMS*: 108–11
TMS, Text 22	*TMS*: 111–14
TMS, Text 23	*TMS*: 114–17
TMS, Text 24	*TMS*: 117–22
TMS, Text 25	*TMS*: 122–24
TMS, Text 26	*TMS*: 124–29
UET 3: no. 1386	Vaiman 1961: 244–46
UET 5: no. 121	Vaiman 1961: 248–50
UET 5: no. 858	Vaiman 1961: 250–54
UET 5: no. 859	Vaiman 1961: 254–56
UET 5: no. 864	Kilmer 1964: 140–42
UM 29-15-192	Neugebauer and Sachs 1984: 248
UMB 16/2 pl. 7	Steele 1951: 25: pl. 7
VAT 672	*MKT* I: 267–68

VAT 6469	*MKT* I: 268–69
VAT 6505	*MKT* I: 270–73
VAT 6546	*MKT* I: 268–69
VAT 6597	*MKT* I: 274–77
VAT 6598 + BM 96957[1]	*MKT* I: 277–87
VAT 7528	*MKT* I: 508–13
VAT 7530	*MKT* I: 287–89
VAT 7531	*MKT* I: 289–94
VAT 7532	*MKT* I: 294–303
VAT 7535	*MKT* I: 303–10
VAT 7537	*MKT* I: 466–78
VAT 7620	*MKT* I: 314–17
VAT 7621	*MKT* I: 290–94
VAT 7848	*MCT*: 141–45
VAT 8389	*MKT* I: 317–35
VAT 8390	*MKT* I: 335–40
VAT 8391	*MKT* I: 319–35
VAT 8512	*MKT* I: 340–45
VAT 8520	*MKT* I: 346–51
VAT 8521	*MKT* I: 351–67
VAT 8522	*MKT* I: 367–73
VAT 8523	*MKT* I: 373–80
VAT 8528	*MKT* I: 353–67; *MKT* III: 59–61
VAT 8547	Sachs 1952: 153–56
YBC 4186	*MCT*: 91
YBC 4607	*MCT*: 91–94
YBC 4608	*MCT*: 49–53
YBC 4612	*MCT*: 103–05
YBC 4652	*MCT*: 100–03
YBC 4657	*MCT*: 66–76
YBC 4662	*MCT*: 71–76
YBC 4663	*MCT*: 69–76
YBC 4666	*MCT*: 76–81
YBC 4668	*MKT* I: 420–66
YBC 4669	*MKT* I: 514–16; *MKT* III: 26–29
YBC 4673	*MKT* I: 506–08; *MKT* III: 29–34
YBC 4675	*MCT*: 44–48
YBC 4695	*MKT* I: 501–02; *MKT* III: 34–39
YBC 4696	*MKT* I: 504–06; *MKT* II: 60–64
YBC 4697	*MKT* I: 485–86; *MKT* III: 40–41

1. Christopher Walker (Letter to author 1986), enclosed a copy of his join, BM 96957 to VAT 6598. Because Mr. Walker and Dr. Michael Roaf plan to publish this text, it will not be included here.

YBC 4698	*MKT* I: 513; *MKT* III: 42–45
YBC 4708	*MKT* I: 389-402
YBC 4709	*MKT* I: 412–20
YBC 4710	*MKT* I: 402–12
YBC 4711	*MKT* I: 503–04; *MKT* III: 45-48
YBC 4712	*MKT* I: 420-66
YBC 4713	*MKT* I: 420-66
YBC 4714	*MKT* I: 487-501
YBC 4715	*MKT* I: 478-85
YBC 5022	*MCT*: 132–36
YBC 5037	*MCT*: 59-66
YBC 6295	*MCT*: 42
YBC 6492	*MCT*: 105–06
YBC 6504	*MKT* III: 22–25
YBC 6967	*MCT*: 129–30
YBC 7164	*MCT*: 81–90
YBC 7243	*MCT*: 136–37
YBC 7273	*MCT*: 131
YBC 7284	*MCT*: 97–98
YBC 7289	*MCT*: 42–43
YBC 7290	*MCT*: 44
YBC 7302	*MCT*: 44
YBC 7326	*MCT*: 130–31
YBC 7997	*MCT*: 98-99
YBC 8588	*MCT*: 75–76
YBC 8600	*MCT*: 57–59
YBC 8633	*MCT*: 53–55
YBC 9852	*MCT*: 44–49
YBC 9856	*MCT*: 99–100
YBC 9874	*MCT*: 90-91
YBC 10522	*MCT*: 131
YBC 10722	*MCT*: 98
YBC 11120	*MCT*: 44
YBC 11126	*MCT*: 44
2N-T 30	Neugebauer and Sachs 1984: 247
2N-T 116	Neugebauer and Sachs 1984: 247
2N-T 174	Neugebauer and Sachs 1984: 245
2N-T 472	Neugebauer and Sachs 1984: 246–47
No Museum Number	*MCT*: 6–10

PROBLEMS: LIST OF TABLETS CATEGORIZED AS CONTAINING ALGEBRAIC/GEOMETRIC, PRACTICAL, OR MIXED PROBLEMS

TABLET	ALGEBRAIC/ GEOMETRIC	PRACTICAL	MIXED
A 24194	x		
A 24195	x		
AO 6484			x
AO 6770			x
A0 8862			x
A0 10642	x		
AO 10822		x	
AO 17264		x	
AOT 304		x[2]	
AOT 305		x[3]	
BM 13901	x		
BM 15285	x		
BM 34568			x
BM 80209	x		
BM 85194			x
BM 85196		x	
BM 85200 + VAT 6599	x		
BM 85210		x	
Böhl Coll. 1821		x	
CBM 12648	x		
CBS 11318	x		
Const. 236		x[4]	
CUA 34		x	
Db_2 146	x		
Erm 15073			x
Erm 15188	x		
Erm 015189	x		
Haddad 104		x	
IM 31210		x	
IM 31247	x		

2. AOT 304 is an Ur III tablet and not included in the tally in Chapter 3.

3. AOT 305 is an Ur III tablet and not included in the tally in Chapter 3.

4. Const. 236 may be dated to the Old Akkadian period and, therefore, is not included in the tally in Chapter 3.

TABLET	ALGEBRAIC/ GEOMETRIC	PRACTICAL	MIXED
IM 31248	x		
IM 43996	x		
IM 52301	x		
IM 52672	x		
IM 53953	x		
IM 53957		x	
IM 53961		x	
IM 53963	x		
IM 53965		x	
IM 54010		x	
IM 54011		x	
IM 54464		x	
IM 54472	x		
IM 54478	x		
IM 54538		x	
IM 54559	x		
IM 55347	x		
Ist 04360	x		
Ist 04552	x		
K 8705		x	
K 11097		x	
LBAT: no. 1647		x	
LBAT: no. 1648		x	
M 10	x		
M 6829		x	
M 6861		x	
M 8613		x	
M 15192 + M 15236		x	
MAH 16055	x		
MLC 1354	x		
MLC 1731	x		
MLC 1842		x	
MLC 1950	x		
MM 86.11.404		x	
NBC 7934		x	
NBC 8082	x		
NCBT 1913	x		
Plimpton 322	x		
Strssbg 362			x
Strssbg 363	x		
Strssbg 364	x		

TABLET	ALGEBRAIC/ GEOMETRIC	PRACTICAL	MIXED
Strssbg 366			x
Strssbg 367	x		
Strssbg 368		x	
TMS, Text 1	x		
TMS, Text 2	x		
TMS, Text 5	x		
TMS, Text 6	x		
TMS, Text 7	x		
TMS, Text 8	x		
TMS, Text 9	x		
TMS, Text 10		x	
TMS, Text 11	x		
TMS, Text 12	x		
TMS, Text 13		x	
TMS, Text 14		x	
TMS, Text 15		x	
TMS, Text 16	x		
TMS, Text 17	x		
TMS, Text 18	x		
TMS, Text 19	x		
TMS, Text 20	x		
TMS, Text 21	x		
TMS, Text 22		x	
TMS, Text 23	x		
TMS, Text 24		x	
TMS, Text 25		x	
TMS, Text 26		x	
UET 3: no. 1386		x[5]	
UET 5: no. 121		x	
UET 5: no. 858	x		
UET 5: no. 859	x		
UET 5: no. 864		x	
UM 29-15-192	x		
UMB 16/2	x		
VAT 672	x		
VAT 6469		x	
VAT 6505	x		
VAT 6546		x	

5. *UET* 3: no. 1386 is an Ur III tablet and not included in the tally in Chapter 3.

TABLET	ALGEBRAIC/ GEOMETRIC	PRACTICAL	MIXED
VAT 6597		x	
VAT 6598 + BM 96957		x	
VAT 7528		x	
VAT 7530		x	
VAT 7531		x	
VAT 7532		x	
VAT 7535		x	
VAT 7537	x		
VAT 7620	x		
VAT 7621		x	
VAT 7848			x
VAT 8389		x	
VAT 8390	x		
VAT 8391		x	
VAT 8512	x		
VAT 8520	x		
VAT 8521		x	
VAT 8522			x
VAT 8523		x	
VAT 8528		x	
VAT 8547	x		
YBC 4186		x	
YBC 4607		x	
YBC 4608			x
YBC 4612	x		
YBC 4652		x	
YBC 4657		x	
YBC 4662		x	
YBC 4663		x	
YBC 4666		x	
YBC 4668	x		
YBC 4669			x
YBC 4673		x	
YBC 4675	x		
YBC 4695	x		
YBC 4696	x		
YBC 4697	x		
YBC 4698		x	
YBC 4708		x	
YBC 4709	x		
YBC 4710	x		

TABLET	ALGEBRAIC/ GEOMETRIC	PRACTICAL	MIXED
YBC 4711	x		
YBC 4712	x		
YBC 4713	x		
YBC 4714	x		
YBC 4715	x		
YBC 5037		x	
YBC 6295	x		
YBC 6492	x		
YBC 6504	x		
YBC 6967	x		
YBC 7164		x	
YBC 7273		x	
YBC 7284		x	
YBC 7289	x		
YBC 7290	x		
YBC 7302	x		
YBC 7326		x	
YBC 7997		x	
YBC 8588		x	
YBC 8600		x	
YBC 8633	x		
YBC 9852	x		
YBC 9856		x	
YBC 9874		x	
YBC 10522	x		
YBC 10722		x	
YBC 11120	x		
YBC 11126	x		
2N-T 30	x		
2N-T 116	x		
2N-T 174	x		
2N-T 472	x		
No Museum Number (*MCT*: 6-10)	x[6]		

6. This notation refers to two tablets, which contain 97 metrological examples of areas of rectangles, squares and circles.

ALGEBRAIC/GEOMETRIC PROBLEMS

TABLET	NO. OF PROBLEMS	SUBJECT	SOLUTION WORKED YES	SOLUTION WORKED NO	ANSWER GIVEN YES	ANSWER GIVEN NO	DRAWING YES	DRAWING NO	DATE
A 24194[7]	247[8]	#1–247: equations		x		x		x	OB
A 24195	177	#1–177: equations		x		x		x	OB
AO 6484*	17	#1: geometrical series	x		x			x	Seleucid
		#2: geometrical series	x		x			x	
		#6: plane geometric figure		x	x			x	
		#8: plane geometric figure	x		[x]			x	
		#9: solid geometric figure	x		[x]			x	
		#10: solid geometric figure	x		[x]			x	
		#11: solid geometric figure	x		[x]			x	
		#12: solid geometric figure	x		[x]			x	
		#13: solid geometric figure	x		[x]			x	
		#14: numbers and reciprocals	x		x			x	
		#15: numbers and reciprocals	x		x			x	
		#16: numbers and reciprocals	x		x			x	

7. The subject is listed for each problem unless more than 10 consecutive examples of the same subject appear. The only exceptions to this rule are BM 85200 + VAT 6599 and Strssbg 364, in which thirty and fourteen problems, respectively, are listed separately to clarify the alternation of broken text with defined problems.

8. A 24914 contains 7 principal problems and 240 variations; the text refers only to 240 items, that is, in rev. v 6-7, 4 **šu-si im-šu** '4 sixties (= 4,0) sections.'

TABLET	NO. OF PROBLEMS	SUBJECT	SOLUTION WORKED YES	SOLUTION WORKED NO	ANSWER GIVEN YES	ANSWER GIVEN NO	DRAWING YES	DRAWING NO	DATE
		#17: numbers and reciprocals	x		x			x	
AO 6770*	5	#1: equations	x			x		x	OB
AO 8862*	8	#1: equations	x		x			x	OB
		#2: equations	x		x			x	
		#3: equations	x		x			x	
		#4: equations		x		x		x	
AO 10642	2 extant	#1: plane geometric figure		x		x	x		OB
		#2: plane geometric figure		x		x	x		
BM 13901	24 (21 extant)	equations	x		x			x	OB
BM 15285	50 (not all extant)	plane geometric figures		x		x	x		OB
BM 34568*	20(+) (19 extant)	#1–11: equations	x		x			x	Seleucid
		#13–15: equations	x		x			x	
		#17–19: equations	x		x			x	
BM 80209	24	#1–24: plane geometric figures		x		x		x	OB
BM 85194*	35	#20: plane geometric figure	x		x		x		Kassite or older†
		#21: plane geometric figure	x		x		x		
		#29: plane geometric figure	x		x		x		
		#33: plane geometric figure	x		x			x	

TABLET	NO. OF PROBLEMS	SUBJECT	SOLUTION WORKED YES	SOLUTION WORKED NO	ANSWER GIVEN YES	ANSWER GIVEN NO	DRAWING YES	DRAWING NO	DATE
BM 85200 +VAT 6599	30	#1: broken	?	?	?	?	?	?	OB
		#2: broken	[x]		[x]			[x]	
		#3: broken	[x]		[x]			[x]	
		#4: broken	[x]		[x]			[x]	
		#5: equations	x		x			x	
		#6: equations	x		x			x	
		#7: equations	x		x			x	
		#8: equations	x		x			x	
		#9: equations	x		x			x	
		#10: broken	?	?	?	?	?	?	
		#11: broken	[x]		[x]			[x]	
		#12: equations	x		x			x	
		#13: equations	x		x			x	
		#14: equations	x		x			x	
		#15: equations	x		x			x	
		#16: equations	x		x			x	
		#17: equations	x		x			x	
		#18: equations	x		x			x	
		#19: equations	x		x			x	
		#20: equations	x		x			x	
		#21: equations	x		x			x	
		#22: equations	x		x			x	
		#23: equations	x		x			x	
		#24: equations	x		x			x	
		#25: equations	x		x			x	
		#26: equations	x		x			x	
		#27: equations	x		x			x	
		#28: broken	?	?	?	?	?	?	
		#29: equations	x		x			x	
		#30: equations	x		x			x	
CBM 12648	?	solid geometric figure; fragmentary	x		x			x	OB
CBS 11318	1	plane geometric figure/metrology	x		x			x	OB

TABLET	NO. OF PROBLEMS	SUBJECT	SOLUTION WORKED YES	SOLUTION WORKED NO	ANSWER GIVEN YES	ANSWER GIVEN NO	DRAWING YES	DRAWING NO	DATE
Db_2 146	1	plane geometric figure	x		x		x		OB
Erm 15073*	8	#1: plane geometric figure	x		x			x	OB
		#2: plane geometric figure	x		[x]		x		
		#3: plane geometric figure	x		[x]			x	
Erm 15188	1	plane geometric figure		x		x	x		OB
Erm 015189	2	obv.: plane geometric figure[9]		x		x	x		OB
		rev.: plane geometric figure		x		x	x		
IM 31247	8	#1: equations	x		x			x	OB
		#2: equations	x		x			x	
		#3: equations	x		x			x	
		#4: equations	x		x			x	
		#5: equations	x		x			x	
		#6: equations	x		x			x	
		#7: equations	x		x			x	
		#8: equations	x		x			x	
IM 31248	2	#1: plane geometric figure	x		x		x		OB
		#2: plane geometric figure	x		x			x	

9. A series of drawings appear on both the obverse and reverse of the tablet. The data is the same for each group of problems.

TABLET	NO. OF PROBLEMS	SUBJECT	SOLUTION WORKED YES	SOLUTION WORKED NO	ANSWER GIVEN YES	ANSWER GIVEN NO	DRAWING YES	DRAWING NO	DATE
IM 43996	1[10]	plane geometric figures	x			x	x		OB
IM 52301	3	#1: plane geometric figure	x		x			x	OB
		#2: plane geometric figure	x		x			x	
		#3: plane geometric figure	x		x			x	
IM 52672	1	equations	x		x			x	OB
IM 53953	1	plane geometric figure	x		x			x	OB
IM 53963	1	plane geometric figure	x		x			x	OB
IM 54472	1	square roots[11]	x		x			x	OB
IM 54478	1	solid geometric figure	x		x			x	OB
IM 54559	1	plane geometric figure	x		x			x	OB
IM 55357	1	plane geometric figure	x		x		x		OB
Ist 04360	16 extant	plane geometric figures		x		x	x		OB

10. The obverse contains a series of triangles with transversals and inscribed numbers. On the reverse the same figure is copied by the pupil but with calculations.

11. The method of solution is similar to YBC 6295.

TABLET	NO. OF PROBLEMS	SUBJECT	SOLUTION WORKED YES	SOLUTION WORKED NO	ANSWER GIVEN YES	ANSWER GIVEN NO	DRAWING YES	DRAWING NO	DATE
Ist 04552	?	plane geometric figures; fragmentary		x		x	x	x[12]	OB
M 10	6	#1: numbers and reciprocals		x	x			x	OB
		#2: numbers and reciprocals		x	x			x	
		#3: numbers and reciprocals		x	x			x	
		#4: numbers and reciprocals		x	x			x	
		#5: numbers and reciprocals		x	x			x	
		#6: numbers and reciprocals		x	x			x	
MAH 16055	2	obv.: series of plane geometric figures		x		x	x		OB
		rev.: series of plane geometric figures		x		x	x		
MLC 1354	1	plane geometric figures	x		?	?	x		OB
MLC 1731	9	#1: fractional expressions		x	x			x	OB
		#2: fractional expressions		x	x			x	
		#3: fractional expressions		x	x			x	
		#4: fractional expressions		x	x			x	
		#5: fractional expressions		x	x			x	

12. Ist 04552 is a fragmentary text. The obverse contains three fragmentary problems; drawings of circles with inscribed numbers appear in an adjoining column. The reverse is completely destroyed.

TABLET	NO. OF PROBLEMS	SUBJECT	SOLUTION WORKED YES	SOLUTION WORKED NO	ANSWER GIVEN YES	ANSWER GIVEN NO	DRAWING YES	DRAWING NO	DATE
		#6: fractional expressions		x	x			x	
		#7: fractional expressions		x	x			x	
		#8: fractional expressions		x	x			x	
		#9: fractional expressions		x	x			x	
MLC 1950	1(+?)	obv.: plane geometric figure	x		[x]		x		OB
		rev.: broken	?	?	?	?	?	?	
NBC 8082	1	plane geometric figure	x		x			x	OB
NCBT 1913	1	plane geometric figure/metrology	x		x			x	OB
Plimpton 322	17 extant	Pythagorean numbers	x		x			x	OB
Strssbg 362*	6	#3: equations		x	x			x	OB
		#4: equations		x	x			x	
Strssbg 363	3	#1: equations	x		x			x	OB
		#2: equations	x		x			x	
		#3: equations	x		x			x	
Strssbg 364	14	#1: plane geometric figure		x		x	x		OB
		#2: plane geometric figure		x		x	x		
		#3: plane geometric figure		x		x	x		
		#4: broken		[x]		[x]	[x]		
		#5: broken		[x]		[x]	[x]		
		#6: plane geometric figure		x		x	x		

TABLET	NO. OF PROBLEMS	SUBJECT	SOLUTION WORKED YES	SOLUTION WORKED NO	ANSWER GIVEN YES	ANSWER GIVEN NO	DRAWING YES	DRAWING NO	DATE
		#7: plane geometric figure		x		x	x		
		#8: plane geometric figure		x		x	x		
		#9: plane geometric figure		x		x	x		
		#10: plane geometric figure		x		x	x		
		#11: plane geometric figure		x		x	x		
		#12: plane geometric figure		x		x	x		
		#13: plane geometric figure		x		x	x		
		#14: plane geometric figure		x		x	x		
Strssbg 366*	4	#1: numbers and reciprocals	x		x			x	OB
		#2: numbers and reciprocals	x		x			x	
		#3: numbers and reciprocals	x		x			x	
Strssbg 367	1	plane geometric figure	x		x		x		OB
TMS, Text 1	1	plane geometric figure		x	x		x		OB
TMS, Text 2	2	#1: plane geometric figure		x	x		x		OB
		#2: plane geometric figure	x			x	x		
TMS, Text 5	262	equations		x		x		x	OB

TABLET	NO. OF PROBLEMS	SUBJECT	SOLUTION WORKED YES	SOLUTION WORKED NO	ANSWER GIVEN YES	ANSWER GIVEN NO	DRAWING YES	DRAWING NO	DATE
TMS, Text 6	59(+)	equations		x		x		x	OB
TMS, Text 7	2	#1: equations	x		x			x	OB
		#2: equations	x		x			x	
TMS, Text 8	3 (2 extant)	#1: equations	x		x			x	OB
		#2: equations	x		x			x	
TMS, Text 9	3	#1: equations	x		x			x	OB
		#2: equations	x		x			x	
		#3: equations	x		x			x	
TMS, Text 11	2	#1: equations	x		[x]			x	OB
		#2: equations	x		x			x	
TMS Text 12,	1	equations	x		[x]			x	OB
TMS, Text 16	2	#1: equations	x		x			x	OB
		#2: equations	x		x			x	
TMS, Text 17	1	equations	x		x			x	OB
TMS, Text 18	1	equations	x		x			x	OB
TMS, Text 19	2	#1: equations	x		x			x	OB
		#2: equations	x		x			x	
TMS, Text 20	2	#1: equations	x		x			x	OB
		#2: equations	x		x			x	
TMS, Text 21	1	plane geometric figure	x		x			x	OB
TMS, Text 23	1	plane geometric figure	x		x			x	OB

TABLET	NO. OF PROBLEMS	SUBJECT	SOLUTION WORKED YES	SOLUTION WORKED NO	ANSWER GIVEN YES	ANSWER GIVEN NO	DRAWING YES	DRAWING NO	DATE
UET 5: no. 858	1	plane geometric figure	x		x			x	OB
UET 5: no. 859	2	#1: equations	x		x			x	OB
		#2: equations		x		x		x	
UM 29-15-192	1	plane geometric figure/metrology	x		x			x	OB
UMB 16/2	1	plane geometric figure/metrology	x		x			x	OB
VAT 672	2	#1: solid geometric figure	x		[x]			x	OB
		#2: solid geometric figure	x		[x]			x	
VAT 6505	12 (6 extant)	#1: numbers and reciprocals	x		[x]			x	OB
		#2: numbers and reciprocals	x		[x]			x	
		#3: numbers and reciprocals	x		x			x	
		#4: broken	?	?	?	?	?	?	
		#5: broken	?	?	?	?	?	?	
		#6: numbers and reciprocals	x		x			x	
VAT 7537	44 (42 extant)	equations		x		x		x	Kassite or older†
VAT 7620	1	plane geometric figure	x		?	?		x	OB
VAT 7848*	7(+)	#2: plane geometric figure	x		x			x	Seleucid
		#5: solid geometric figure	x		x			x	
VAT 8390	2	#1: equations	x		x			x	OB

TABLET	NO. OF PROBLEMS	SUBJECT	SOLUTION WORKED YES	NO	ANSWER GIVEN YES	NO	DRAWING YES	NO	DATE
		#2: equations	x		x			x	
VAT 8512	1	plane geometric figure	x		x			x	OB
VAT 8520	2	#1: numbers and reciprocals	x		x			x	OB
		#2: numbers and reciprocals	x		x			x	
VAT 8522*	4	#4: equations	x		x			x	OB
VAT 8547[13]	4	#1: cube roots	x		x			x	OB
		#2: cube roots	x		x			x	
		#3: cube roots	x		x			x	
		#4: cube roots	x		x			x	
YBC 4608*	5	#1: plane geometric figure	x		x			x	OB
		#2: plane geometric figure	?	?	?	?		x	
		#3: plane geometric figure	x		?	?		x	
		#4: plane geometric figure	x		x			x	
YBC 4612	15	#1–15: plane geometric figures		x	x			x	OB
YBC 4668[14]	286 (181 extant)	equations		x	x[15]	x		x	Kassite or older†

13. VAT 8547 contains four cube root problems, which assume the student is familiar with tables of cube roots.

14. YBC 4668 is treated as the main text and YBC 4713 and YBC 4712 as containing variants to it.

15. Answers are given in only five cases: problems #1, #11 (var: YBC 4712) #C85, #C89 and #C95.

TABLET	NO. OF PROBLEMS	SUBJECT	SOLUTION WORKED YES	SOLUTION WORKED NO	ANSWER GIVEN YES	ANSWER GIVEN NO	DRAWING YES	DRAWING NO	DATE
YBC 4669*	26 (25 extant)	#B2: equations		x		x		x	
YBC 4675	1	plane geometric figure	x		x		x		OB
YBC 4695	97	#1–97: equations		x		x		x	Kassite or older†
YBC 4696[16]	52	#1–52: plane geometric figures		x		x		x	Kassite or older†
YBC 4697	25	#1–25: equations		x		x		x	Kassite or older†
YBC 4709	55	#1–55: equations		x		x		x	Kassite or older†
YBC 4710	35	#1–35: equations		x		x		x	Kassite or older†
YBC 4711	131 (100 extant)	equations		x		x		x	Kassite or older†
YBC 4712		See YBC 4668							
YBC 4713		See YBC 4668							
YBC 4714	43 (39 extant)	equations		x	x	x[17]		x	Kassite or older†

16. The problems in YBC 4696 contain first and second degree equations, the elements of which are transferred to a triangle, which is divided into two parts by a transversal parallel to the base.

17. Answers given to all problems in YBC 4714 except problems #34 and #37.

TABLET	NO. OF PROBLEMS	SUBJECT	SOLUTION WORKED YES	SOLUTION WORKED NO	ANSWER GIVEN YES	ANSWER GIVEN NO	DRAWING YES	DRAWING NO	DATE
YBC 4715	50–60 (45 extant)	equations		x	x[18]	x		x	Kassite or older†
YBC 6295	1	cube root	x		x			x	OB
YBC 6492	24	equations		x		x		x	OB
YBC 6504	4	#1: equations	x		x			x	OB
		#2: equations	x		x			x	
		#3: equations	x		x			x	
		#4: equations	x		x			x	
YBC 6967	1	numbers and reciprocals	x		x			x	OB
YBC 7289	1	plane geometric figure		x	x		x		OB
YBC 7290	2	#1: plane geometric figure		x	x		x		OB
		#2: plane geometric figure[19]		x		x	x		
YBC 7302	1	plane geometric figure		x	x		x		OB
YBC 8633	3	#1: plane geometric figure	x		x		x[20]		OB
		#2: plane geometric figure	x		x		<x>		

18. Answers are given in only problems #1 and #1' in YBC 4715.

19. YBC 7290 rev. contains the drawing of a trapezoid without inscribed numbers.

20. The three problems of this text are related by paramaters and by the drawing in problem #1. Problem #1 involves **sag-du** 'triangle'; problems #2 and #3 concern **sag-ki-gu$_4$** 'trapezoid,' which refers to the auxiliary figure of the illustration, which is a triangle subdivided into three triangles.

TABLET	NO. OF PROBLEMS	SUBJECT	SOLUTION WORKED YES	SOLUTION WORKED NO	ANSWER GIVEN YES	ANSWER GIVEN NO	DRAWING YES	DRAWING NO	DATE
		#3: plane geometric figure	x		x		<x>		
YBC 9852	1	plane geometric figure	x	(part)	x		x		OB
YBC 10522	1	broken; numerical calculations	x	(part)	?	?		x	OB
YBC 11120	1	plane geometric figure		x	x		x		OB
YBC 11126	1	plane geometric figure		x	x		x		OB
2N-T 30	1	plane geometric figure/metrology	x		x			x	OB
2N-T 116	1	plane geometric figure/metrology	x		x			x	OB
2N-T 174	1	solid geometric figure	x		x		x		OB
2N-T 472	1	plane geometric figure/metrology	x		x			x	OB
No Museum Number (*MCT*: 6–10)	97	plane geometric figures/ metrology	x		x			x	OB

PRACTICAL PROBLEMS

TABLET	NO. OF PROBLEMS	SUBJECT	SOLUTION WORKED YES	SOLUTION WORKED NO	ANSWER GIVEN YES	ANSWER GIVEN NO	DRAWING YES	DRAWING NO	DATE
AO 6484*[21]	17	#3: wall with wood on top	x		x			x	Seleucid
		#4: wall with wood on top	x		x			x	
		#5: field	x		x			x	
		#7: field	x		x			x	
AO 6770*	5	#2: interest	x		x			x	OB
		#3: original quantity	x			x		x	
		#4: bitumen	x		x			x	
		#5: field plans	x			x		x	
AO 8862*	8	#5: bricks	x		x			x	OB
		#6: bricks	x		x			x	
		#7: bricks	x		x			x	
		#8: bricks	?	?	?	?		x	
AO 10822	10 extant	#1: bricks		x		x		x	OB
		#2: bricks		x		x		x	
		#3: bricks		x		x		x	
		#4: bricks		x		x		x	
		#5: bricks		x		x		x	
		#6: bricks		x		x		x	
		#7: bricks		x		x		x	
		#8: bricks		x		x		x	
		#9: bricks		x		x		x	
		#10: bricks		x		x		x	
AO 17264	1	inheritance	x		x			x	MB, Kassite
AOT 304	1	bricks		x	x			x	Ur III

21. The subject is listed for each problem unless more than 10 consecutive examples of the same subject appear. The only exception is NBC 7934, because the three problems are treated as a group due to their subject matter.

TABLET	NO. OF PROBLEMS	SUBJECT	SOLUTION WORKED YES	SOLUTION WORKED NO	ANSWER GIVEN YES	ANSWER GIVEN NO	DRAWING YES	DRAWING NO	DATE
AOT 305	4	#1: irrigation		x	x			x	Ur III
		#2: irrigation		x	x			x	
		#3: irrigation		x	x			x	
		#4: irrigation		x	x			x	
BM 34568*	20(+) (19 extant)	#12: reed leaning against wall	x		x			x	Seleucid
		#16: metalwork	x		x			x	
BM 85194*	35	#1: construction *(arammu)*	x		x			x	Kassite or older†
		#2: construction *(kikurru)*	x		x			x	
		#3: construction *(kikurru)*	x		x			x	
		#4: defense works	x		x		x		
		#5: construction (**bàd** = *dūru*)	x		x			x	
		#6: clepsydra	x		x			x	
		#7: clepsydra	x		x			x	
		#8: clepsydra	x		x			x	
		#9: field plan	x		x			x	
		#10: original quantity	x		x			x	
		#11: original quantity	x		x			x	
		#12: original quantity	x		x			x	
		#13: original quantity	x		x			x	
		#14: *kiššû* 'reed bundle'	x		x			x	
		#15: *kiššû* 'reed bundle'	x		x			x	
		#16: brickworks	x		x			x	
		#17: construction (**bàd** = *dūru*)	x		x			x	
		#18: construction (**bàd** = *dūru*)	x		x			x	
		#19: brickworks	x		x			x	

TABLET	NO. OF PROBLEMS	SUBJECT	SOLUTION WORKED YES	SOLUTION WORKED NO	ANSWER GIVEN YES	ANSWER GIVEN NO	DRAWING YES	DRAWING NO	DATE
		#22: standard measuring vessels	x		x			x	
		#23: work	x		x			x	
		#24: standard measuring vessels	x		x			x	
		#25: defense works	x		x			x	
		#26: defense works	x		x			x	
		#27: field	x		x			x	
		#28: construction (*ḫirītu*)	x		x			x	
		#30: bricks	x		x			x	
		#31: herding and herd growth	x		x			x	
		#34: clepsydra	x		x			x	
		#35: clepsydra	x		x			x	
BM 85196	18	#1: construction (*appu*)	x		x		x		Kassite or older[†]
		#2: *kiššû*	x		x			x	
		#3: construction (*tamlâtu*)	x		x			x	
		#4: construction (*tamlâtu*)	x		x			x	
		#5: *kiššû*	x		x			x	
		#6: construction	x		x			x	
		#7: construction (^{giš}*sikkatu*)	x		[x]			x	
		#8: construction	x		x			x	
		#9: wood leaning against wall	x		x			x	
		#10: construction (^{giš}*sussulu*)	x		x			x	
		#11: *nalbantu*	x		x			x	
		#12: work	x		[x]			x	
		#13: work	x		x			x	
		#14: standard measuring vessels	x		x			x	
		#15: canals	x		x		x		
		#16: canals	x		x			x	

TABLET	NO. OF PROBLEMS	SUBJECT	SOLUTION WORKED YES	NO	ANSWER GIVEN YES	NO	DRAWING YES	NO	DATE
		#17: construction (*arammu*)	x		x		x		
		#18: metalwork	x		x			x	
BM 85210	13 (12 extant)	#1: defense works	x		x			x	Kassite or older[†]
		#2: defense works	x		x			x	
		#3: defense works	x		x			x	
		#4: defense works	x		x			x	
		#5: defense works	x		x			x	
		#6: defense works	x		x			x	
		#7: defense works	x		x			x	
		#8: construction (**ki-gub** = *manzazu*)	x		x			x	
		#9: *nalbantu*	x		[x]			x	
		#11: original quantity	x		x			x	
		#12: original quantity	x		x			x	
		#13: clepsydra	x		x			x	
Böhl Coll. 1821	1	constructions (**uru**^{ki} = *ālu*)	x		x		x		OB
Const. 236	1	construction (**iz-zi** = *igāru*)		x	x			x	OAkk
CUA 34	4	#1: fields		x	x			x	OB
		#2: fields		x	x			x	
		#3: fields		x	x			x	
		#4: fields		x	x			x	
Erm 15073*	8	#4: canals	x		x			x	OB
		#5: canals	x		x		x		
		#6: construction (*iku*)	x		x		x		
		#7: construction (*iku*)	x		x		x		
		#8: construction (*pitiqtu*)	x		x			x	

TABLET	NO. OF PROBLEMS	SUBJECT	SOLUTION WORKED YES	SOLUTION WORKED NO	ANSWER GIVEN YES	ANSWER GIVEN NO	DRAWING YES	DRAWING NO	DATE
Haddad 104	10	#1: log	x		x			x	OB
		#2: log	x		x			x	
		#3: log	x		x			x	
		#4: standard measuring vessel	x		x			x	
		#5: standard measuring vessel	x		x			x	
		#6: *maḫīru* 'price equivalency'	x		x			x	
		#7: mud plaster	x		x			x	
		#8: mud plaster	x		x			x	
		#9: bricks	x		x			x	
		#10: bricks	x		x			x	
IM 31210	8	#1: inheritance	x		x			x	OB
		#2: inheritance	x		x			x	
		#3: inheritance	x		x			x	
		#4: inheritance	x		x			x	
		#5: inheritance	x		x			x	
		#6: inheritance	x		x			x	
		#7: inheritance	x		x			x	
		#8: inheritance	x		x			x	
IM 53957	1	original quantity	x		x			x	OB
IM 53961	1	construction (*pitiqtu*)	x		x			x	OB
IM 53965	1	fields	x		x			x	OB
IM 54010	1	fields	x		x			x	OB
IM 54011	2	#1: construction (*pitiqtu*)	x		x			x	OB
		#2: excavation	x		x			x	
IM 54464	1	*maḫīru* 'price equivalency'	x		x			x	OB
IM 54538	1	bricks	x		x			x	OB

TABLET	NO. OF PROBLEMS	SUBJECT	SOLUTION WORKED YES	NO	ANSWER GIVEN YES	NO	DRAWING YES	NO	DATE
K 8705	1	inheritance[22]	x	?	?			x	Late Assyrian
K 11097	1	inheritance[23]	x	?	?			x	Late Assyrian
LBAT: no. 1647	2 (1 extant)	#1: construction (**gur**$_7$ = *karû*)	x		x?			x	LB
LBAT: no. 1648	6 (3 extant)	#1: inheritance	x		x			x	LB
		#2: inheritance	x		x			x	
		#4: standard measuring vessels	x		x			x	
M 6829	19	tabulation of the number of sheep consecrated for an *ikribu* offering and the number of sheep set free		x	x			x	OB
M 6861	2	#1: accumulation of quantity of grain according to an official fixed tariff		x	x			x	OB
		#2: accumulation of quantity of grain according to an official fixed tariff		x	x			x	

22. K 8705 and K 11097 contain instructions and calculations concerning the lore of the *bārû*-priests, who examined the sheep livers. Like the Mari tabular texts, they form a bridge between pure mathematical problem texts and economic/administrative tablets.

23. See note 22.

TABLET	NO. OF PROBLEMS	SUBJECT	SOLUTION WORKED YES	SOLUTION WORKED NO	ANSWER GIVEN YES	ANSWER GIVEN NO	DRAWING YES	DRAWING NO	DATE
M 8613	30 entries	amount of grain accumulated each day of a month of 30 days		x	x			x	OB
M 15192 + M 15236	13/14	amount of precious metal lost from masses of gold and silver during a particular treatment		x	x			x	OB
MLC 1842	1	*maḫīru* 'price equivalency'	x		[x]			x	OB
MM 86.11.404	4 (1 extant)	#2: original quantity	x		[x]			x	Seleucid
NBC 7934	3	#1–#3: foundation for a wall and pillar		x	x			x	OB
Strssbg 362*	6	#1: inheritance	x		x			x	OB
		#2: *maḫīru* 'price equivalency'	?	?	x			x	
		#5: field plan		x	x			x	
		#6: construction (*arammu*)		x		x		x	
Strssbg 366*	4	#4: field plan	x		x			x	OB
Strssbg 368	1	field plan	x		x			x	OB
TMS, Text 10	1	inheritance	x		x			x	OB
TMS, Text 13	1	*maḫīru* 'price equivalency'	x		x			x	OB

TABLET	NO. OF PROBLEMS	SUBJECT	SOLUTION WORKED YES	SOLUTION WORKED NO	ANSWER GIVEN YES	ANSWER GIVEN NO	DRAWING YES	DRAWING NO	DATE
TMS, Text 14	1	construction (**gur**$_7$ = *karû*)	x		x			x	OB
TMS, Text 15	1	doors	x		x			x	OB
TMS, Text 22	1	interest	x		x			x	OB
TMS, Text 24	2	#1: canals/ excavation	x		x			x	OB
		#2: canals/ excavation	x		x			x	
TMS, Text 25	2	#1: canals	x		x			x	OB
		#2: forest and reed thicket	x		x			x	
TMS, Text 26	3(+)	#1: inheritance	x		x			x	OB
		#2: inheritance	x		x			x	
		#3: inheritance	x		x			x	
UET 3: no. 1386	1	construction (**iz-zi** = *igāru*)		x	x			x	Ur III
UET 5: no. 121	4	#1: inheritance		x	x			x	OB
		#2: herding and herd growth		x	x			x	
		#3: herding and herd growth		x	x			x	
		#4: herding and herd growth		x	x			x	
UET 5: no. 864	1	field	x		x			x	OB
VAT 6469	?	commerce	x		?	?		x	OB
VAT 6546	?	commerce	x		?	?		x	OB
VAT 6597	5	#1: inheritance	x		[x]			x	OB

TABLET	NO. OF PROBLEMS	SUBJECT	SOLUTION WORKED YES	SOLUTION WORKED NO	ANSWER GIVEN YES	ANSWER GIVEN NO	DRAWING YES	DRAWING NO	DATE
		#2: inheritance	x		x			x	
		#3: inheritance	x		x			x	
		#4: inheritance	x		x			x	
		#5: inheritance	x		[x]			x	
VAT 6598	7(+)	#1: bricks	[x]		[x]			x	OB
+ BM 96957	(6 extant)	#2: bricks	x		x			x	
		#3: bricks	x		x			x	
		#4: bricks	x		x			x	
		#5: doors	x		x		x		
		#6: doors	x		x		x		
VAT 7528	24/25 (6 extant)	#1–19: canals[24]		x	x			x	Kassite or older[†]
VAT 7530	6	#1: *maḫīru* 'price equivalency'		x		x		x	OB
		#2: *maḫīru* 'price equivalency'		x		x		x	
		#3: *maḫīru* 'price equivalency'		x		x		x	
		#4: *maḫīru* 'price equivalency'		x		x		x	
		#5: *maḫīru* 'price equivalency'		x		x		x	
		#6: *maḫīru* 'price equivalency'		x		x		x	
VAT 7531	4	#1: inheritance		x		x		x	OB
		#2: inheritance		x		x		x	
		#3: inheritance		x		x		x	
		#4: inheritance		x		x		x	
VAT 7532	1	field plan	x		x		x		OB
VAT 7535	3	#1: field plan	x		x		x		OB
		#2: field plan	x		x		x		

24. The answers to problems #6, #7, #9, #10 and #19 are found on the tablet; the answers to problems #5, #8 and #11–18 can be restored.

TABLET	NO. OF PROBLEMS	SUBJECT	SOLUTION WORKED YES	NO	ANSWER GIVEN YES	NO	DRAWING YES	NO	DATE
		#3: field plan		x		x	x		
VAT 7621	2 extant	#1: inheritance		x		x	x		OB
		#2: inheritance		x		x		x	
VAT 7848*	7(+)	#1: field	x		x			x	Seleucid
		#3: field	x		x			x	
		#4: field	x		x			x	
		#6: field	x		x			x	
		#7: field	x		x			x	
VAT 8389	4	#1: fields	x		x			x	OB
		#2: fields	x		x			x	
		#3: fields	x		x			x	
		#4: fields	x		x			x	
VAT 8391	6	#1: fields	x		x			x	OB
		#2: fields	x		x			x	
		#3: fields	x		x			x	
		#4: fields	x		x			x	
		#5: fields	x		x			x	
		#6: fields	x		x			x	
VAT 8521	4	#1: interest	x		x			x	OB
		#2: interest	x		x			x	
		#3: interest	x		x			x	
		#4: interest	x		x			x	
VAT 8522*	4	#1: logs	x		x		x		OB
		#2: inheritance	x		x			x	
		#3: herding and herd growth		x	x			x	
VAT 8523	6	#1: canals	x		x			x	OB
		#2: canals	x		x			x	
		#3: canals	x		x			x	
		#4: canals	x		x			x	
		#5: canals	x		x			x	
		#6: canals	x		x			x	
VAT 8528	2	#1: interest	x		x			x	OB

TABLET	NO. OF PROBLEMS	SUBJECT	SOLUTION WORKED YES	NO	ANSWER GIVEN YES	NO	DRAWING YES	NO	DATE
		#2: interest	x		x			x	
YBC 4186	1	irrigation	x		x			x	OB
YBC 4607	10	#1: bricks		x	x			x	OB
		#2: bricks		x	x			x	
		#3: bricks		x	x			x	
		#4: bricks		x	x			x	
		#5: bricks		x	x			x	
		#6: bricks		x	x			x	
		#7: bricks		x	x			x	
		#8: bricks		x	x			x	
		#9: bricks		x	x			x	
		#10: bricks		x	x			x	
YBC 4608*	5	#5: inheritance	x		x			x	OB
YBC 4652	11[25]	#3: original quantity		x	[x]			x	OB
		#4: original quantity		x	[x]			x	
		#5: original quantity		x	[x]			x	
		#6: original quantity		x	[x]			x	
		#7: original quantity		x	x			x	
		#8: original quantity		x	x			x	
		#9: original quantity		x	x			x	
		#10: original quantity		x	x			x	
		#19: original quantity		x	x			x	
		#20: original quantity		x	x			x	

25. The colophon notes 22 problems but only 12 are partly preserved, of which 6 can be fully restored.

TABLET	NO. OF PROBLEMS	SUBJECT	SOLUTION WORKED YES	SOLUTION WORKED NO	ANSWER GIVEN YES	ANSWER GIVEN NO	DRAWING YES	DRAWING NO	DATE
		#21: original quantity		x	x			x	
		#22: original quantity		x	x			x	
YBC 4657[26]	31	#1–3: excavation (**ki-lá** = *kalakku*)		x	x			x	OB
		#4–6: excavation (**ki-lá** = *kalakku*)		x	[x]			x	
		#7: excavation (**ki-lá** = *kalakku*)		x		x		x	
		#8–16: excavation (**ki-lá** = *kalakku*)		x	x			x	
		#17–19: excavation (**ki-lá** = *kalakku*)		x		x		x	
		#20–31: excavation (**ki-lá** = *kalakku*)		x	x			x	
YBC 4662	10	#19–22: excavation (**ki-lá** = *kalakku*)	x		x			x	OB
		#23: excavation (**ki-lá** = *kalakku*)	[x]		[x]			x	
		#24–26: excavation (**ki-lá** = *kalakku*)	x		x			x	
		#27: excavation (**ki-lá** = *kalakku*)	x		[x]			x	
		#28: excavation (**ki-lá** = *kalakku*) See YBC 4657	x		x			x	

26. YBC 4657, YBC 4662 and YBC 4663 form a group containing 31 problems, systematically arranged, dealing with **ki-lá** = *kalakku* 'excavations (of volumes of earth).' YBC 4657 gives the data and most of the answers to 31 problems, which are worked out in three other tablets of which two are preserved, YBC 4662 and YBC 4663. YBC 4663 treats problems #1–#8, and YBC 4662 treats problems #19–28. The parameters are the same for all problems (with the exception of #31, which may be the first example of another tablet which followed.

TABLET	NO. OF PROBLEMS	SUBJECT	SOLUTION WORKED YES	SOLUTION WORKED NO	ANSWER GIVEN YES	ANSWER GIVEN NO	DRAWING YES	DRAWING NO	DATE
YBC 4663	8	#1–8: excavation (**ki-lá** = *kalakku*) See YBC 4657	x		x			x	OB
YBC 4666	23 extant[27]	#1–22: canal		x	x			x	OB
		#23: canal		x		x		x	
YBC 4669*	26 (25 extant)	#1: standard measuring vessels		x	x			x	Kassite or older†
		#2: standard measuring vessels		x	x			x	
		#3: standard measuring vessels		x	x			x	
		#4: standard measuring vessels		x	x			x	
		#5: standard measuring vessels		x	x			x	
		#6: standard measuring vessels		x	x			x	
		#7: standard measuring vessels		x	x			x	
		#8: standard measuring vessels		x	x			x	
		#9: standard measuring vessels		x	x			x	
		#10: bricks		x	x			x	
		#11: bricks		x	[x]			x	
		#12: bricks		x	x			x	
		#13: bricks		x	x			x	
		#14: bricks		x	[x]			x	
		#B1: work	x		x			x	
		#B3: original quantity		x	x			x	
		#B4: original quantity		x	x			x	
		#B5: original quantity		x	[x]			x	
		#B6: metalwork		x	x			x	

27. According to the colophon the text includes 26 problems; however, only 23 are extant.

TABLET	NO. OF PROBLEMS	SUBJECT	SOLUTION WORKED		ANSWER GIVEN		DRAWING		DATE
			YES	NO	YES	NO	YES	NO	
		#B7: herding and herd growth		x		x		x	
		#B8: herding and herd growth		x		x		x	
		#B9: standard measuring vessels		x	[x]			x	
		#B10: work		x	x			x	
		#B11: interest		x	x			x	
YBC 4673	23	#1: bricks		x		x		x	Kassite
	(21 extant)	#2: bricks		x	x			x	or older†
		#3: bricks		x		x		x	
		#4: bricks		x	x			x	
		#5: bricks		x	x	(part)		x	
		#6: bricks		x		x		x	
		#7: bricks		x	?	?		x	
		#8: bricks		x	x?	x		x	
		#9: construction (*pitiqtu*)		x	x			x	
		#10: construction (*pitiqtu*)		x	x			x	
		#11: construction (*pitiqtu*)		x	x			x	
		#12: construction (*pitiqtu*)		x	?	?		x	
		#13: construction (e = *iku*)		x		x		x	
		#14: construction (e = *iku*)		x	[x]			x	
		#15: construction (e = *iku*)		x	x			x	
		#16: work		x		x		x	
		#17: work	x		x			x	
		#18: work	x		x			x	
		#19: work	x		x			x	
		#20: work	x		x			x	
		#21: work	x		x			x	
YBC 4698	17	#1: interest		x		x		x	Kassite
		#2: interest		x		x		x	or older†
		#3: commerce		x		x		x	

TABLET	NO. OF PROBLEMS	SUBJECT	SOLUTION WORKED YES	SOLUTION WORKED NO	ANSWER GIVEN YES	ANSWER GIVEN NO	DRAWING YES	DRAWING NO	DATE
		#4: commerce		x		x		x	
		#5: commerce		x		x		x	
		#6: commerce		x	x			x	
		#7: commerce		x	x			x	
		#8: commerce		x	x			x	
		#9: commerce		x	x			x	
		#10: commerce		x	x			x	
		#11: commerce		x	x			x	
		#12: commerce		x		x		x	
		#13: commerce		x		x		x	
		#14: *maḫīru* 'price equivalency'		x		x		x	
		#15: *maḫīru* 'price equivalency'		x		x		x	
		#16: *maḫīru* 'price equivalency'		x		x		x	
		#17: original quantity		x	x			x	
YBC 4708	60	#1–60: bricks							Kassite
		#1–5,10,16,34, 37,40,47,49–55:		x	x			x	or older[†]
		#21:		x	[x]			x	
		#6,11–15,17–20, 22–33,35,36:		x		x		x	
		#38,39,41–46,48, 56–60:		x		x		x	
		#7–9:		x	?	?		x	
YBC 5037	44	#1–44: excavation (**ki-lá** = *kalakku*)							OB
		#1,4,10,13–24, 26,28–30,32,33:		x		x		x	
		#2,3,5–9,11,12, 25,27,31,34–44:		x	x			x	
YBC 7164	18	#1–18: canal		x	x			x	OB

TABLET	NO. OF PROBLEMS	SUBJECT	SOLUTION WORKED YES	SOLUTION WORKED NO	ANSWER GIVEN YES	ANSWER GIVEN NO	DRAWING YES	DRAWING NO	DATE
YBC 7273	1	herding and herd growth		x	x			x	OB
YBC 7284	1	bricks	x		x			x	OB
YBC 7326	1	herding and herd growth		x	x			x	OB
YBC 7997	1	brickworks	x			x		x	OB
YBC 8588	1	excavation (**ki-lá** = *kalakku*)	x		x			x	OB
YBC 8600	1	log	x		x			x	OB
YBC 9856	2	#1: work		x	x			x	OB
		#2: inheritance		x	x			x	
YBC 9874	2	#1: canal		x		x	x		OB
		#2: canal	x		x			x	
YBC 10722	1	bricks	?	?	?	?		x	OB

MIXED: ALGEBRAIC/GEOMETRIC AND PRACTICAL PROBLEMS

TABLET	NO. OF PROBLEMS	SUBJECT	SOLUTION WORKED YES	SOLUTION WORKED NO	ANSWER GIVEN YES	ANSWER GIVEN NO	DRAWING YES	DRAWING NO	DATE
AO 6484*	17	#1: geometrical series	x		x			x	Seleucid
		#2: geometrical series	x		x			x	
		#3: wall with wood on top	x		x			x	
		#4: wall with wood on top	x		x			x	
		#5: field	x		x			x	
		#6: plane geometric figure		x	x			x	
		#7: field	x		x			x	
		#8: plane geometric figure	x		[x]			x	
		#9: solid geometric figure	x		[x]			x	
		#10: solid geometric figure	x		[x]			x	
		#11: solid geometric figure	x		[x]			x	
		#12: solid geometric figure	x		[x]			x	
		#13: solid geometric figure	x		[x]			x	
		#14: numbers and reciprocals	x		x			x	
		#15: numbers and reciprocals	x		x			x	
		#16: numbers and reciprocals	x		x			x	
		#17: numbers and reciprocals	x		x			x	
AO 6770*	5	#1: equations	x			x		x	OB
		#2: interest	x		x			x	
		#3: original quantity	x			x		x	

TABLET	NO. OF PROBLEMS	SUBJECT	SOLUTION WORKED YES	SOLUTION WORKED NO	ANSWER GIVEN YES	ANSWER GIVEN NO	DRAWING YES	DRAWING NO	DATE
		#4: bitumen	x		x			x	
		#5: field plans	x			x		x	
AO 8862*	8	#1: equations	x		x			x	OB
		#2: equations	x		x			x	
		#3: equations	x		x			x	
		#4: equations		x		x		x	
		#5: bricks	x		x			x	
		#6: bricks	x		x			x	
		#7: bricks	x		x			x	
		#8: bricks	?	?	?	?	?	x	
BM 34568*	20(+) (19 extant)	#1–11: equations	x		x			x	Seleucid
		#12: reed leaning against wall	x		x			x	
		#13: equations	x		x			x	
		#14: equations	x		x			x	
		#15: equations	x		x			x	
		#16: metalwork	x		x			x	
		#17–19: equations	x		x			x	
BM 85194*	35	#1: construction *(arammu)*	x		x			x	Kassite or older†
		#2: construction *(kikurru)*	x		x			x	
		#3: construction *(kikurru)*	x		x			x	
		#4: defense works	x		x		x		
		#5: construction (**bàd** = *dūru*)	x		x			x	
		#6: clepsydra	x		x			x	
		#7: clepsydra	x		x			x	
		#8: clepsydra	x		x			x	
		#9: field plan	x		x			x	
		#10: original quantity	x		x			x	
		#11: original quantity	x		x			x	

TABLET	NO. OF PROBLEMS	SUBJECT	SOLUTION WORKED YES	SOLUTION WORKED NO	ANSWER GIVEN YES	ANSWER GIVEN NO	DRAWING YES	DRAWING NO	DATE
		#12: original quantity	x		x			x	
		#13: original quantity	x		x			x	
		#14: *kiššû* 'reed bundle'	x		x			x	
		#15: *kiššû* 'reed bundle'	x		x			x	
		#16: brickworks	x		x			x	
		#17: construction (**bàd** = *dūru*)	x		x			x	
		#18: construction (**bàd** = *dūru*)	x		x			x	
		#19: brickworks	x		x			x	
		#20: plane geometric figure	x		x		x		
		#21: plane geometric figure	x		x		x		
		#22: standard measuring vessels	x		x			x	
		#23: work	x		x			x	
		#24: standard measuring vessels	x		x			x	
		#25: defense works	x		x			x	
		#26: defense works	x		x			x	
		#27: field	x		x			x	
		#28: construction (*ḫirītu*)	x		x			x	
		#29: plane geometric figure	x		x		x		
		#30: bricks	x		x			x	
		#31: herding and herd growth	x		x			x	
		#32: fragmentary	x		x			x	
		#33: plane geometric figure	x		x			x	
		#34: clepsydra	x		x			x	
		#35: clepsydra	x		x			x	

TABLET	NO. OF PROBLEMS	SUBJECT	SOLUTION WORKED YES	NO	ANSWER GIVEN YES	NO	DRAWING YES	NO	DATE
Erm 15073*	8	#1: plane geometric figure	x		x			x	OB
		#2: plane geometric figure	x		[x]		x		
		#3: plane geometric figure	x		[x]			x	
		#4: canals	x		x			x	
		#5: canals	x		x		x		
		#6: construction (*iku*)	x		x		x		
		#7: construction (*iku*)	x		x		x		
		#8: construction (*pitiqtu*)	x		x			x	
Strssbg 362*	6	#1: inheritance	x		x			x	OB
		#2: *maḫīru* 'price equivalency'	?	?	x			x	
		#3: equations		x	x			x	
		#4: equations		x	x			x	
		#5: field plan		x	x			x	
		#6: construction (*arammu*)		x		x		x	
Strssbg 366*	4	#1: numbers and reciprocals	x		x			x	OB
		#2: numbers and reciprocals	x		x			x	
		#3: numbers and reciprocals	x		x			x	
		#4: field plan	x		x			x	
VAT 7848*	7(+)	#1: field	x		x			x	Seleucid
		#2: plane geometric figure	x		x			x	
		#3: field	x		x			x	
		#4: field	x		x			x	
		#5: solid geometric figure	x		x			x	
		#6: field	x		x			x	

TABLET	NO. OF PROBLEMS	SUBJECT	SOLUTION WORKED YES	SOLUTION WORKED NO	ANSWER GIVEN YES	ANSWER GIVEN NO	DRAWING YES	DRAWING NO	DATE
		#7: field	x		x			x	
VAT 8522*	4	#1: logs	x		x		x		OB
		#2: inheritance	x		x			x	
		#3: herding and herd growth		x	x			x	
		#4: equations	x		x			x	
YBC 4608*	5	#1: plane geometric figure	x		x			x	OB
		#2: plane geometric figure	?	?	?	?		x	
		#3: plane geometric figure	x		?	?		x	
		#4: plane geometric figure	x		x			x	
		#5: inheritance	x		x			x	
YBC 4669*	26 (25 extant)	#1: standard measuring vessels		x	x			x	
		#2: standard measuring vessels		x	x			x	
		#3: standard measuring vessels		x	x			x	
		#4: standard measuring vessels		x	x			x	
		#5: standard measuring vessels		x	x			x	
		#6: standard measuring vessels		x	x			x	
		#7: standard measuring vessels		x	x			x	
		#8: standard measuring vessels		x	x			x	
		#9: standard measuring vessels		x	x			x	
		#10: bricks		x	x			x	
		#11: bricks		x	[x]			x	
		#12: bricks		x	x			x	
		#13: bricks		x	x			x	
		#14: bricks		x	[x]			x	

TABLET	NO. OF PROBLEMS	SUBJECT	SOLUTION WORKED YES	SOLUTION WORKED NO	ANSWER GIVEN YES	ANSWER GIVEN NO	DRAWING YES	DRAWING NO	DATE
		#B1: work	x		x			x	
		#B2: equations		x		x		x	
		#B3: original quantity		x	x			x	
		#B4: original quantity		x	x			x	
		#B5: original quantity		x	[x]			x	
		#B6: metalwork		x	x			x	
		#B7: herding and herd growth		x		x		x	
		#B8: herding and herd growth		x		x		x	
		#B9: standard measuring vessels		x	[x]			x	
		#B10: work		x	x			x	
		#B11: interest		x	x			x	

Appendix 1

Indices

The glossary lists Akkadian words in the mathematical texts. The Sumerian equivalent or logographic spelling when found in the texts follows the Akkadian word in parenthesis. The translations are standard based on existing dictionaries (*CAD* and *AHw*) and glossaries (*MCT*, *MKT*, *TMB*, Kilmer 1960) unless indicated.

Each entry in the glossary contains a bibliography for the Akkadian forms of the word in the nominative, accusative and/or genitive and in the context of various expressions. The Akkadian forms are followed by the bibliography for the Sumerian equivalent by itself, with various suffixes listed in alphabetic order and in the context of various expressions. Finally, when applicable, the coefficients are listed with references. All bibliographical references are listed in alphabetical order by museum number with the rare exceptions of *LBAT*, *TMS*, *UMB* and *UET*. A standard set of pages for each tablet can be found in the list of tablets in the charts. When a particular word is discussed, the information follows the bibliographic data in a separate paragraph.

A list of words beginning with **gán** appear in the glossary s.v. *eqlu* for convenience. In many cases, the meaning remains to be determined (see Kilmer 1990b). We hesitate to translate these phrases by geometrical terms. *kiššû* and *nalbantu*, formerly translated by 'truncated cone' and 'truncated pyramid', respectively, illustrate the problems in attributing geometrical meanings to words (see Chapter 4 Section XVI).

A reverse list of Sumerian words and their Akkadian equivalents (when known) appears at the end of Appendix 1.

GLOSSARY OF REALIA FROM CUNEIFORM MATHEMATICAL TEXTS

aban šadî (= **na$_4$-kur-ra**) 'rock'
coefficients:
    20 *ša* **na$_4$-kur-ra**: YBC 5022: 57

abāru (= **a-gar$_5$**) 'lead'
coefficients:
    2,24 **igi-gub a-gar$_5$** [x]: A 3553: 23

2,24 *a-[ba-ru]-um*(?): YBC 7243: 38 [Kilmer (1960: 293 n. 1) suggested this reading based on the value of the coefficient and its position in the series of metals; she also proposed a possibile restoration of *a-[na-ku]-um.*]

a-ba-rum 1,[5]2,30 *i-g[i-gu-bu-šu]*: IM 52916, IM 52685, IM 52304 Section I(b): 4

1,52,30 *šà a-ba-ri*: *TMS*, Text 3: 60

abnu (= **na**$_4$) 'stone, stone weight'

ab-ni-ia: AO 6770 rev. 4, 10

na$_4$: AO 6770 obv. 18

YBC 4652 obv. [1–3], 5–18; rev. 1, 3, 4, 6, 7, 9, 11, 13

YBC 4698 obv. i 8, 10; obv. ii 7, 12; rev. i 1, 6, 14; rev. ii 13, 19

na$_4$**-mu**: AO 6770 rev. 3

na$_4$**-ka**: YBC 4698 obv. i 7, 10; rev. i 15

abu (= **ad**) 'father'

ad *ana* **dumu**-*šú ša i-ram-mu*: K 11097: 3

abullu 'city gate, entrance gate of a building or building complex'

i-na pa-ni a-bu-li-im: BM 85194 obv. i 3

i-na pa-ni a-bu-lim: BM 85196 rev. ii 10

agāru 'to hire'

ag-ra-am a-gu-ur: YBC 10722 rev. 1

lú ḫu[n-g]á *a-gu-ur*: YBC 10722 rev. 4

agru (= **lú ḫun-gá**) 'hired man'

ag-ra-am a-gu-ur: YBC 10722 rev. 1

lú ḫun-gá: YBC 4657 obv. 1, 4, 6, 9, 10, 13, 14, 16; rev. 24

YBC 4666 obv. 2, 5, 8, 10; rev. 1, 4, 8, 11, 13, 27

YBC 5037 obv. 11

YBC 7164 obv. 2

YBC 10722 rev. 4

lú ḫun-gá and **erim** appear to be used interchangeabley (Weitemeyer 1962: 37).

agû (= **aga**) 'crown'

aga lugal: BM 15285 rev. 2

agurru (= **sig$_4$-al-ùr-ra**) 'kiln-burnt brick'

a[*-gu-ru-*]*um*: YBC 7243: 38 (see *abāru*)

a-gu-ri-im: IM 49949: 14

^{sig4}*a-gu-ru-um*: YBC 7243: 44

ú-ku!-ur!-ru!-um: YBC 7243: 43

sig$_4$-al-ùr-ra: A 3553: 25, 26

AO 10822 rev. 8, 9

BM 85194 obv. iii 38; rev. i 26

CBS 10996 obv. ii 5'

VAT 6598 obv. 2, 3, [4], 5, 6, 7 (**sig$_4$-al-ùr-ra**-*ka*), [13], 16, 17; rev. i 8

YBC 4607 obv. 16; rev. 1, 13, 15, 16, 18

YBC 5022: 6, 34, 35

é-sig$_4$ sig$_4$-al-ur$_5$-ra: VAT 6598 obv. [3], [13], 15; rev. i 1

coefficients:

2,15 **igi-gub sig$_4$-al-ùr-ra**: VAT 6598 obv. 5; rev. i 8

2,15 **igi-gub é-sig$_4$ sig$_4$-al-ùr-ra**: VAT 6598 obv. 15 [Christopher Walker (Letter to author 1986), enclosed a copy of his join, BM 96957, to VAT 6598, which contains problems dealing with **é-sig$_4$** and **é-sig$_4$ sig$_4$-al-ur$_5$-ra**, to which the term *kuburrû* 'thickness' and the coefficient of kiln-burnt bricks or 2;15 are applied.]

2,24 *a*[*-gu-ru-*]*um*: YBC 7243: 38

2(?),42 ^{sig4}*a-gu-ru-um*: YBC 7243: 44

3,20 *ša* **sig$_4$-al-ùr-ra**: YBC 5022: 34

3,44 **igi-gub sig$_4$-al-ùr-ra**: A 3553: 26

5,24 **igi-gub** PEŠ **sig$_4$-al-ùr-ra**: A 3553: 25 [The coefficient 5,24 is applied to the number of burnt bricks (of ⅔ **kùš** a square side) used for laying 1 **sar** area in the problem text, YBC 4607 rev. 13–15 (Kilmer 1960: 289–90 n. 2).]

1,52,30 *ša* **sig$_4$-al-ùr-ra ki**-2: YBC 5022: 35

48 1,40 1,20 *ú-ku!-ur!-ru!-um*: YBC 7243: 43

[...], 15 ^{giš}**má-lá sig$_4$-al-ùr-ra**: CBS 10996 obv. ii 5'

with *nalbanu*:

2,42 *na-al-ba-an* **sig$_4$-al-ùr-ra**: YBC 5022: 6

with *nazbalu*:

30 **igi-gub** *na-za-ba-al a-gu-ri-im*: IM 49949: 14

with reference to *nazbalša*

1,41,15 *na-az-ba-al-ša*: YBC 5022: 7 (referring to kiln-burnt brick)

YBC 9819 listed two additional types of bricks with different measurements not found in these mathematical problems dealing with bricks, namely, *ukurum*, (possibly a variant of *agurru*) and *zarin* (Dunham 1982: 27–41; Chapter 4, Summary to Section I. BRICKS AND BRICKWORKS).

aḫatu (= **nin**) 'sister'
nin (or NIN): YBC 10522 obv. 3

aḫu (= **šeš**) 'brother'
a-ḫu-um: *a-ḫu-um e-li a-ḫi-im*: YBC 9856 obv. 10
a-ḫi ṣe-eḫ-ri-im: YBC 9856 obv. 9
a-ḫi-im: YBC 9856 obv. 10
šeš: *UET* 5: no. 121: 2
VAT 6597 obv. ii 1, 3, 6, 7, 8, 10, 14
VAT 8522 obv. ii 1c, 3d
šeš-(x)-**kam**: *UET* 5: no. 121: 5, 6, 7, 8
šeš gal: *UET* 5: no. 121: 2
10 **šeš-e-ne**: Strssbg. 362 obv. 1
šeš-meš: AO 17264 obv. 2
LBAT: no. 1648 obv.(?) 8'
TMS, Text 10: 5, 26
TMS, Text 26 obv. i 13; obv. ii 4
VAT 8522 obv. ii 1, 4e
YBC 4608 rev. 14, 20
šeš-a-ni: VAT 7531 obv. 4; rev. 4
YBC 9856 obv. 8
šeš gal: VAT 6597 obv. ii 5, 10; rev. [9], 11
VAT 8522 obv. ii 4, 2e, 3e
šeš tur: VAT 6597 obv. i 11; obv. ii 11, 13; rev. 5, 6
VAT 8522 obv. ii 2e
šeš ús(-sa): VAT 6597 obv. i 6, 7, 10; rev. 8
VAT 8522 obv. ii 4
šeš *ana* **šeš**: K 11097: 6
šeš ugu šeš: Strssbg. 362 obv. 2, 4, 18
VAT 8522 obv. ii 4 (**šeš gal u[gu šeš ús-]sa**), 6
YBC 4608 rev. 14 (**šeš ugu** <**šeš**>), 16

akalu (= ^{giš}**ninda**) receptacle for 1 **ninda**
^{giš}**ninda**: YBC 4669 obv. i 13, 17, 21; obv. ii 1, 5, 9

akkadû 'Akkadian'
ki-a-am ak-ka-du-ú: *TMS*, Text 9: 18
i-na ak-ka-di-i: *TMS*, Text 9: 10

á-kur-ra see *aban šadî*

alamdimmû 'form, figurine'
[*alam-dím-me-e*]: K 8705: 2 (// K 8865 rev. 6 in Borger 1957: 194)

allum (= ^{giš}**al**) 'hoe'
al: MM 86.11.404 obv. 10(?) [*al* may be an abbreviation for *allu* in MM 86.11.404 obv. 7, 10 (see *CAD*, s.v. *allu*).]
al-ia: MM 86.11.404 obv. 7
al-lu-um: IM 52916, IM 52685, IM 52304 Section E: rev. 25'–29
al-lum: Haddad 104 #10 iv 37
(x) *al-lim*: Haddad 104 #9 iv 24; #10 iv 41
iš-ka-ar al-lim: Haddad 104 #9 iv 26; #10 iv 43
10 *al-lu-ú*: MM 86.11.404 obv. 6
al-li ba-la-lim: Haddad 104 #9 iv 17
al-li ḫa-ba-tim: Haddad 104 #9 iv 16–17
al-li la-ba-nim: Haddad 104 #9 iv 15, 17; #10 iv 27
al-li ne-eq-re-bi-im: Haddad 104 #9 iv 20
coefficients:
a-na 40 **ninda** *a-za-bi-il* 2,13,20 *al-lu-um*
a-na 20 **ninda** *a-za-bi-il* 5 *al-lu-um*
a-na 15 **ninda** [*a-za-bi-il* x] *al-lu-um*
[*a-na*] 10 **ninda** *a-z*[*a-bi-il*] 6,40 *al-lu-um*
[*a-na* 5] **ninda** *a-za-*[*bi-il*] 10 al-lu-um: IM 52916, IM 52685, IM 52304 Section E: rev. 25'–29
20 *al-li ḫa-ba-tim* 20 *al-li la-ba-nim* 10 *al-li ba-la-lim lu-pu-ut*: Haddad 104 #9 iv 16–17
5 *al-li ne-eq-re-bi-im*: Haddad 104 #9 iv 20
with *allum* implied:
10 *ba-la-lim lu-pu-tam*: Haddad 104 #10 iv 33
20 *ḫa-ba-tam* 20 *la-ba-nam* 10 *ba-la-lam l*[*u-pu-u*]*t*: Haddad 104 #10 iv 29

Cf. 2,13,20(?) *ù* 1,15 GIŠ-DUB-ÍL: YBC 5022: 45 [Kilmer (1960: 289. n. 1) suggested that the DUB looks like AB but that ^{giš}**al** = *allu* is expected; the coefficient 2,13,20 is found in *TMS*, Text 3: 37m in broken context.]

The *allu* is described as having 1–4 teeth; however, the *allu* of 4 teeth is called *ti-te-nu* or *ra-'-i-zu* (see Sumerian ^{giš}**al-zú** = *šin-nu alli*). Different kinds of *allu* are used for breaking the earth, making bricks, etc. Thureau-Dangin (1924b: 145 and 1926: 31) showed that the *allu* has two or more teeth, resembling the forked Egyptian hoe represented on the mastaba of the Ancient Empire.

alpu (= **gu$_4$**) 'ox'
coefficients:

x x [x] *la* x[...]x *al-pí-im e-pé-ša-am*
[x] x [x] x [...] x *i-gi-gu-bu-šu*: IM 52916, IM 52685, IM 52304 Sections F & G: rev. 30'–31'

For a discussion of terms related to bovines, see Kilmer (1990b: 86, 88).

ālu (= **uru, uru^{ki}**) 'city'

i-na **uru**: *LBAT*: no. 1647 i 2'
uru^{ki}: BM 85194 obv. i 37
BM 85210 obv. i 1, 22; ii 2
Böhl Coll. 1821 1, 4, 5, 6
uru *a-ṣa-ba-at*: BM 85210 obv. ii 15
uru^{ki} *lu-úṣ-ba-at*: BM 85194 rev. ii 10, 24–25
BM 85210 obv. ii 6, 16 (<**uru^{ki}**> *lu<-úṣ-ba-at>*)
uru^{ki} *na-ki-ir ^{d}Marduk a-ṣa-ba-at*: BM 85194 rev. ii 7, 22
BM 85210 obv. ii 1 (<**uru^{ki}**>)
uru^{ki} na-ki-ir *^{d}Marduk lu-úṣ-ba-at*: BM 85210 obv. i 3
uru bil: Böhl Coll. 1821 7, 12
uru libir(-ra): Böhl Coll. 1821 7, 12, 14

alulūtu (a metal)
coefficients:

a-lu-lu-tum 2,24 *i-gi-gu-bu-šu*: IM 52916, IM 52685, IM 52304 Section I(b): 6

amama (meaning unknown)

4,20,6,40 **gán** *am-am-ma* x **[ì]-dub igi-gub**-*ú*: BM 47860 obv. 1

Perhaps, the meaning of *am-am-ma* is elucidated by rev. 4–8 which describes 'the circle of the evening meal of Ekur' and 'every one of the shrines... all of the cult platforms'. Livingstone 1986: 21 translated 'volume of the Amamma'.

amaru (= **sig$_4$-anše**) 'brick pile'

a-ma-ri-im: *TMS*, Text 3: 48

IM 49949: 6 (*a*!-*ma-ri-im*)
sig$_4$-anše: IM 52301 rev. 22
A 3553: 24
VAT 8523 rev. 6, 11
YBC 4673 obv. ii 19, 27
YBC 5022: 16
YBC 7243: 40
sig$_4$-anše du$_8$-du$_8$: YBC 4673 obv. ii 13, 17; obv. iii 7 **sig$_4$-a[nše(?)] du$_8$-du$_8$**
sig$_4$-anše gub: YBC 4669 obv. iii 2, 5
sig$_4$-anše sig$_4$: AO 10822 rev. 13, 14, 16
sig$_4$-anše sig$_4$-áb: AO 10822 rev. 18, 19, 21
sig$_4$ sig$_4$-anše: YBC 4708 obv. i [1], 5, 8, 11, 14; obv. ii 1, 14; obv. iii 1, 21, [26]; rev. i 1, 5, 13, 21, 28; rev. ii 1, 6, 11, 17, 22, 27 (**sig$_4$ <sig$_4$-> anše**); rev. iii [1], 13

coefficients:
4,10 *ša* **sig$_4$(?)-anše(?)**: IM 52301 rev. 22
5 *šà a-ma-ri-im*: *TMS*, Text 3: 48
7,12 **igi-gub(?)** BI *a*!-*ma-ri-im*: IM 49949: 6
7,12 **igi-gub sig$_4$-anše**: A 3553: 24
7,12 *ša* **sig$_4$-anše**: YBC 5022: 16
7,12 **sig$_4$-[an]še**: YBC 7243: 40

amatu 'utterance, spoken word, thing, matter'
i-na a-mat [d]*Anu u An-tum liš-lim*: AO 6484 u.e.
iš-ka-ar a-wa-ti: YBC 9874 rev. 7

amēlum (= **lú**) 'man'
iš-te-en a-wi-lu-ú: AO 8862 iii 28–29
iš-te-en a-wi-lu-um: AO 8862 iv 5
iš-ka-ar iš-te-en a-wi-li-im: IM 53961 obv. 4; rev. 4–5
[iš]-ka-ar a-wi-lim iš-te-en: IM 54011 rev. 4
6 *a-wi-lu-ka*: IM 54538 rev. 2 [In obv. 4 *ṣābu* is used as a synonym for *awīlūka*.]
lú: Haddad 104 #10 iv 28
Erm 15073 rev. iii 22, 25
M 6829: [1''-3''] [*i-nu-ma* **lú** *ik-ri-bi-šu* **udu-ḫi-a** *ma-da-tim a-na* [d]*Il-tim i-ka-ar-ra-ab*...] (restored by // text *ARM* 24 #260: 1–3)
UET 3: no. 1386: 12

VAT 8523 rev. 7 [Thureau-Dangin (*TMB*: 124–25) corrected **uš** to **lú** in VAT 8523 rev. i 1, 19, 20, 22, 25.]

lú-1: *TMS*, Text 3: 46
YBC 4662 obv. 40
YBC 5022: 43

lú 1-e: VAT 7528 obv. i [4](?)
YBC 4657 rev. 7
YBC 4666 obv. 12, 20, 22, 24; rev. 15
YBC 4669 rev. i 1; rev. iii 7
YBC 4673 obv. i 6; obv. ii 10, 26; obv. iii 13, 20; rev. i 6
YBC 7164 obv. 5, 14, 16, 22; rev. 8, 16

lú-1-**kam**: A 3553: 27

1 **lú**: BM 85194 obv. i 3, 11, 14, 21, 24, 30, 33; obv. ii 20, 25
BM 85196 rev. i 2, 23, [33], 34, 37, 39
Haddad 104 #7 iv 5; #8 iv 9, 14; #9 iv 16, 25; #10 iv 42
VAT 8523 rev. 14, 16, 23, 24

coefficients:

10 **gín éš-kàr lú**-1: YBC 5022: 43

10 *ša* **saḫar** *za-ba-li* **éš-kàr lú**-1: *TMS*, Text 3: 46

6 **igi-gub kàr lú**-1-**kam**: A 3553: 27 [Kilmer (1960: 290 n. 3–4) suggested that the value 6 may refer to the reciprocal of the 10 **gín** volume assignment used in the solution of excavation problems and that the value 45 (45 **igi-gub a-rá**: A 3553: 28) is the number by which to multiply (**a-rá**) the reciprocal of a standard daily work assignment.]

3,45 **éš-kàr** 1 **lú** (*ša* **íd** *sé-ke-ri-im*): VAT 8523 rev. 14–15, 16, 23, 24

amūtu (= **bà**) 'liver (examined by the haruspex), liver model, omen'

bà sig$_5$-at: K 8705: 5

[6 **ki-meš**] **bà**: K 8705: 6 (// K 8865 rev. 10 in Borger 1957: 194)

siḫir (= **nigin**)-ti *amūti*: K 11097: 13 (see Borger 1957: 192)

annaku (= **an-na**) 'tin'

⌜*a-na-ku-um*⌝: IM 52916, IM 52685, IM 52304 Section I(b): 5

an-na: A 3553: 20
MM 86.11.404 obv. 4 (4½ **gín an-na**)
TMS, Text 3: 57
YBC 4698 rev. ii 7, 9, 10, 11
YBC 5022: 27

coefficients:

1,24 *šà* **a[n]-na**: *TMS*, Text 3: 57

2,24 *ru-uq-qú ša* **an-na**: YBC 5022: 27

1,20,21 **igi-gub an-n[a]**: A 3553: 20
⌜*a-na-ku-um*⌝ 2,36(?),15 *i-gi-[gu-bu-šu]*: IM 52916, IM 52685, IM 52304 Section I(b): 5 [Kilmer 1985]

Goetze (1948: 35) suggested translating **an-na** as 'tin' rather than 'lead', basing his translation on recipes for making bronze in Ur III tablets. He proposed the following development: Sumerian **(kug) an-a(k)** > Akk. *annakum* > Sum. ***annagga** > **nagga**.

ANŠE (economic term) estimated weight loss?
ANŠE: YBC 4698 obv. i 11, 16, 18; obv. ii 2, 8 13, 16, 18; rev. i 2, 7, 9, 10, 12, 13, 16; rev. ii 9, 15, 18

apkal šamni (*apkallu* = **nun-me**) 'diviner (lit.: expert in lecanomancy?, see *CAD* s.v. *apkallu* 3b)
[*mār*] **lúḫal nun-me** [*šamni*]: K 11097: 7

aplu (= **ibila**) 'heir, oldest son, son'
ibila(-bi): *UET* 5: no. 121: 2
AO 6484 colophon

appu 'spur of land (made artifically), causeway, bund'
ap-pu-um: BM 85196 obv. i 1

appu 'top of a tree'
i-na ap-pi-šu: VAT 8522 obv. i 3, 8a, [8]

apu (= **gišgi**) 'reed thicket, cane brake'
[g]išgi *u* **gištir** *ak-ba-ru*: *TMS*, Text 25 obv. ii 1 [Von Soden (1964: 50) suggested the reading [*ša i-na a*]*pim u qištim ak-sú!-ru*.]

arammu 'ramp before a city gate'
a-ra-am-mu(-um): BM 85194 obv. i 1
BM 85196 rev. ii 8
a-ra-mu: Strssbg. 362 rev. 15
a-ra-mi: Strssbg. 362 rev. 20

Thureau-Dangin (*TMB*: 21) suggested that *arammu* is an inclined plane which gives access to a city and probably crosses an enclosed canal. The ramp is elevated against the rampart and serves as a road of approach (see Heb.

s^e^lalah), as described in sieges in Thucydides, Josephus and Herodotus (Thureau-Dangin 1933e: 126).

Assyrian inscriptions describe seige operations that often involve the construction of *arammu* of beaten earth. The ramp is not only used for military purposes; in a Neo-Babylonian contract from Nippur, it is used as a synonym for a beaten earth jetty which serves as a pier, and in lexical texts it is explained by the term *titurru* 'earth dam' (Thureau-Dangin 1932h: 114–115).

Salonen (1965: 26) suggested *arammu* is a '(siege) dam'.

arḫu (= **sig$_4$-áb**) 'half-brick'
ar-ḫu-um: IM 49949: 15
sig$_4$-áb: AO 10822 rev. 7, 8, 9, 18, 19, 20, 21, 22
CBS 10996 obv. ii 4'
YBC 4607 obv. 10; rev. 10, 12
YBC 5022: 4
with **sig$_4$-anše** (= *amaru*):
sig$_4$-anše sig$_4$-áb: AO 10822 rev. 18, 19, 21
with *tarkibūtu*:
5 *tar-ki-bu-ta* **sig$_4$**: AO 10822 rev. 7
coefficients:
with *malallû*:
[...], 10 **^giš^má-lá sig$_4$-ab**: CBS 10996 obv. ii 4'
with *nalbanu*:
5,24 *na-al-ba-an* **sig$_4$-áb**: YBC 5022: 4
with *nazbalu*:
3,0 **igi-gub** *na-az-ba-al*(!) *ar-ḫu-um*: IM 49949: 15
cf. 3 *ša* ½ **kùš**: YBC 7243: 47
with reference to *nazbalša*:
3,22,30 *na-az-ba-al-ša*: YBC 5022: 5 (referring to ½-brick)

arḫu (= **itu**) 'month'
itu: K 11097: 6
itu (x) **kam**: BM 85194 rev. i 50
BM 85196 rev. i 7
VAT 7528 obv. iii 6 (?)
itu-1 7 ½ **kam** (mng. 37½ days): YBC 7164 obv. 9, 11
itu ^d^igi-kur: M 6861: 15
itu *ṣí-li-li-tum*: *TMS*, Text 5 colophon

ašbatu 'residence, dwelling place'

a-aš-ba-tum: BM 85194 obv. iii 1 [Note alternative readings which are problematic, namely, *a-aš-ku-tum*, referring to *aškuttu* 'the plaque which anchors the peg(s) or doorknob to the doorjamb' (Leichty 1987: 193), and *sà!-an!-ku!-tum* (von Soden 1937: 194).]

ašlu 'rope'

aš-lam **bil**: BM 85196 obv. ii 29
aš-li ú-sà-an-ni-iq: BM 85196 obv. ii 31

aširtu (*ešertu*) 'sanctuary, socle'

eš-re-e-ti gi-mir-[*ši-na* (x x x)]: BM 47860 rev. 5

ašru (= **ki**) 'place, location (e.g., of an ominous feature)'

[6 **ki-meš**] **bà**: K 8705: 6

atappu (= **pa$_5$-sig**) '(small) branch canal'

a-ta-ap ta-ra-aḫ-ḫi: BM 85196 rev. i 19
[*a-*]*ta-a*[*p*] *ta-ra-aḫ-ḫi-im-mi*: YBC 9874 rev. 1
a-ta-pa[*-a*]*m*: BM 85196 rev. i 21
a-ta-ap-pi: *TMS*, Text 24: 34
pa$_5$-sig: *TMS*, Text 3: 33
TMS, Text 24: 54
TMS, Text 25 obv. i 7
VAT 7528 obv. i 1, 6, 12, 18, 24; obv. ii 1, 7, 13, 19, 23, 27, 31; obv. iii 1, 7, 17; rev. ii 4, 7, 13
YBC 4666 obv. 1, 5, 7, 9, 11, 13, 15, 17, 19, 21, 23; rev. 1, 4, 7, 10, 12, 14, 16, 18, 20, 22, 24, 26; left edge
YBC 7164 obv. 1, 4, 6, 8, 10, 12, 15; rev. 17, 19
pa$_5$-sig libir-ra: YBC 7164 obv. 19, 21, 23, 26; rev. 1, 3, 6, 10, 13

coefficients:

48 *šà* **pa$_5$-sig**: *TMS*, Text 3: 33
igi 48 **igi-gub pa$_5$** *pu-ṭú-ú*[*r*]: *TMS*, Text 25 obv. i 7

See 48 **a-meš** *pat-ti* CBS 10996 obv. iii 24'
See 48 *me-e* **a-šà**: YBC 7243: 26

Powell (1988:166) showed *atappu* is 'a holding ditch into which water is forced by blocking a larger canal that feeds it', based on *TMS*, Text 25 #1, and derived from Sumerian **a tab** 'to hold or block water'.

atartu 'excess'

[*a-tar-tum za-k*]*ar šu-mi*: K 8705: 5 [*CAD* s.v. *zakar šumi* noted that *atartu* 'excess' in the protasis means *zakar šumi* 'fame' in the apodosis.]

[*a*]-*tar-ti*: K 8705: 7

atḫû 'members of a group of persons of equal status and age' (see *CAD* s.v *atḫû*)

at-ḫu-u: IM 31210 §3: 2; §6: 1; §7: 2

ki-pa-tam a-na ši-na at-ḫi za-za-am e-pé-ša-am: IM 52916, IM 52685, IM 52304 Section D: rev. 19'

[...] *ka-ma-ra-am at-ḫi-i* [...]: IM 52916, IM 52685, IM 52304 Section F: 53'

at-ḫi-ka: IM 31210 §1: 23; §6: 5, 7

atkuppu (= **ad**-KID) a craftsman making objects of reeds
coefficients:

10,48,6 *ša* **gir$_4$-ad**-KID: YBC 5022: 55 [Neugebauer and Sachs (*MCT*: 135) read [2]0,48,6 but Kilmer (1960: 291) read the numbers cited and suggested 10,48,6 as the coefficient for *kupru ša ina kīrim* 'asphalt in the oven' in YBC 7243: 24.]

This craft to include weaving mats and manufacturing reed boats and reed containers, which were then coated with bitumen to make them water-tight (*CAD* s.v. *atkuppu*).

bābu (= **ká**) 'door, gate, entrance'

[...] *i-ru-um ba-ba-am ṣí-it* [...]-*im*: IM 52916, IM 52685, IM 52304 Section F: 52'

ká: *TMS*, Text 15: 1
VAT 6598 rev. i 19
YBC 4673 rev. i 13

balālum 'to mix'
coefficients:

20 *al-li ḫa-ba-tim* 20 *al-li la-ba-nim* 10 *al-li ba-la-lim lu-pu-ut*: Haddad 104 #9 iv 16–17

with *allum* implied:

10 *ba-la-lim lu-pu-tam*: Haddad 104 #10 iv 33

20 *ḫa-ba-tam* 20 *la-ba-nam* 10 *ba-la-lam l*[*u-pu-u*]*t*: Haddad 104 #10 iv 29

banû ‘to build, make’
 e *ab-ni*: BM 85194 obv. i 42
 ḫi-ri-tam ab-ni: BM 85194 obv. i 39
See **a-šà** *ab-ni*: *TMS*, Text 17: 1

bārû (= ^{lú}**ḫal**) ‘diviner’
 ^{lú}**ḫal**: K 11097: 1
 [*mār*] ^{lú}**ḫal nun-me** [*šamni*]: K 11097: 7

bārûtu (= ^{lú}**ḫal**) ‘divination’; N.B. *niṣirti bārûti* are a series of extispicy texts
 ba-ru-ti: K 8705: 2
 sal-šeš ^{lú}**ḫal**-*t*[*i*]: K 11097: 1

biltu (= **gún**) ‘load’
coefficients:
 3 **bán še gún [gi]š**-KA[K+LI]Š-**ra**
 1 (**gur**) 1 **bariga** *ḫa-a-mu* **gún** ^{giš}**mar-gíd-da**
 4,30 *za-ba-lu* 6 (**gur**) **in-nu gún ki[-min]**
 12,30 *za-ba-lu* 1 (**gur**) **in-nu gún [ki-min]**: CBS 10996 obv. iii 29’–32’

bīt kunukki (= **é-kišib**) ‘storehouse’
coefficients:
 15 *ša* **é-kišib a**: YBC 5022: 33
 3,36 *ša* **é-kišib in-nu-da**: YBC 5022: 31
 3,45 *ša* **é-kišib suḫ$_4$**: YBC 5022: 32
 4,48 **é-kišib še**: YBC 5022: 61
 4,48 **igi-gub é-kišib in-nu-d[a]**: A 3553: 37
 6,40 **igi-gub é-kišib** [x]: A 3553: 11
 6,40 *ša* **é-kišib še**: YBC 5022: 30
Cf. 6,40 **gán ì-dub** še-im **igi-gub**-*ú*: BM 47860 obv. 7

 For a discussion of the factor 6,40, see Powell (1984a: 45).

bītu (= **é**) ‘house, dwelling place, temple, palace, etc.’
 [*ša i*]-*na* **é** *ša-ap-*[*ku*]: M 6861: 11–12
 [...] **kù-babbar é**: *LBAT*: no. 1648 obv.(?) 14’

bītu epšu (= **é-dù-a**) ‘house plot’
coefficient:
 20 **igi-gub é-dù-a**: A 3553: 10

Walker (Letter to author 1986) enclosed a copy of his join, BM 96957, to VAT 6598, which contains a problem dealing with **é-dù-a** and reflects the use of the coefficient 0;20.

būrtu (= **túl**) 'well, cistern'
túl: BM 85194 obv. iii 38, 39, 45, [46]; rev. i 26
YBC 4186 obv. 1, 5, 8

daltu (= ^{giš}**ig**) 'door'
^{giš}**ig**: IM 52916, IM 52685, IM 52304 Sections F and G: rev. 32'–33'
TMS, Text 15: 1, 2, 3 [von Soden (1964: 49) suggested reading **ba ig**-*ši as di!-ik-šu/ši* in *TMS*, Text 15.]
coefficients:
^{giš}**ig** x [...] x [...] 5(?),40 *i-gi-gu-bu-šu*
^{giš}**ig** x [...] 5 *i-gi-gu-bu-šu*: IM 52916, IM 52685, IM 52304 Sections F and G: rev. 32–33'

dibdibbu (= ^{giš}**dib-dib**) 'water clock'
^{giš}**dib-dib**: BM 85194 obv. ii 27, 30, 34, 36, 41
BM 85210 rev. ii 10

dilūtu (= ^{giš}**a-lá**, ^{giš}**a-bal**) water-hoisting device
coefficients:
45(?) ^{giš}**a-bal**: YBC 7243: 12
9,39 **a-meš** ^{giš}**a-lá**: CBS 10996 obv. iii 25'

dišpu (= **làl**) 'honey'
coefficients:
20 **lál**: YBC 7243: 29

DUG-GU-ZI (meaning uncertain); possibly, (^{dug})**gu-zi** 'cup?'
DUG-GU-ZI: BM 34568 rev. i 12

dūru (= **bàd**) '(city) wall'
bàd: BM 85194 obv. ii 19; rev. 1, 20; rev. ii 8, 17, 30, 31
BM 85210 obv. i 1, 5, 9, 11, 13, 15, 17, 18, 22; obv. ii 4 (*iš-tu ḫu-ur-ḫur-i*[*m*] *a-na* **bàd** *ka-ša-di-im*); obv. ii 5, 9, 16, 24, 25

Salonen (1965: 26) suggested the meaning '(siege) wall'.

ᵈEa-banû (= *ᵈé-a-***dù**) 'Ea-creator', refers to the fourth string of an instrument
coefficients:

sa *ᵈé-a-***dù** *ù* **sa** *qud-mu-ú* 4,1 **sa šub-múrub**: CBS 10996 obv. i 17'
sa *šá-***ge₆** *ù* **sa** *ᵈé-a-***dù** 2,4 **sa** ⌜*titur*⌝ **murub₄**-*tú*: CBS 10996 obv. i 20'
sa *ᵈé-a-***dù** [*ù* **sa** 4,x **sa**...]: CBS 10996 obv. i 24' [Kilmer's (1984: 70) collation of U.3011 obv. i 4 **sa**-4-**tur** a-*ba-nu*-[*ú*] Sumerian: 'fourth small'; Akkadian: 'Ea-creator' may provide the restoration for CBS 10996 obv. i 24'.]

eleppu (= ᵍⁱˢ**má**) 'boat'

e-le-pu-um: IM 52916, IM 52685, IM 52304 Section I(d): 1
e-le-pí-im: IM 49949: 12
ᵍⁱˢ**má**: BM 85194 rev. iii 7
IM 49949: 12
YBC 7243: 19

coefficients:

5 ᵍⁱˢ**má**: YBC 7243: 19
4,[x],8 **igi-gub** ᵍⁱˢ**má** *e-le-pí-im*: IM 49949: 12
e-le-pu-um 7,12 *i-gi-gu-bu-ša*: IM 52916, IM 52685, IM 52304 Section I(d): 1

See Salonen (1939: 11–12) for a discussion of *eleppu*.

elmešu (= **sud-ág**) a precious stone; a kind of metal(?)
coefficients:

2,8 *ru-uq-qú ša* **sud-ág**: YBC 5022: 28

Lewy (1949b: 164 n. 3) suggested that *elmešu* is a mixture of gold and silver called electrum. Such mixtures of gold and silver are known from decorations of objects from the Royal Cemetery at Ur and add to the range of colors at the disposal of the goldsmith. Muhly (1973: 178) suggested both Akk. *ešmaru* and Heb. ḫašmal *may be translated as 'electrum'*.

Neugebauer and Sachs (*MCT*: 134) proposed the translation 'brass' for *elmešu*.

embūbu 'reed pipe'
coefficients:

sa 3(!)-*šú* **sig** *ù* **sa** 3-*šú uḫ-ri* 3,4 **sa** *em-bu-bu*: CBS 10996 obv. i 15'

eperu (= **saḫar**) 'earth, soil, dirt'
? *e-pi-ri-ia*: AO 8862 iv 22, 26
saḫar: A 3553: 32–36

IM 49949: 4
IM 52916, IM 52685, IM 52304 Section E: rev. 23'
TMS, Text 3: 40, 42, 46, 47, 71
YBC 4673 obv. ii 12, 15
YBC 5022: 3, 8, 42
YBC 7243: 32

coefficients:

5 *ša* **saḫar**: *TMS*, Text 3: 47
10 *šà* **saḫar** *ka-ru*: *TMS*, Text 3: 40
10 *šà* **saḫar** *za-ba-li* **éš-kàr lú**-1: *TMS*, Text 3: 46
20 *šà* **saḫar** *za-ba-li*: *TMS*, Text 3: 42
[x],30 **saḫar** *i-za-ab-bi-il*!: *TMS*, Text 3: 71 [See von Soden 1964: 47 for this reading.]
3,45 **igi-gub saḫar kù šu-ba-an-te**
16 **igi-gub kù saḫar ba-an-te**
10(?),1,30(?) **igi-gub saḫar ki-lá**
10(?),1,20 **igi-gub saḫar íl-íl** $\textbf{sig}_4$ **du**
10(?),1,6,40(?) **igi-gub saḫar íl-íl im-dù-a**: A 3553: 32–36 [Kilmer (1960: 290 n. 7) suggested that lines 32–33 deal with the relation of volume (**saḫar**) to wages (**kù**). The coefficients in these two lines, 3,45 and its reciprocal 16, are used in solving a problem concerning damming up (*kāsiru*) a canal by using a dike comprised of ⅔ brick and ⅓ soil in VAT 8523 #5.]

with *maššû*:

1,40 *ma-aš-šu-ú-um ša* **saḫar**: YBC 5022: 42 [Note a coefficient of 1,40 in broken context in BM 47680 rev. 2.]

with *nalbanu*:

7,12 *na-al-ba-an* **saḫar**: YBC 5022: 3

with *nazbalu*:

1,40 *na-az-ba-al* **saḫar**: YBC 5022: 8
1,40 **igi-gub saḫar** *na-az-ba-al* **saḫar**: IM 49949: 4
na-az-ba-al **saḫar** 1,40 *i-gi-gu-bu*: IM 52916, IM 52685, IM 52304 Section E: rev. 23' [Note a coefficient of 1,40 in broken context in BM 47680 rev. 2.]
6,40 *na-az-ba-al* **saḫar**: YBC 7243: 32

epēru 'to supply, provide'

i-tì-nu-um uš-te-pi-ra-an-ni: AO 8862 iv 7–8
i-tì-nu-um uš-te-pi-ra-an-ni-i-ma: AO 8862 iii 32–33

eqlu (= **a-šà, gán**) 'field'

a-šà: CBS 10996 obv. iii 26'–28'

TMS, Text 3: 16–18
YBC 4186 obv. 3, 6, 7, 12
YBC 7243 obv. 26
a-šà giš-kár: *UET* 5: no. 864: 1
a-šà *Gu-la*: *UET* 5: no. 864: 16
a-šà-*im*: BM 85194 rev. ii 35, 38
a-šà-bi *ma-la ma-ṣú*: VAT 7531 obv. 3, 8, 17; rev. 3
bu-ur **a-šà**-*ka*: IM 54010 rev. 2
gán: BM 47860 obv. 1–13, [14], [15]
YBC 7243: 2, 3, 9
GÁN(?)-GA ŠE(?): IM 52301 rev. 22
GÁN GI[Š...]: BM 15285: Text XII
gán ^{giš}**ban**: BM 15285: Texts L, M
gán ^{giš}**illar**: *TMS*, Text 3: 13–15
gán giš-ku(?) [...]: BM 15285: Text Q
gán ^{giš}**má-gur$_8$**: BM 15285: Text Q
gán má-gur$_8$: YBC 5022: 47
YBC 7243: 6
GÁN-MAN: YBC 5022: 20, 53
gán gi[**š**(?)**-sa**]**r**(?): BM 15285: Text M
gán sa-li: A 3553: 7
gán ud-dà-sar: BM 15285: Text K
gán-u$_4$-sakar$_x$: A 3553: 4
BM 15285: Texts K, O, P
BM 85210 rev. i 5, 9, 16, 17
MLC 1354 obv.1
YBC 5022: 21, 52, 54
YBC 7243: 8
gán-ud-a(?)**-sar**: YBC 7243: 7
See **ud-gán-sar**: BM 85194 rev. iii 1
gán zà-mí: YBC 5022: 48
gán ^{giš}**zà-mí**: BM 15285: Text J
gán zarà: A 3553: 5
YBC 5022: 49, 51
YBC 7243: 5

coefficients:

a-šà:

48 *me-e* **a-šà**: YBC 7243: 26 [See *mû* regarding this coefficient.]
1,30 **a-šà a-meš** *ša-qu-ú* [...]
18 **a-šà a-meš** *ša-qu-ú* **ki-min**
2,46,40 **a-šà a-meš** *ša-qu-ú* **ki-min**: CBS 10996 obv. iii 26’–28’

13,20(?) **igi-gub** *šà* **a-šà še**
56,40 **ri** *šà* **a-šà še**
28,20 [*p*]*i-ir-ku šà* **a-šà še**: *TMS*, Text 3: 16–18

gán:

15 **gán** [...]: YBC 7243: 2
10 **gán** [...]: YBC 7243: 3
45(?) **gán** [...]: YBC 7243: 9
30 *ša* GÁN(?)-GA ŠE(?): IM 52301 rev. 22
13,7,30 **igi-gub** *šà* **gán** ^{giš}**illar**: *TMS*, Text 3: 13
52,30 **ri** *šà* **gán** ^{giš}**illar**: *TMS*, Text 3: 14
30 *pi-ir-ku šà* **gán** ^{giš}**illar**: *TMS*, Text 3: 15
12,30 **gán má-gur$_8$**: YBC 7243: 6
12,30 *šá* **gán má-gur$_8$**: YBC 5022: 47
6(?) *ša* GÁN-MAN: YBC 5022: 20
4,36,10 *ša* GÁN-MAN **ki**-2: YBC 5022: 53
4[5] **igi-gub gán sa-li**: A 3553: 7
10,15,36 **gán u$_4$-sakar$_x$**: A 3553:4
45(?) *ù* 10 *ša* **gán u$_4$-sakar$_x$**: YBC 5022: 21
5,24(?) **gán-ud-a**(?)**-sar**: YBC 7243: 7
4,20 *ša* **gán u$_4$-sakar$_x$ ki**-2: YBC 5022: 52
45 *ša* **gán u$_4$-sakar$_x$** ki-3: YBC 5022: 54
14,36,10 **gán u$_4$-sakar$_x$** *ki*(?) *in*-ŠUB: YBC 7243: 8

refers to **gán u$_4$-sakar$_x$**:

15 **igi-gub-ba**: BM 85210 rev. i 18 (?) (see *TMB* no. 98: 18)
MLC 1354 obv. 6

33,20 *ša* **gán zà-mí**: YBC 5022: 48
26,40 **gán zà-mí**: YBC 7243: 4
26,40 *ša* **geštu zà-mí**: YBC 5022: 50
26,40 **igi-gub geštu zà-mí**: A 3553: 6
12,30 **gán zarà**: YBC 7243: 5
12,30 **igi-gub gán** ^{giš}**zarà**: A 3553: 5
22,30 *ù* 12,30 *ša* **gán zarà**: YBC 5022: 49
3,30 *ša* **gán zarà ki**-2: YBC 5022: 51

Goetze (1948: 34) suggested **gán** indexes areas in the form characteristic of the objects mentioned; he also proposed that if **gán** denotes the 'body' of these objects then the sign should be read PISAN. For a disucssion of geometric shapes and forms based on known realia and indexed by **gán** 'plane', see Kilmer (1990b: 84–86).

with *našpaku*:

4,20,6,40 **gán** *am-am-ma* x [**ì**]**-dub igi-gub**-*ú*: BM 47860 obv. 1
20,6,16 **gán** *ur-ru-uḫ* **sag**-3 **ki-min igi-gub**-*ú*: BM 47860 obv. 2
30,3,46 **gán** *ki-ir-ṣi* **ki-min igi-gub**-*ú*: BM 47860 obv. 3
16,40,2,30 **gán** *ka-re-e* **ki-min igi-gub**-*ú*: BM 47860 obv. 4
4,10,3,7,30 **gán** *ka-ṣal* **ki-min ki-min igi-gub**-*ú*: BM 47860 obv. 5
12,30,6,22,30 **gán gaba-ri ki-min igi-gub**-*ú*: BM 47860 obv. 6
6,40 **gán ì-dub** še-im **igi-gub**-*ú*: BM 47860 obv. 7
6,0 **gán ì-dub** *me-*⌜*e*⌝ **igi-gub**-*ú*: BM 47860 obv. 8
6,0 **gán ì-dub** x DIŠ **igi-gu**[**b**-*ú*]: BM 47860 obv. 9
6,12 **gán ì-dub** x ŠE DU_8 NU **igi-g**[**ub**-*ú*]: BM 47860 obv. 10
6,12 **gán ì-dub** [x] x DIŠ [**igi-gub**-*ú*]: BM 47860 obv. 11
6,24 **gán ì-**[**dub** ... **igi-gub**-*ú*]: BM 47860 obv. 12
6,24 **g**[**án ì-dub** ... **igi-gub**-*ú*]: BM 47860 obv. 13
6,45 [**gán** ... **igi-gub**-*ú*]: BM 47860 obv. 14
7,30 [**gán**...**igi-gub**-*ú*]: BM 47860 obv. 15

Livingstone (1986: 20) transcribed **gán** as *ikû*; however, the term is explained by the summary in rev. 4–8 and the synonym **a-šà-meš** in rev. 6.

ereqqu (= ^{giš}**mar-gíd-da**) 'wagon'
coefficients:

1 (**gur**) 1 **bariga** *ḫa-a-mu* **gún** ^{giš}**mar-gíd-da**: CBS 10996 obv. iii 30'

erinnu (= ^{giš}**erin**) 'cedar'

^{giš}**erin**: VAT 8522 obv. i 1, 5

erṣetu 'earth'

sal-šeš ^{lú}**ḫal**-*t*[*i pirišti šamê u erṣeti*]: K 11097: 1

eršu (= ^{giš}**ná**) 'couch'
coefficients:

[x] **igi-gub** ^{giš}**ná** ^{giš}**gu-za**: A 3553: 43

erû (= **urudu**) 'copper'
coefficients:

1,12 *šà* **urudu**: *TMS*, Text 3: 54
1,12 **igi-gub urudu**: A 3553: 18
1,12 *ra-ṭù-um ša* **urudu**: YBC 5022: 22
1,12 *ra*(?)-*ṭù*(?)-*um ša* **urudu**: YBC 7243: 36
9 **ki-lá**(?)**-bi**(?) *ša* **urudu**: YBC 7243: 33

esēḫu 'to assign' (used of fields, persons, provisions or goods)
e-si-ḫa: K 11097: 6

esittu 'mortar'
coefficients:
e-si-it-tum 3,36 *ig-gi-gu-bu-ša*: IM 52916, IM 52685, IM 52304 Section I(d): 2

eṣēdu 'harvest'; meaning uncertain in IM 54010
e-ṣí-da: IM 54010 obv. 3, 4 [von Soden (1952: 52) suggested reading *e-ṣí-id*.]
[*iš*]*-ka-ar e-ṣ*[*e-di*] *e-pé-ša-am*: IM 52916, IM 52685, IM 52304 Section I(d): 4
[*e*]*-ṣ*[*i-id š*]*a* **bùr gán a-šà**-*im* **še**-*um* **en-nam**: BM 85194 rev. ii 35

etpešu 'wise, clever'
et-pe-šú ḫa-as-si mu-du-ú: K 11097: 7

gabarû (= **gaba-ri**) 'copy (of a tablet)'
ki pī **dub gaba-ri** *la-ba-ri-šú*: BM 47860 rev. 12

GÁ-DUB ŠE (meaning unknown)
4,48 GÁ-DUB ŠE: YBC 5022: 61

GÁN(?)-GÁ ŠE(?)
30 *ša* GÁN(?)-GÁ ŠE(?): IM 52301 rev. 22

GÁN GI[Š...]
GÁN GI[Š...]: BM 15285: Text XII

GÁN-MAN (meaning unknown)
coefficients:
6(?) *ša* GÁN-MAN: YBC 5022: 20
4,36,10 *ša* GÁN-MAN **ki**-2: YBC 5022: 53

gaṣṣu (= **im-babbar**) 'gypsum, whitewash'
coefficients:
26 **im-babbar**: YBC 5022: 56
...8(?) **im-babbar**: YBC 7243: 15

GI BUL IR (meaning unknown)
coefficients:
7,12 **igi-gub** GI BUL IR: A 3553: 40

7,12 is also found as a coefficient in the following:
7,12 **igi-gub**(?) BI *a!-ma-ri-im*: IM 49949: 6
7,12 **igi-gub** $\mathbf{sig_4}$**-anše**: A 3553: 24
7,12 *ša* $\mathbf{sig_4}$**-anše**: YBC 5022: 16
7,12 $\mathbf{sig_4}$**-[an]še**: YBC 7243: 40
e-le-pu-um 7,12 *i-gi-gu-bu-ša*: IM 52916, IM 52685, IM 52304 Section I(d): 1
7,12 *na-al-ba-an* **saḫar**: YBC 5022: 3
7,12 *ša na-aš-pa-ak* $\mathbf{gur_7}$: *TMS*, Text 3: 51

GI-NÍGIN-NA (meaning unknown)
coefficients:
4,58 **igi-gub** GI-NÍGIN-NA: A 3553: 38

GIŠ-DUB-ÍL (meaning unknown)
coefficients:
2,13,20(?) *ù* 1,15 *ša* GIŠ-DUB-ÍL: YBC 5022: 45 [Kilmer (1960: 289 n. 1) suggested that the DUB looks like AB but that $^{\mathbf{giš}}$**al** = *allu* is expected.]

giš-kár (no known Akkadian equivalent) 'task'
a-šà giš-kár: *UET* 5: no. 864: 1

GIŠ-LAGAB (meaning uncertain)
2[0] GIŠ-LAGAB: A 3553: 47(a)

The reading is ambiguous; if *ḫūratu* 'plant dye' is the equivalent of GIŠ-LAGAB, it is not attested for OB (Kilmer 1960: 301).

gubbu (meaning uncertain)
coefficients:
6,40 *i-gi-gu-ub-bi-im gu-ub-bi-im*: IM 52301 rev. 20

6,40 is also found as a coefficient in the following:
6,40 *ša* **é-kišib še** YBC 5022: 30
6,40 **im-dub**(?) YBC 7243: 17
6,40 *na-az-ba-al* **saḫar** YBC 7243: 32

Possible readings for this word are:
gubbu 'well, cistern' (NA, West Semitic lw.)
kuppu 'catchwater at a well, well' (RS, SB)
quppu 'chest, receptacle' (from OB on).

gurunnu 'heap (of sesame)'
coefficients:
1,20 *gu-ru-un* **še-giš-ì**
20 **kimin**
1,20 **kimin**
1 **kimin**: CBS 10996 obv. iii 4'–7'

(**dug**)**gu-zi** see DUG-GU-ZI

ḫabātum
coefficients:
20 *al-li ḫa-ba-tim* 20 *al-li la-ba-nim* 10 *al-li ba-la-lim lu-pu-ut*: Haddad 104 #9 iv 16–17
with *allum* implied:
20 *ḫa-ba-tam* 20 *la-ba-nam* 10 *ba-la-lam l*[*u-pu-u*]*t*: Haddad 104 #10 iv 29

ḫad/ṭālum (work at the third level of canal digging)
ḫa-d/ṭa-lum iḫ-d/ṭi-il: BM 85196 rev. i 37

Goetze (1948: 35–36) suggested that *ḫadālum iḫdil* 'he/it backed up (the water)', be construed as an absolute infinitive with the *-um* ending being adverbial, not nominative. Oppenheim (1947: 128) translated *ḫad/ṭālum* as 'to stretch a rope (for demarcation)'.

ḫammu 'litter of grain'
coefficient:
1 (**gur**) 1 **bariga** *ḫa-a-mu* **gún** **giš****mar-gíd-da**: CBS 10996 obv. iii 30'

ḫasīs sammî (= **geštú zà-mí**) See *sammû*.

ḫassu 'wise'
et-pe-šú ḫa-as-si mu-du-ú: K 11097: 7

ḫasû (= **ḫar**) 'lung'
la-'-ab **ḫar**: K 8705: 4

ḫaṣābu (= **kud**) 'to cut or break off (reeds, branches)'
ḫaṣbu:
gi-kud: VAT 7532 obv. 1
VAT 7535 obv. 1; rev. 1, 25
gi-kud-a: A 3553: 45
YBC 5022: 64
ḫaṣābu:
igi (x) **gál**(-**là**)-*šu ḫu-ṣú-ub*(-*ma*): VAT 7532 obv. 9, 12
VAT 7535 obv. 10, 12; rev. 9, 11
aḫ-ṣú-ub-šu-ma: IM 53965 obv. 4
aḫ-ṣu-ub-šu-ú-ma: Strssbg. 368 obv. 2
aḫ-ṣú-bu-ú-šu ú-te-er-šu-um-ma: Strssbg. 368 obv. 3
ta-aḫ-ṣú-bu: IM 53965 rev. 2
iḫ-ḫa-aṣ-bu: Strssbg. 366 rev. 5 (½ **ninda** *ri-iš* **gi** 5 **kùš** *ša iḫ-ḫa-aṣ-bu*)
Strssbg. 368 rev. 2
VAT 7532 obv. 17; rev. 6
VAT 7535 obv. 17, 24; rev. 21, 30–31
ša iḫ-ḫa-aṣ-ba-an-ni(-ma) ú-te-er-šum-ma: VAT 7532 obv. 6
VAT 7535 rev. 6
igi (x) **gál**(-*šu*) *iḫ-ḫa-aṣ-ba-an-ni-ma*: VAT 7532 obv. 2–3
VAT 7535 obv. 3; rev. 3, 27
igi (x) **gál**(-**là**-*šu*) *ù* (y) *iḫ-ḫa-aṣ-ba-an-ni*(-*ma*): VAT 7532 obv. 4
VAT 7535 obv. 4–5, 6–7; rev. 4–5, 28–29
coefficients:
[1] **igi-gub gi-kud-⌜a⌝**: A 3553: 45
1 **gi-kud-a**: YBC 5022: 64

ḫaṣbu '(formed) clay, pot sherd, (small pot)'
coefficients:
12 *šà ḫa-aṣ-bi*: *TMS*, Text 3: 65

ḫerû 'to dig'
eḫ(?)-*re-e*: YBC 9874 rev. 5
e[*ḫ*(?)-*r*]*e*(?)-*a*-[*m*]*a*: YBC 9874 rev. 3
iḫ-re: YBC 7164 obv. 16, 17, 18

ḫimmatu 'faggot of reeds'
coefficients:
72 ^{giš}**má-lá** *ḫi-im-ma-ti*: CBS 10996 obv. ii 10'

ḫirītu ‘ditch, canal, moat’
ḫi-ri-tum: BM 85194 rev. ii 41
ḫi-ri-tam: BM 85194 obv. i 39
ḫi-ri-tim: BM 85194 obv. i 41, 48; ii 4

ḪI-*rum* ‘field(?)’
ḪI {*uš*} *rum*: AO 17264 obv. 1

Thureau-Dangin (1934c: 62) suggested that this word is the equivalent of **sag-ki-gud** ‘trapezoid’.

ḫubullu (= **ur$_5$-ra**) ‘debt with interest’
a-na **ur$_5$-ra** *ad-di-in-ma*: VAT 8528 rev. i 10

ḫulu ‘road’ → here: ‘method (of solution)’; synonymn for *nēpešum*
ḫu-li: *TMS*, Text 16: 7

ḫuluḫtu ‘light colored frit’; *CAD* s.v. *ḫuluḫtu* assigned this meaning on the basis of the Sumerian **an-zaḫa** ‘heated or fused lead’.
2 *šà ḫu-lu-uḫ-tim*: *TMS*, Text 3: 61 [von Soden (1964: 47) emended the text to *ḫu-lu-uḫ-ḫi!*]

ḫuluqqû ‘commercial losses, lost merchandise’
[...] *ma-la ur-ta-ba-bu-ú ḫu-lu-q*[*a-am*...]: IM 52916, IM 52685, IM 52304 Section F: 54’

ḫurāṣu (= **guškin**) ‘gold’
guškin: A 3553: 22
BM 34568 rev. i i 12, 14, 15, 17
IM 52916, IM 52685, IM 52304 Section I(b): 3
M 15192 + M 15236 rev. 11–13, [14], 15–19 // *ARM* 24: no. 146
TMS, Text 3: 59
YBC 4698 obv. i 13, 15; rev. ii 16, 17
YBC 5022: 24, 25
YBC 7243: 34, 37

coefficients:
1,20 *r*[*a-ṭ*]*ù-um* <*ša*> **guškin**: YBC 7243: 34
guškin 1,48 [*i-gi-gu-bu-šu*]: IM 52916, IM 52685, IM 52304 Section I(b): 3
1,48 **igi-gub guški[n]**: A 3553: 22
[1],48 *šà* **guškin**: *TMS*, Text 3: 59

1,48 *ra-ṭù-um ša* **guškin**: YBC 5022: 24
1,48 [*ra-ṭù-um*] *ša* [**guš**]**kin**: YBC 7243: 37:
2,15 *ru-uq-qú ša* **guškin**: YBC 5022: 25

ḫurḫuru 'city moat'(?)
ḫur-ḫu-ru: BM 85194 rev. ii 25
ḫur-ḫ[*u-ri*]: BM 85194 rev. ii 11
iš-tu ḫu-ur-ḫur-i[*m*] *a-na* **bàd** *ka-ša-di-im*: BM 85210 obv. ii 4

idū (= **á**) 'wages' (per man per day)
saḫar du$_8$ **á iz-zi**(sic!): *UET* 3: no. 1386: 7 [Vaiman read **á-giš-gi**.]
á-bi: YBC 4663 rev. 3
á-bi [*i*]*-na-ad-di-kum*: YBC 4663 obv. 39
á-bi (^{lú}**ḫun-gá**) **en-nam**: YBC 4657 obv. 13
YBC 4663 obv. 34
6 (**še**) **á**(**-bi**)**-am**$_5$: YBC 4663 obv. 20, 26, 39, rev. 14
6 **še á** ^{lú}**ḫun-gá**: YBC 4666 rev. 8, 11, 13
YBC 7164 obv. 2
6 (**še**) **á-bi** (^{lú}**ḫun-gá**): YBC 4657 obv. 1, 4, 6, 9, 10, 13, 14, 16
YBC 4663 obv. 1, 7, 14; rev. 2
YBC 4666 rev. 27
YBC 5037 obv. 2, 3, 5, 6, 7, 9, 11, 12, 14, 16, 17, 19, 21, 22, 23, 24
1 (**bán**) (**še**) **á**(**-bi**) ^{lú}**ḫun-gá**: YBC 4666 obv. 2, 5, 8, 10
1 (**bán**) **še-ta-àm á-bi** ^{lú}**ḫu**[**n-g**]**á**: YBC 4657 rev. 24
1 ^{lú}**ḫun-gá á-bi** 1 (**bán**) 5 **sìla**: YBC 4666 rev. 1, 4
(x) *a-na i-di i-ši*: YBC 4663 obv. 6, 11, 18, 23, 29
igi *i-di* **du**$_8$: YBC 4663 rev. 16
[...(**bán**)] 5 **sìla** *še-a-am i-di-šu il-qé-ma*: YBC 10722 rev. 3

igāru (= **é-sig**$_4$, **iz-zi**, var. **i-zi**) 'wall (of a building)'
i-gar: AO 6484 obv. 6, 7
i-ga-ri-im: IM 49949: 7
é-sig$_4$: VAT 6598 obv. 3, 4 (**é-sig**$_4$*-ya*), 8, 14, 17, 19, 22, [23]; rev. i [1], 3 (**é-sig**$_4$*-ya*), 10 (**é-sig**$_4$*-ka*), 18
é-sig$_4$ **sig**$_4$**-al-ur**$_5$**-ra**: VAT 6598 obv. [3], [13], 15; rev. i 1
iz-zi: A 3553: 41
AO 10822 rev. 10, 11, 12
Const 236: 1, 4
UET 3: no. 1386: 7 [**saḫar du**$_8$ **á iz-zi**(sic!)]

with *kuburrû*:

ku-bu-ur-re-e **iz-zi**: AO 10822 rev. 12

ku-bu-ur-re **iz-zi**: AO 10822 rev. 10

coefficients:

2,15 **igi-gub é-sig$_4$ sig$_4$-al-ùr-ra**: VAT 6598 obv. 15

See 2,15 **igi-gub sig$_4$-al-ùr-ra** in VAT 6598 obv. 5; rev. i 8

6 **igi-gub** *i-ga-ri-im*: IM 49949: 7

6 **igi-gub iz-zi**: A 3553: 41

Walker (Letter to author 1986) enclosed a copy of his join, BM 96957, to VAT 6598, which contains problems dealing with **é-sig$_4$** 'wall', to which the coefficient 6, the coefficient of a wall, is applied.

Cf. *kal* 1-**še u$_4$**-60-**kam** 3⅔ <**sar**> 5 **gín**: *UET* 3: no. 1386: 9 [This is a volume text in which 60 man-days translates into 3,45 **sar** excavated or constructed; the work assignment of 0;3,45 **sar** is used as the coefficient for the earth wall or *pitiqtu*; the section of the text which transforms the volume of earth into man-days is introduced in line 7 by **saḫar du$_8$ á iz-zi**(!).]

igru 'hire, wages, rent' (pl. tant.)

coefficients:

30 **igi-gub** *šà pu-ut i-gi-ri-im*: *TMS*, Text 3: 35 [The construction is problematic; the norm is *ana igrim*. *CAD* s.v. *igru* notes that *igru* seems to be replaced by *idū* in OB with the exception of Codex Hammurabi and Mari references. It occurs only rarely in literary texts of that period.]

ikribu 'votive offering'

[*i-nu-ma* **lú** *ik-ri-bi-šu* **udu-ḫi-a** *ma-da-tim a-na* ^{d}*Il-tim i-ka-ar-ra-ab*...]: M 6829: [1''–3''] (restored by // text *ARM* 24 #260: 1–3)

iku (= **e**) 'dike'

i-ku-um: Erm 15073 rev. ii 7

i-ki-im: BM 85194 obv. ii 17

i-ki-ia: Erm 15073 rev. ii 9

e: BM 85194 obv. i 42, 44; ii 7, 13

Strssbg. 364 rev. 7

YBC 4673 rev. i 16

e libir-ra: YBC 4673 rev. i 21; rev. ii 1

In discussing BM 85194 #4, Thureau-Dangin (1932c: 24–25) suggested that the dike (**e** = *iku*) is an earth embankment, which is probably constructed by using the dirt excavated from digging a moat (*ḫirītu*).

Waetzoldt (1990) listed three kinds of dams or dikes: 1) a dam to block a watercourse or canal on both sides, 2) a protective dam to prevent flooding an area far from the main dam and 3) raised dikes which brought water from a canal or reservoir to a field.

iku u palgu are complementary terms. *palgu* 'irrigation ditch' is bordered by *iku*.

imgiddû (= **im-gíd-da**) 'oblong tablet'

im-gíd-da: IM 52916, IM 52685, IM 52304 Sections F and G: rev. 43'

im-gíd-da da PN_1 *ù* PN_2: *TMS*, Text 5 colophon

5,20 **im-gíd-da**: BM 47860 rev. 4

coefficients:

im-gíd-da [...] *te u* (coefficient list; context broken): IM 52916, IM 52685, IM 52304 Sections F and G: rev. 43'

im-gíd-da is read *imgiddû*, *liginnu*, *u'iltu*, and *nibzu* before the NB period; the term describes the tablet as oblong as opposed to wide (*CAD* s.v. *giṭṭu*).

The following definitions apply:

giṭṭu 'one column tablet with literary content, tablet containing receipt or certificate, document or deed written on parchment' (NB, LB, Sum. lw) (*CAD* s.v. *giṭṭu*)

imgiddû 'one column tablet' (SB, Sum. lw) (*CAD* s.v. *imgiddû*)

nibzu 'tablet, document' (NA, NB, Sum. lw) (*CAD* s.v. *nibzu*)

liginnu 'tablet (containing texts or excerpts, especially for teaching purposes)' (OB, SB, NA, NB) (*CAD* s.v. *liginnu*)

u'iltu 'in Assyria a literary tablet containing excerpts' (LB, SB); 'promissory note, obligation' (NB, LB) (*AHw* s.v. *u'iltu*)

im-KI(?)-**a** meaning unknown; reading uncertain

coefficients:

44 **im**-KI(?)-**a**: YBC 7243: 16 [Kilmer (1960: 301) lists this item with clay and clay objects.]

imlû (= **im-lá**) meaning unknown

i-im-lu-ú[*-um*]: YBC 5022: 40

im-li-im: *TMS*, Text 3: 63

im-lá: A 3553: 17

BM 85194 obv. iii 5, 6, 11, 18, 20, 21

BM 85210 rev. ii 4, 7, 8

YBC 7243: 14

im-lá-bi: BM 85194 obv. iii 14

coefficients:

18 *i-im-lu-ú*[*-um*]: YBC 5022: 40

18 *šà im-li-im*: *TMS*, Text 3: 63

18 **im-lá**: BM 85194 obv. iii 20; obv. iii 6 (**igi** 18 **im-lá** <*pu-ṭur*>); obv. iii 11 (18 **im<-lá>-bi**); see BM 85194 obv. iii 15 (**igi** 18 **du$_8$-a**)

YBC 7243: 14

18 **igi-gub im-lá**: A 3553: 17

imlû occurs between the coefficients for clay and bitumen in *TMS*, Text 3; however, in A 3553 *imlû* occurs between coefficient for spices and metals.

immeru (= **udu**, **udu-níta**) 'rams'

udu...: VAT 8522 rev. 2

ša 10 **udu** *ik-ru-bu* 1 **udu** *ba-al-ṭà-*(var. *ṭú*)*-tim ú-wa-aš-ša-ar* etc.:

M 6829: 1'–38' (see // text *ARM* 24: no. 260 *ba-al-ṭa-am*, *ba-al-ṭa-tim*)

udu-... **-ḫi-a sila$_4$-gub**: VAT 8522 rev. 3, 4

udu-ḫi-a: UET 5: no. 121: 9 (sic!), 12, 15

a-di **udu-ḫi-a** ᵈ*I*[*l-tim la ú-wa-š*]*e-ru*: M 6829: 39' // *ARM* 24: no. 260

[*i-nu-ma* **lú** *ik-ri-bi-šu* **udu-ḫi-a** *ma-da-tim a-na* ᵈ*Il-tim i-ka-ar-ra-ab*...]:

M 6829: [1''–3''] (restored by // text *ARM* 24 #260: 1–3)

udu-níta: K 8705: 8, 9, 10, 15

im-šu 'section, paragraph' (separated off by lines)

(x) **im-šu**: VAT 7528 colophon

YBC 4657 bottom edge

YBC 4666 left edge

YBC 4668 colophon

YBC 4673 colophon

YBC 4695 colophon

YBC 4696 colophon

YBC 4697 colophon

YBC 4708 colophon

YBC 4709 colophon

YBC 4710 colophon

YBC 4711 colophon

YBC 4712 colophon

YBC 4713 colophon

YBC 4714 colophon

YBC 4715 [colophon]
YBC 5037 rev. 28
(x) **im-šu-meš**: A 24194 rev. v 6–7
YBC 4607 left edge
YBC 4652 rev. [14]

īn alpim (= **igi gu$_4$**) lit.:'ox eye'; meaning uncertain
coefficients:
i-in al-pí-im 16,52,30 *i-g[i-gu-bu-šu]*: IM 52916, IM 52685, IM 52304 Section C: rev. 12'
16,52,30 **igi-gub** *šà* **igi gu$_4$**: *TMS*, Text 3: 19
52,30 **ri** *šà* **igi gu$_4$**: *TMS*, Text 3: 20
30 *pi-ir-ku šà* **igi gu$_4$**: *TMS*, Text 3: 21

Kilmer (1990: 86) classified **igi gu$_4$** among 'terms for specific geometric constructions that are based on highly stylized art forms and artifacts related to bovine faces and bovine lyres' and agreed with Vaiman (1963: 80) that the 'ox eye' is a biconvex figure constructed from 120° arcs.

ipru (= **še-ba**) 'barley ration distributed by administration of temple, palace, etc.'
giš 2(0 **sìla**) **še-ba**: M 6861: 11

irbu 'income, import duty, offering' meaning uncertain in IM 54010 due to poor state of text
i-gi ir-bi: IM 54010 obv. 5
ir-bé-e i-na qa-na: IM 54010 obv. 2

irtu 'chest'
[*ir*]-*ti*: K 8705: 12

isiqtu 'mark'
a-na i-si-iq-tim: BM 85194 obv. ii 35
a-na i-si-iq-tim ú-ul i-sa-an-ni-iq: BM 85194 obv. ii 28-29 (<-*iq*>), 39, 43

isqu (= **giš-šub-ba**) 'lot', (interval of tuning)
coefficients:
sa *qud-mu-ú ù* **sa** 3-*šú* **sig** 1,3 **sa giš-šub(!)-ba**: CBS 10996 obv. i 18'

iṣu (= **giš**) 'wood, tree'
i-ṣu-um: Haddad 104 #1 i 9; #2 i 10

i-ṣú-um: Haddad 104 #2 i 18; #3 i 19, 30, 37; ii 30
i-ṣa-am: Haddad 104 #3 i 22
i-ṣi-im: Haddad 104 #1 i 1, 4; #2 i 12
i-ṣí-im: Haddad 104 #2 i 14, 16; #3 i 23, 25 (*ku-bu-úr i-ṣí-im*), 26, 28, 41; ii 4, 7, 9, 14, 16, 20, 25, 26, 28
YBC 7243: 35
YBC 8600 obv. 1
iṣ-ṣi: AO 6484 obv. 6, 9
ki(-ip)-pat **giš**: YBC 8600 obv. 7; rev. 3
ku-bu-ur i-ṣí-im: YBC 7243: 35
ku-bu-ur **giš**: YBC 8600 obv. 5, 6
ku-bu-ur-re-e **giš**: YBC 5022: 58
coefficients:
4,48 *ša ku-bu-ur i-ṣí-im*: YBC 7243: 35
4,48 *ša ku-bu-ur-re-e* **giš**: YBC 5022: 58
referring the the thickness of a log:
4,48 **igi-gub-ba**: YBC 8600 obv. 4

išartu 'normal interval on a musical scale, normal tuning'
coefficients:
3,6 **sa** *ti*-x *i-šar-tum*: CBS 10996 obv. i 8'
sa *ša*-**ge**$_6$ *ù* **sa** 4 *uḫ-ri* 2,6 **sa** *i-šar-tum*: CBS 10996 obv. i 13'

išdu (= **suḫuš, úr**) 'foundation'
iš-di: VAT 7848 obv. 14
i-na iš-di-šu: VAT 8522 obv. i 2 (refers to base of a tree, antonymn of *appu*)
i-na iš-di-šu-ú: Haddad 104 #3 i 21
suḫuš: AO 6484 obv. 7 (**suḫuš** *i-gar*)
BM 85194 obv. i 2, 7 (**suḫuš saḫar-ḫi-a**)
úr: BM 85196 obv. i 18, 20; rev. ii 9

Cf. *iš-tu* **umbin saḫar-ḫi-a**: BM 85210 obv. ii 3 (**umbin** may be a scribal error for **suḫuš**.)

iškaru (= **éš-kàr**) 'assignment of work' (per man per day)
iš-ka-ra-am: YBC 9856 obv. 1
iš-ka-ar x [...]: IM 52916, IM 52685, IM 52304 Section I(e): 1
iš-ka-ar a-wa-ti: YBC 9874 rev. 7
iš-ka-ar al-lim: Haddad 104 #9 iv 26; #10 iv 43
[*iš*]-*ka-ar a-wi-lim iš-te-en*: IM 54011 rev. 4

[*iš*]-*ka-ar iš-te-en a-wi-li-im*: IM 53961 obv. 4; rev. 4–5
iš-ka-ar 1 **lú**: Haddad 104 #9 iv 16, 25; #10 iv 42
[*i*]*š-ka-ar* **lú** *m*[*i*]-*nu-um*: Haddad 104 #10 iv 28
[*iš*]-*ka-ar e-ṣ*[*e-di*] *e-pé-ša-am*
[*iš*]-*ka-ar za-ri-i ša še-im e-pé-š*[*a-a*]*m*: IM 52916, IM 52685, IM 52304
Section I(d): 4–5 (assignments in connection with agricultural work)
iš-ka-ar si-ri-im: Haddad 104 #7 iii 39
éš-kàr: A 3553: 27 (**kàr**)
AO 8862 iv 21 tabulation
BM 85194 obv. i 9, 20, 31, 34, 35; obv. ii 1, 5, 7, 22
BM 85196 obv. ii [33]; rev. i 33, 38
Erm 15073 rev. i 7, 18
TMS, Text 3: 46 (**éš-kàr lú**-1)
VAT 7528 obv. i 3, 9, 14, 20; rev. ii 9
VAT 8523 rev. 7 (**éš-kàr** *šá* **lú**); 14, 16, 23, 24 (**éš-kàr** 1 **lú**)
YBC 4657 obv. 1, 4, 6, 8, 11, 12, 14, 16; rev. 7, 9, 11, 13, 15 16, 19, 20, 22, 24
YBC 4662 obv. [36], 39, 41; rev. [3], 7, 11, 14, 18, 21, 25, 27, 31, [36]
YBC 4663 obv. 1, 5, 7, 9, 14, 17, 20, 22 (**éš<-kàr>**), 27, 32, 33, 36; rev. 2, 5, 14, 17
YBC 4666 obv. 2, 5, 7, 9, 11, 13, 15, 17, 19, 21, 23; rev. 1, 4, 8, 11, 13, 15, 17, 19, 21, 23, 24, [26]
YBC 4669 rev. i 1, 3, 6
YBC 4673 obv. iii 12, 19; rev. i 5; rev. iii 17
YBC 5022: 43 (**éš-kàr lú**-1)
YBC 5037 obv. 6, 7 (<**éš**->), 8, 10, 12, 14, 16, 17, 19, 21, 22–24 (<**éš**->)
YBC 7164 obv. 1, 24, [27]; rev. <1>, 8, 15
YBC 7243: 20
YBC 8588 obv. 8, 9, 12; edge 2; rev. 5
éš-kàr *ša* $\mathbf{u_4}$: AO 8862 iv 21 tabulation
2 (**bán**) **éš-kàr**: YBC 4673 obv. i 24; obv. ii 2

coefficients:
⅓ **sar saḫar éš-kàr**: VAT 7528 obv. i 3, 9, 14, 20; rev. ii 9
⅓ **gín** (**saḫar**) **éš-kàr**: YBC 4666 obv. 2, 5, 7, 9, 11, 13, 15, 17, 19, 21, 23; rev. 1, 4
YBC 7164 obv. 24, [27]; rev. <1>, 8, 15
10 **gín éš-kàr lú**-1: YBC 5022: 43
10 **éš-kàr**: BM 85194 obv. i 9, 20, 31, 34; obv. ii 1, 5, 22

BM 85196 obv. ii [33]
Erm 15073 rev. i 7, 18
YBC 4673 rev. iii 17
10 … **éš-kàr**: BM 85194 obv. i 35; obv. ii 7
10 **éš-kàr** <**ki-**>**lá**: YBC 7243: 20
10 (**gín**) (**saḫar**) **éš-kàr**: BM 85196 rev. i 33
YBC 4657 obv. 1, 4, 6, 8, 11, 12, 14, 16; rev. 7, 9, 11, 13, 15, 16, 19, 20, 22, 24
YBC 4662 obv. 36, 41; rev. 3, 11, 18, 25, 36
YBC 4663 obv. 1, 7, 14, 20, 32, 33; rev. 2, 14
YBC 4666 rev. 8, 11, 13, 15, 17, 19, 21, 23, 24, 26
YBC 4669 rev. i 3
YBC 5037 obv. 6, 7 (<**éš-**>), 8, 10, 12, 14, 16, 17, 19, 21, 22–24 (<**éš-**>)
YBC 7164 obv. [1]
10 *ša* **saḫar** *za-ba-li* **éš-kàr lú**-1: *TMS*, Text 3: 46
6 **igi-gub kàr lú**-1-**kam**: A 3553: 27
referring to *pitiqtu*:
3,45 **éš-kàr**: YBC 4673 obv. iii 12, 19; rev. i 5

itinnu (= **šitim**) 'housebuilder'
i-tì-nu-um: AO 8862 iii 32; iv 6
YBC 4673 obv. i 14

ittû 'omen, ominous sign'
it(?)-*ta*-[*a-šú-nu*]: K 8705: 8

ittû (= **esir**, **esír**) 'crude bitumen'
[*ma-ḫi-ir*…] *ma-ḫi-ir* **esir** *ma-ḫi-ir* […]: IM 52916, IM 52685, IM 52304 Section F: 46'
coefficients:
15 *šà* **esir**: *TMS*, Text 3: 66
16 **igi-gub esír**: A 3553: 14
30 *ša* **esír**(?): YBC 5022: 57 [See Kilmer 1960: 291]

ka-[…]
coefficients:
5 *ka*-[…]: YBC 7243: 1

kabāru 'to become fat, heavy, thick, strong' (said of animals, people, parts of the body, parts of the exta, timber and trees, ears of barley, etc.)

ka-ba-ar: VAT 6598 rev. i 2

$^{\text{[g]iš}}$**gi** *u* ^{giš}**tir** *ak-ba-ru*: *TMS*, Text 25 obv. ii 1 [Von Soden (1964: 50) suggested the reading [*ša i-na a*]*pim u qištim ak-sú!-ru.*]

24 **saḫar** *ak-ba-ru*: *TMS*, Text 25 obv. ii 3

ik-bi-ir: VAT 8522 obv. i 2, 3, 7a, 7

YBC 8600 obv. 2

kalakku (= **ki-lá**) 'excavation'

ka-la-ak-ku: *TMS*, Text 24: 33, 37 (*ka-la-ak-ku* **ugu** *ka-la-ak-*[*ki*])

ka-la-ak-kum: YBC 8588 obv. 8

ka-la-ak-ki: *TMS*, Text 24: 37 (*ka-la-ak-ku* **ugu** *ka-la-ak-*[*ki*]; see *TMS*: 117–22)

ka-la-ak-ki-im: *i-na iš-te-en ka-la-ak-ki-im* (x) *ka-la-ak-ku*: YBC 8588 obv. 2–3, 6–7

(x) *ka-la-ak-ku*: YBC 8588 obv. 3, 7, 10

ki-lá: A 3553: 34

YBC 4657 obv. 1, 4, 6, 8, 10, 12, 14, 16, 18–22, 24, 26, 27, 29, 31, rev. 1, 3, 5, 7, 9, 11, 13, 15, 16, 18, 20, 22, 24, bottom edge

YBC 4662 obv. 1, 12, 24, 36, 41 , rev. [3], 11, 18, 25, 31

YBC 4663 obv. 1, 7, 14, 20, 26, 33, 40, rev. 1, 13

YBC 5037 obv. 1, 3, 4, 6, 7, 8, 10, 12, 14, 16, 17, 19, 21-33, rev. 1–10, 12–14, 16, 18, 20, 22, 24, 26

YBC 7243: 20 (<**ki-**>**lá**)

YBC 8588 obv. 1

coefficients:

10 **éš-kàr** <**ki-**>**lá**: YBC 7243: 20

10(?),1,30(?) **igi-gub saḫar ki-lá**: A 3553: 34

kalû 'yellow ochre' (referring to any earth or ore of a specific yellow color, regardless of its chemical composition)

coefficients:

54 *ka-lu-ú-um*: YBC 5022: 56

karābu 'to dedicate an offering by pronouncing the relevant formulas'

ša 10 **udu** *ik-ru-bu* 1 **udu** *ba-al-ṭà-*(var. *ṭú*)*-tim ú-wa-aš-ša-ar* etc.: M 6829: 1'–38' (see // text *ARM* 24: no. 260 *ba-al-ṭa-am, ba-al-ṭa-tim*)

[*i-nu-ma* **lú** *ik-ri-bi-šu* **udu-ḫi-a** *ma-da-tim a-na* ^{d}*Il-tim i-ka-ar-ra-ab*...]: M 6829: [1''–3''] (restored by // text *ARM* 24 #260: 1–3)

karpatu (= **dug**) 'earthen container'
dug an-bar: YBC 4698 obv. i 12
dug guškin: YBC 4698 obv. i 13
dug x [...]: BM 47860 rev. 3

kartappu (= **kir$_4$-dib/dab**) 'groom (for leading donkeys and horses and as a title of a court official), a high administrative official'
kir$_4$(?)-**dib**-*šu*: VAT 8522 obv. i 5a, 7a

kāru 'quay, market'
ka-ru: *TMS*, Text 3: 40
ka-ru-um: IM 52301 rev. 22
[*i-*]*na ka-a-ri*: MM 86.11.404 obv. 5
ka-ar-šu: Haddad 104 #3 i 20
coefficients:
10 *sà* **saḫar** *ka-ru*: *TMS*, Text 3: 40
7,30 *ka-ru-um*: IM 52301 rev. 22
Cf. 7,30 [**gán** ....**igi-gub**-*ú*]: BM 47860 obv. 15

karû (= **gur$_7$**) 'pile of barley prepared for storage'
ka-re-e: BM 47860 obv. 4
gur$_7$: *LBAT*: no. 1647 i 3', 11'
TMS, Text 3: 51
TMS, Text 14 obv. 1, 8, 9; rev. 4
coefficients:
7,12 *šà na-aš-pa-ak* **gur$_7$**: *TMS*, Text 3: 51
a-na 8 *na-aš-pa-ak* **gur$_7$** *i-*[*ší-ma*]: *TMS*, Text 14: rev. 4
16,40,2,30 **gán** *ka-re-e* **ki-min igi-gub**-*ú*: BM 47860 obv. 4

kasāmu 'to cut (down) trees, wood, weeds, herbs'
coefficients:
10 *šà šu-ri ka-sà-mi*: *TMS*, Text 3: 68

kāsiru 'dam'(in problems dealing with damming up canals)
ka-sí-ri-im: VAT 8523 obv. i 3, 7, 9, 19, 23, 28; obv. ii 4, 7, 12, 19, 28, 30; rev. 4

kaṣlu (meaning unknown)
coefficient:
4,10,3,7,30 **gán** *ka-ṣal* **ki-min ki-min igi-gub**-*ú*: BM 47860 obv. 5

The phrase appears in a list of granary coefficients and may be explained by the summary in rev. 4-8, which describes 'the circle of the evening meal of Ekur' and 'every one of the shrines... all of the cult platforms'. Livingstone 1986: 21 translated 'volume of the Kaṣlu'.

kaspu (= **kù-babbar** or **kù**) 'silver'

kù: A 3553: 32–33
VAT 8528 obv. i [4], 12, 16, 26; obv. ii [-2], 6, 9, 14, 16, 20; rev. i 1, 3, 6, 8
YBC 4657 obv. 4, 6, 8, 10, 12, 14, 16,
YBC 4663 obv. 12, 18, 20, 25, 33; rev. 1
YBC 4698 rev. i 12; rev. ii 8, 10, 11, 19
YBC 5037 obv. 4, 6–8, 10, 12, 14, 17, 19, 21–24

kù-babbar: A 3553: 21
BM 85196 rev. ii 18, 20, 23, 24
Haddad 104 #3:i 20, 21, 31–34, 36; ii 5, 11–13, 15–17, 24
IM 31210 §2: 10; §6: 1, 6; §7: 2, 4, 18
IM 52916, IM 52685, IM 52304 Section F: 48' (*ne-mé-l*[*a-am*...] -*am* **kù-babbar** ša-qa-la-am), Section F: 49' (**⸢ì-giš** *ù* [**⸢ì-ša**]**ḫ** *ša-ma-am* **kù-babbar ì-giš** u **ì-š**[**aḫ** *k*]*a-ma-r*[*a-am*])
IM 54464 obv. 11; rev. 9
LBAT: no. 1648 obv.(?) 8', 12', 14'
M 15192 + M 15236 obv. 1, 3, 5, 6; rev. 8, 9 // *ARM* 24: no. 146
MLC 1842 obv. 4, 8
Strssbg. 362 obv. 1, 7
TMS, Text 3: 59
TMS, Text 10: [1], 5, 26
TMS, Text 13: 1, 3, 13, 14*, 16 [*See von Soden 1964: 49.]
UET 5: no. 121: 1
VAT 6597 obv. ii 10; rev. 11
VAT 7530 obv. 2, 5, 9, 10, [13], 19, 20; rev. [6]
VAT 8521 obv. [1]
VAT 8522 obv. i 4, 5, 3b, 4b; obv. ii 3, 7, 4d
VAT 8528 obv. i 1, 7, 9, 10; obv. ii [-4], 5, 11
YBC 4657 obv. 2, 3,
YBC 4663 2, 7, 14, 26, 30, 38, 40; rev. 4, 13, 16
YBC 4666 rev. 8, 9
YBC 4669 rev. ii 5, 9; rev. iii 18, 22
YBC 4698 obv. i 9, 14, 17
YBC 5022: 23, 26

YBC 5037 obv. 2, 3, 16
YBC 7164 obv. 2, 3
YBC 7243: 51
YBC 9856 obv. 8

kù dirig: YBC 4698 obv. ii 10, 11, 15; rev. i 4, 5, 8
kù-babbar dirig: YBC 4698 obv. ii 5
sag kù-babbar 'initial capital'
VAT 8521 obv. 7, 8, 11, 19, 20, [23]; rev. [5], 6, 9, 17, 18, 21
VAT 8528 obv. i 14
YBC 4669 rev. iii 23

sag kù-bi: YBC 4698 obv. ii 6

coefficients:
1,36 *ša* **kù-babbar**: *TMS*, Text 3: 59
1,30 *ra-ṭù-um ša* **kù-babbar**: YBC 5022: 23
4 *ru-uq-qú ša* **kù-babbar**: YBC 5022: 26
26,40 *ru-qú-um ša* **kù-babbar**: YBC 7243: 51
2,36 **igi-gub kù-b[abbar]**: A 3553: 21
3,45 **igi-gub saḫar kù šu-ba-an-te**
16 **igi-gub kù saḫar ba-an-te**: A 3553: 32–33

kibsu 'example (math.)', lit.: 'path, way of acting, rite, behavior'
3[0] *ki-ib-su*: BM 85200 + VAT 6599 colophon
šu-nigin (x) *ki-bi-su*: BM 85196 colophon
VAT 6505 colophon
šu-nigin 35 *ki-bi-is* **šid-meš**: BM 85194 colophon
20 [*k*]*i-ib-*<*sa-*>*tum*: BM 85210 colophon

kīdu 'region outside the city, open country'
i-na ki-di(?)*-im*: IM 54538 obv. 3 [von Soden (1952: 51) suggested reading **ki-su**$_7$ = *maškanum*.]

kikurrû 'chapel or cult socle'
ki-ku$_8$*-ru-*[*u*]*m*: BM 85194 obv. i 13, 23 ([*ki*]*-ku*$_8$*-rum-um*)

ki-lá-bi (see **im-lá**) meaning unknown
ki-lá-bi: BM 85194 obv. iii 10

kilzappu (= **gìr-gub(-ba)**) 'footstool, pedestal'
gìr-gub: BM 85210 obv. i 13, 15, 17, 19 (**gìr<-gub>**)
YBC 4669 obv. iii 3, 4, 7
gìr-gub-ba: BM 85210 obv. i 2, 4, 8

kimṣu 'knee, shin calf, support (mathematical texts)'
i-na ki-im-ṣí-im: BM 85194 obv. i 14
saḫar-ḫi-a ki-im-ṣí-im: BM 85194 obv. i 16, 25–26

kīr atkuppi (= **gir$_4$-ad-**KID) 'wicker-worker's oven'
coefficients:
10,48,6 *ša* **gir$_4$-ad-**KID: YBC 5022: 55 [Neugebauer and Sachs (*MCT*: 135) read [2]0,48,6 but Kilmer (1960: 291) read the numbers cited.]

kirbānu (= **im-lag**) 'clump of clay'
coefficients:
14,24 *šà ki-ir-ba-ni*: *TMS*, Text 3: 64
6,40 **im-lá**: YBC 7243: 17
4,16 **[im(?)]-lag(?)**: YBC 7243: 18

kirṣu 'preparation of cereal; pinched off, unformed matter (clay, dough, etc.)'
30,3,46 **gán** *ki-ir-ṣi* **ki-min igi-gub**-*ú*: BM 47860 obv. 3

The phrase appears in a list of granary coefficients and may explained by the summary in rev. 4–8, which describes 'the circle of the evening meal of Ekur' and 'every one of the shrines... all of the cult platforms'. Livingstone 1986: 21 translated 'volume of the Kirṣu'.

kīru (= **gir$_4$**) 'oven'
coefficients:
10,45,6 **esír-ḫi-a** *ša i-na ki-ri-im*: YBC 7243: 24 [Kilmer (1960: 291) read 10,48,6 here and in YBC 5022: 55 as the coefficient of the wicker-worker's oven.]
10,48,6 *ša* **gir$_4$-ad-**KID: YBC 5022: 55 [Neugebauer and Sachs (*MCT*: 135) read [2]0,48,6 but Kilmer (1960: 291) read the numbers cited.]

kirû (**kiri$_6$** = **giš-sar**) 'garden, orchard, palm grove'
kiri$_6$ *a-na qa-ti zi-ki-ri iš-ša-ka-*[*a*]*n*: *TMS*, Text 3: 70

kisibirru (= **še-lú**^{sar}) 'coriander'
coefficients:
16 **igi-gub še-lú**^{sar}: A 3553: 16

kisû (= **kí-sa**) 'abutment or platform' protected by a spur of land (*appum*) to create a triangular (underwater) prism in BM 85196 #1
kí-sa kin: BM 85196 obv. i 7

For discussion of this term, see Thureau-Dangin (1935a: 5) and Dunham (1982: 37–39).

kiššû (= **gi-níg-sa-ḫi-a**, **gi-sa**, **gi-sa-ḫi-a**) 'bundle of reeds'
  gi-sa: A 3553: 39
    BM 85194 obv. iii 23, 31
    BM 85196 obv. i 8
  gi-sa-ḫi-a: BM 85194 obv. iii 29
    BM 85196 obv. i 24, 26, 27
  gi-níg-sa-ḫi-a: CBS 10996 obv. ii 8'–9'
coefficients:
  6 **igi-gub gi-sa**: A 3553: 39
  30 ^{giš}**ma-lá gi-níg-sa-ḫi-a**
  75 ^{giš}**ma-lá gi-níg-sa-ḫi-a e**: CBS 10996 ii 8'–9'

See Sachs (1944: 33–36) for a discussion of this term.

kitmu 'closed interval or tuning on a musical instrument'
coefficients:
  4,3 **sa** *kit-mu*: CBS 10996 obv. i 7'

kubru 'thickness'
  ku-bu-ur i-ṣí-im: YBC 7243: 35
  ku-bu-úr i-ṣí-im: Haddad 104 #3 i 25
  ku-bu-ur **giš**: YBC 8600 obv. 5, 6
coefficients:
  4,48 *sa ku-bu-ur i-ṣí-im*: YBC 7243: 35
See 4,48 *sa ku-bu-ur-re-e* **giš**: YBC 5022: 58

kubtu (= **im-dugud**) 'lump of earth or metal'
coefficients:
  8,20 *na-al-ba-an* **im-dugud**: YBC 5022: 14
  7,38 *na-az-ba-al-ša*: YBC 5022: 15 (referring to *kubtu*)

kuburrû 'thickness (of a building, wall, etc.)'
  ku-bu-re: VAT 6598 obv. 12; rev. i 9 (*ku*<*-bu-re*>), 10 (*ku*<*-bu-re*>)
  ku-bu-re **é-sig**$_4$: VAT 6598 obv. 3-4, 4 (**é-sig**$_4$-*ya*), 8, [14], 18–19, 23
    (**é-sig**$_4$-*ya*)
  ku-bu-ur-re-e **giš**: YBC 5022: 58
  ku-bu-ur-re **iz-zi**: AO 10822 rev. 10

ku-bu-ur-re-e **iz-zi**: AO 10822 rev. 12 [*TMB* 229: 10 restored *ku-bu-ru-ú* without **iz-zi**.]
coefficients:
4,48 *sa ku-bu-ur-re-e* **giš**: YBC 5022: 58
See 4,48 *sa ku-bu-ur i-ṣí-im*: YBC 7243: 35
referrring to the thickness of a log:
4,48 **igi-gub-ba**: YBC 8600 obv. 4
a-na 5,0 **igi-gub**: VAT 8522 obv. i 4a, 3c

Walker (Letter to author 1986) enclosed a copy of his join, BM 96957, to VAT 6598, which contains problems dealing with **é-sig$_4$** 'wall', to which the term *kuburrû* 'thickness' and the coefficient 6, the coefficient of a wall, is applied. In three problems the **é-sig$_4$** is made of *agurru* 'kiln-burnt bricks' and the coefficient of kiln-burnt bricks or 2;15 are applied.

kullumu 'to show an ominous sign'
gim *ú-kal-li-mu-ka*: K 8705: 10
with *sikkatu*: 'to show somebody the peg driven in the field as a legal act accompanying the transfer of a field'
uku-ús *sí-ik-ka-as-su ku-ul-li-im-šu*: VAT 7531 obv. 5, 10, 12–13, 19–20; rev. 5–6
VAT 7621 obv. 4, 6

kunāšu (= **zíz**) 'emmer'
zíz àra: BM 47860 rev. 11

Livingstone (1986: 20–21) suggested ḪAR = **àra** = *ṭênu* 'to grind, mill'.

kupru [= **esír(-è-a)**] 'asphalt'
ku-up-ri: *TMS*, Text 3: 67
esír-*šu*: AO 6770 rev. 11
esír-è: A 3553: 13
esír-è-a: AO 6770 rev. 11
YBC 5022: 18
esír-ḫi-a: YBC 7243: 24, 31
coefficients:
12 *ša ku-up-ri*: *TMS*, Text 3: 67
15 **igi-gub esír-è**: A 3553: 13
15 *ša* **esír-è-a**: YBC 5022: 18
15 **esír-ḫi-a**: YBC 7243: 31

referring to *kupru*:

igi-gub-bu-um 15: AO 6770 rev. 13

10,45,6 **esír-ḫi-a** *ša i-na ki-ri-im*: YBC 7243: 24 [Kilmer (1960: 291) read the coefficient as 10,48,6 here and in YBC 5022: 55 as the coefficient of *atkuppu*).

kūru 'crucible (for metal and glass), kiln, brazier'
coefficients:

9 **igi-gub** *šà ku-ri-im*: *TMS*, Text 3: 53

kurullu 'shock, pile of sheaves (on a harvested field)'

ku-ru-li-šu: BM 85196 rev. i 2

kurummatu (= **šuk, šuku, kurum$_6$**) 'ration'

sag šuk-*ia*: YBC 4669 rev. i 14

šuk-*ia*: YBC 4669 rev. i 11

kussû (= **^{giš}gu-za**) 'chair'
coefficients:

[x] **igi-gub ^{giš}ná ^{giš}gu-za**: A 3553: 43

kušāru meaning unknown
coefficients:

[x] **igi-gub** x **gal** *ku-ša-ri*: IM 49949: 11

kutabšu(?) meaning uncertain; cf. *kutabsu* (synonym list), mistake for *kubšu* 'headdress'
coefficients:

ku-ta-ab-ši [...]-*ar*: IM 52916, IM 52685, IM 52304 Sections F and G: rev. 45'

ku-ZI-*im* (meaning unknown)

iš-ka-ra-am i-na ku-ZI-*im id-di-nu-nim-ma*: YBC 9856 obv.1

la/i'bu a kind of inflammation (see Borger 1957: 192-93)

la-'-ab **ḫar**: K 8705: 4

labānum 'to fashion bricks'

la-ba-nam: Haddad 104 #10 iv 29

al-li la-ba-nim: Haddad 104 #9 iv 15, 16–17; #10 iv 27

a-la-bi-in: Haddad 104 #10 iv 28

coefficients:
with *allu*:

20 *al-li ḫa-ba-tim* 20 *al-li la-ba-nim* 10 *al-li ba-la-lim lu-pu-ut*: Haddad 104 #9 iv 16–17

20 *ḫa-ba-tam* 20 *la-ba-nam* 10 *ba-la-lam l*[*u-pu-u*]*t*: Haddad 104 #10 iv 29

Walker (Letter to author 1986) enclosed a copy of his join, BM 96957, to VAT 6598, which contains problems dealing with **é-sig**$_4$ 'wall', to which the verb *labānu* (III/1) is applied in relation to making bricks for this edifice.

latāku 'to check measurements, calculate work in progress'

^{gi}gur *pa-nu-um i-na* **[e]gir** *al-tu-uk-ma*: BM 85196 rev. i 9
al-tu-uk: BM 85196 rev. i 11
[*al*]-*t*[*u-ku*]: BM 85196 rev. i 13

libittu (= **sig**$_4$) 'brick'

li-bi-tu-um: IM 54538 obv. 3
li-<*bi*>*-ta-am*: AO 8862 iv 2
li-bi-ta-am: AO 8862 iv 14
li-bi-it-ti: AO 8862 iv 23
li-bi-ti: AO 8862 iv 18, 21
li-bi-ti-ka: IM 54538 rev. 1
sig$_4$: A 3553: 29-31, 35
AO 8862 iii 29; iv 5; 21 tabulation
AO 10822 rev. 6, 7, 8, 10, 13, 14, 15, 16, 17
BM 85194 obv. iii 49; rev. iii 7, 10
CBS 10996 obv. ii 3', 6'
Haddad 104 #9 iv 21, 22, 23, 25; #10 iv 38, 39, 40, 42
IM 52916, IM 52685, IM 52304 Section E: rev. 22'
IM 49949: 5
TMS, Text 3: 41
YBC 4607 obv. 1, 6; rev. 4, 7, 9
YBC 4669 obv. iii 8, 9, 14 (N.B. l. 12 should read 10⅓ **s[ar** 5 **gín**...**]**)
YBC 4673 obv. i 1, 7, 10, 11 (**sig**$_4$ **al-til-la**), 17, 21, 22 (**sig**$_4$ **al-til-la**), 25; obv. ii 1, 6, 8, 18; obv. iii 1(?), 4(?)
YBC 5022: 2, 9, 10, 11, 12, 13, 36, 38, 41
YBC 7284 rev. 1
YBC 7997 rev. 8
YBC 10722 rev. 2

sig$_4$-bi: AOT 304 obv. 5
YBC 4708 obv. i 3, 4; rev. ii 30, 31
sig$_4$-ḫi-a: BM 85194 rev. iii 9
with *amaru* (**sig$_4$-anše**):
sig$_4$-anše sig$_4$: AO 10822 rev. 13, 14, 16
sig$_4$ sig$_4$-anše: YBC 4708 obv. i [1], 5, 8, 11, 14; obv. ii 1, 14; obv. iii 1, 21, [26]; rev. i 1, 5, 13, 21, 28; rev. ii 1, 6, 11, 17, 22, 27 (**sig$_4$ <sig$_4$>-anše**); rev. iii [1], 13
coefficients:
4,30 **igi-gub sig$_4$ íl-í[l]**
3,22,30 **igi-gub sig$_4$ in**-[x]
1,41,15 **igi-gub sig$_4$ íb**-[x]: A 3553: 29–31 [Kilmer (1960: 288 n. 3) demonstrated that the numerical sequence in A 3553: 29–31 is identical to the order of coefficients for *nazbalum* 'load, carrying operation' of **sig$_4$**, **sig$_4$-áb**, and **sig$_4$-al-ùr-ra**, respectively. The text is broken so that the descriptions remain unclear in A 3553: 30–31.]
10(?),1,20 **igi-gub saḫar íl-il sig$_4$ dù**: A 3553: 35
24 *ša* **sig$_4$** *šà ḫa-a-ti*: *TMS*, Text 3: 41 [von Soden (1964: 47) suggested reading *šà-ḫa-a-ti*.]
with *malallû*:
[...^{giš}**má-lá**] **sig$_4$**: CBS 10996 obv. ii 3'
4,20 ^{giš}**má-lá sig$_4$**-⅔-*ti*: CBS 10996 obv. ii 6'
Cf. Mogens Weitemeyer (1962: 52–53) for dockets involving the transportation of bricks by boat (= ^{giš}**má**).
with *mašqaltum*:
4,24 *ma-aš-qá-al-tum ša* **sig$_4$**: YBC 5022: 36
with *maššû*:
6 *ma-aš-šu-ú-u*[*m ša* **si]g$_4$**: YBC 5022: 41
with *nalbanu*:
1,12 *na-al-ba-an* 1 **kùš sig$_4$**: YBC 5022: 9
4,48 *na-al-ba-an* ½ **kùš sig$_4$**: YBC 5022: 11
9 *na-al-ba-an* ⅓ **kùš sig$_4$**: YBC 5022: 13
with *nazbalu*:
na-az-bal **sig$_4$** 45 *i-gi-gu-bu*: IM 52916, IM 52685, IM 52304 Section E: rev. 22'
4,30 *na-az-ba-al* **sig$_4$**: YBC 5022: 2
4,30 **igi-gub sig$_4$** *na-az-ba-al* **sig$_4$**: IM 49949: 5
with reference to *nazbalša*:
45 *na-az-ba-al-ša*: YBC 5022: 10 (referring to 1-**kùš** brick)
4,23,36 *na-az-ba-al-ša*: YBC 5022: 12 (referring to ½-**kùš** brick)

with *taddītum*:

14,24 *ta-ad-di-tum ša* **sig$_4$**: YBC 5022: 38

with *tarkibūtu*:

7 *tar-ki-bu-ta* **sig$_4$**: AO 10822 rev. 7

with reference to a **sig$_4$** of 8;20 **ma-na** weight:

igi-gub-ba-bi 12: YBC 7284 obv. 3

cf. **sig$_4$** used as a determinative:

2(?),42 ^{sig4}*a-gu-ru-um*: YBC 7243: 44

lillidu (= **sila$_4$-gub**) 'offspring (of children and flocks), mature young of animals'

li-li-di: YBC 7326 obv. 3

li-li-du: *li-li-du e-li li-li-di*: YBC 7326 obv. 3

udu-... -ḫi-a sila$_4$-gub: VAT 8522 rev. 3, 4

u$_8$-udu-ḫi-a *ù* **sila$_4$-gub**: YBC 4669 rev. ii 12, 14, 24

sila$_4$-gub ugu sila$_4$-gub: YBC 4669 rev. ii 22

li-LUM (work at the first level of digging a canal)

li-LUM: BM 85196 rev. i 36

lipû (= **ì-udu**) 'sheep fat'

coefficients:

20 **ì-udu**: YBC 7243: 51

lulû 'antimony'

coefficients:

1,15 *šà lu-li-im*: *TMS*, Text 3: 55

maḫāru 'to buy, receive'

i-na ša am(?)*-ḫu-ru*: MLC 1842 left edge

maḫāṣu (= **tag**) 'to stamp'

ki-lá nu-na-tag: YBC 4652 obv. [1], [3], [6], [8], ⌜10⌝, 12, 14, 17; rev. 1, 4, 7, 11

Neugebauer and Sachs translated **ki-lá nu-na-tag** 'I did not weigh it'; however, in Ur III economic texts and passages from *ana ittišu* the Sume 'an ink stamp or a stamp applied to a lump of clay attached to cloth' or 'stamping of various symbols on measures of capacity' (*MCT*: 102–103).

maḫāṣu specialized meaning: 'to weave'
  ^túg^*ṣu-ba-tum* 48 **uš** *i-na* **u$_4$-1-kam** 20 *im-ḫa-aṣ*: BM 85194 rev. i 47

maḫīru (= **ganba**) 'price equivalent'
  ma-ḫi-rum: Strssbg. 362 rev. 5
    VAT 7530 obv. 10, 12
  ma-ḫi-ra-am e-pé-ša-am ma-ḫi-ir **še**-*im ma-ḫi-ir* [...]
  ma-ḫi-ir **še-ì-giš** *ma-ḫi-ir* **zú-lum** *ma-ḫi-ir* **síg**
  [*ma-ḫi-ir*] **ì-giš** [*ma-ḫi-i*]*r* **ì-sag-dug-ga** *ma-ḫi*[-*ir* **ì-šaḫ**]
  [*ma-ḫi-ir*...] *ma-ḫi-ir* **esír** *ma-ḫi-ir* [...]
  [*m*]*a-ḫ*[*i-ra-tim na*]*-sa-ḫa-am ša-ma-am ù k*[*a-ma-ra-*]*am*: IM 52916, IM 52685, IM 52304 Section F: 43'–47'
  ma-ḫi-ir: VAT 7530 obv. 5, 14, 15, 16
  i-na ma-ḫi-ir 1 **bán** 5 **sìla ì-šaḫ** 1 **bán ì-giš**: IM 54464 obv. 2
  ma-ḫi-ir na-ḫi-im: IM 54464 obv. 3
  ma-ḫi-ir **ì-giš** *ù* **ì-šaḫ** *ka-ma-ra-a*[*m*]: IM 52916, IM 52685, IM 52304 Section F: 50'
  ma-ḫi-ri-ia(/*-ka*): MLC 1842 obv. 3, 4, 11; rev. 3
  ganba(-e): MLC 1842 obv. 1, 2, 6
    VAT 7530 obv. 21; rev. 7
    YBC 4698 rev. i 17, 21; rev. ii 1
  ganba-ḫi-a: YBC 4698 rev. ii 7

māḫiru (= **gaba-ri**) 'rival, withstanding attack (used as an adjective describing a door or weapon)
coefficients:
  12,30,6,22,30 **gán gaba-ri ki-min igi-gub**-*ú*: BM 47860 obv. 6

The phrase appears in a list of granary coefficients and may be explained by the summary in rev. 4-8, which describes 'the circle of the evening meal of Ekur' and 'every one of the shrines... all of the cult platforms'. The reading of **gaba-ri** as *gabarû* 'copy, duplicate' does not apply here (cf. rev. 12). Livingstone 1986: 21 read *gab-ri* 'strong' and translated 'volume of the strong one'.

maḫrû (= **igi**) 'front, fore leg'
  [igi]-*ù šá* **udu-níta**: K 8705: 8

majāru 'ploughed'
  ma-ia-ru: CUA 34 rev. (d)

makāru 'to irrigate'

a-šà (*ki ma-ṣi*) *am-qú-ur*: YBC 4186 obv. 3, 12
1 **sìla** *pí<-sa-nu-um>* 10 **šu-si** *am-qú*(?)-*ur*: BM 85194 rev. iii 42
10 *šu-pu-ul me-e ša* **a-šà** *im-qú-ru*: YBC 4186 obv. 6, 7

makāsu 'to collect a share from a rented field (from a tenant or respective shares of owner and tenant)'

am-ku-us: VAT 8389 obv. i 1, 2; rev. i 1, [2]
VAT 8391 obv. i 1, 2, 21, 22; rev. i [3], 4; rev. ii 10, 11
im-ku-sú: VAT 8389 obv. i 6, 8, 15, 19; obv. ii 18, 22; rev. i 6, 7, 11, 17; rev. ii 3, 7
VAT 8391 obv. i 6, 7, 11, 14, 26, 27; obv. ii 1, 6, 26, 30; rev. i 8, 9, 13, 20, 24; rev. ii 4, 7, 15 (*im-ku<-sú>*), 16, 20, 26

makurru (= ^{giš}**má-gur**$_8$) 'cargo boat'

gán má-gur$_8$: YBC 5022: 47
YBC 7243: 6
gán ^{giš}**má-gur**$_8$: BM 15285 Text Q

coefficients:

[12,3]0 *ša* **gán má-gur**$_8$: YBC 5022: 47
12,30 **gán má-gur**$_8$: YBC 7243: 6

Salonen (1939: 12–14) described ^{giš}**má-gur**$_8$ as a ship with a high sternpost or sea boat. Kilmer (1985; 1990b: 85) showed that the 'cargo boat shape' is a biconvex figure constructed from 90° arcs, as illustrated in the drawing in BM 15285.

malallû (= ^{giš}**má-lá**) 'cargo boat'

^{giš}**má-lá**: BM 85196 obv. i 24
CBS 10996 obv. ii 3'–10'
YBC 5022: 29

coefficients:

6 *ša* ^{giš}**má-lá**: YBC 5022: 29
[... ^{giš}**má-lá**] **sig**$_4$
[...], 10 ^{giš}**má-lá sig**$_4$**-áb**
[...], 15 ^{giš}**má-lá sig**$_4$**-al-ùr-ra**
4,20 ^{giš}**má-lá sig**$_4$-⅔-*ti*
50 ^{giš}**má-lá gi-šu-kin**
30 ^{giš}**má-lá gi-níg-sa-ḫi-a**

75 ^{giš}**má-lá gi-níg-sa-ḫi-a è**
72 ^{giš}**má-lá** *ḫi-im-ma-ti*: CBS 10996 obv. ii 3'–10'

maltaktu 'water clock'
coefficients:
12 *ma-al-ta-ak-tum*: YBC 7243: 50
12 *ma-al-ta-ak-tum ša mu-ši-tim*: YBC 5022: 60

manzazu (= **ki-gub**) 'socle'
ki-gub: BM 85210 rev. i 2

marru (= **mar**) 'spade'
mar: *LBAT*: no. 1648 rev. 8'(?)

māru (= **dumu**) 'son'
ad *ana* **dumu**-*šú ša i-ram-mu*: K 11097: 3
[*mār*] ^{lú}**ḫal nun-me** [*šamni*]: K 11097: 7
ma-ru-šu-nu: VAT 7621 obv. 1, 7

mašqaltum 'weight'
coefficients:
4,24 *ma-aš-qá-al-tum ša* **sig$_4$**: YBC 5022: 36

maššû a basket or other container used for carrying
coefficients:
6 *ma-aš-šu-ú-u*[*m ša* **si**]**g$_4$**: YBC 5022: 41
1,40 *ma-aš-šu-ú-um ša* **saḫar**: YBC 5022: 42

maṭû verb used to explain what is lost from a mass of precious metal during a particular treatment
i-na x **ma-na** y **gín kù-babbar/guškin** z **gín** *im-ṭi*: M 15192 + M 15236: 1–19 // *ARM* 24: no. 146

middatu (= **ága**) measure (of area, length, time, capacity)
mi-in-da-as-sú **nu-zu**: Strssbg. 368 obv. 1
VAT 7535 obv. 1; rev. [1], 25–26
ága-bi: YBC 4668 obv. i 30, 36, 43, 47 (= YBC 4713 obv. ii 15, 21; obv. iii 1, 8)
YBC 4713 obv. ii 15, 21; obv. iii 1, 8

middu term applied to a coefficient used in calculating the capacity in **bariga** from the volume in **sar** in relation to spreading mud plaster over a surface
coefficients:
5 *mi-id-di-im*: Haddad 104 #7 iv 4; #8 iv 13

miksu 'tax'
mi-ik-sa-am e-pé-ša-[*am*]: IM 52916, IM 52685, IM 52304 Section F: 51'

*mi-*LU*-ti* (reading uncertain)
coefficients:
1,36 **a-meš** *mi-*LU*-ti*: CBS 10996 obv. iii 21'

MÌN (meaning uncertain, pertains to string of stringed instrument)
coefficients:
sa MÌN 3-*šú ù* **sa** *šá*-**ge**$_6$ 5,2 **sa murub**$_4$-*tu*: CBS 10996 obv. i 19'

minâtu 'limbs, body (sum of all parts of the body), size, proportions, health' (pl.)
mi-na-a-te: K 8705: 12

mû (= **a**) 'water'
mu-ú: *TMS*, Text 25 obv. i 4 (*mu-ú a-na* 40 **erim-ḫi-a**), 6 (*mu-ú šà* **erim**), 7
me-⸢*e*⸣: BM 47860 obv. 8
me-e **a-šá**: YBC 7243: 26
me-e **k[i-t]a**: *TMS*, Text 25 obv. i 8
me-e pa-ta-a-tim: YBC 7164 rev. 17, 19
me-e-sa az-zu-ul-ma i-na me-e-ša a-na 1 **šu-si** *šu-up-lim* **a-šà** *ki ma-ṣi am-qú-ur*: YBC 4186 obv. 2–3
i-na pa-ni me-e: BM 85196 obv. i 2
10 *šu-pu-ul me-e ša* **a-šà** *im-qú-ru*: YBC 4186 obv. 6, 7
a: YBC 7243: 30
é-kišib a: YBC 5022: 33
na-az-ba-al **a**: YBC 5022: 44
a-meš: CBS 10996 obv. iii 21'–28'
coefficients:
6,0 **gán ì-dub** *me-*⸢*e*⸣ **igi-gub**-*ú*: BM 47860 obv. 8
15 **a**: YBC 7243: 30
15 *ša* **é-kišib a**: YBC 5022: 33
1,20 *ù* 15 *na-az-ba-al* **a**: YBC 5022: 44
48 *me-e* **a-šà**: YBC 7243: 26
1,36 **a-meš** *mi-*LU*-ti*

1,12 **a-meš** [giš]*zi-rí-qum*
1,18 **a-meš** *pat-ti*
48 **a-meš** *pat-ti*
9,39 **a-meš** [giš]**a-lá**
1,30 **a-šà a-meš** *ša-qu-ú* [...]
18 **a-šà a-meš** *ša-qu-ú* **ki-min**
2,46,40 **a-šà a-meš** *ša-qu-ú* **ki-min**: CBS 10996 obv. iii 21'–28'

For the coefficient 48, compare the following:
48 **a-meš** *pat-ti* CBS 10996 obv. iii 24'
48 *me-e* **a-šà**: YBC 7243: 26
48 *šà* **pa$_5$-sig**: *TMS*, Text 3: 33
igi 48 **igi-gub pa$_5$** *pu-ṭú-ú*[*r*]: *TMS*, Text 25 obv. (i 7. See Powell 1988: 165).

mūdû 'informed'
et-pe-šú ha-as-si mu-du-ú: K 11097: 7

mušarû 'garden'
mušarîsam 'garden by garden'
coefficients:
iš-x [...] *mu-ša-ri-ša-am* (coefficient list; broken context): IM 52916, IM 52685, IM 52304 Sections F and G: rev. 37'

AHw s.v. *mušarišam* referred to *ugārišam* 'meadow by meadow'; however, *CAD* s.v. *mušarû* suggested the meaning of *mušarišam* is uncertain.

mūšu, *mūšītu* (= **ge$_6$**) 'night, nighttime'
mu-ši-tim: YBC 5022: 60
ge$_6$: CBS 10996 obv. iii 12'–13'
u$_4$-*ma* **ge$_6$** mng. 'always': K 11097: 6
coefficients:
12 *ma-al-ta-ak-tum ša mu-ši-tim*: YBC 5022: 60
See: 12 *ma-al-ta-ak-tum*: YBC 7243: 50
40 **u$_4$**-*mu a-na* **ge$_6$** *na-pa-lu*
40 **ge$_6$** *a-na* **u$_4$**-*mu na-pa-lu*: CBS 10996 obv. iii 12'–13'

muttallītu see *muttalliktum*

muttalliktum type of pavement
45 **igi-gub** *mu-ut-ta-al-li-i*[*k*]-*tum*: YBC 5022: 39

45 **igi-gub** *mu-ta-al-li-ik-tum*: IM 49949 rev. 16
[4]5 *mu-ta-lik-tim*: Haddad 104 #10 iv 30

See 45 *ta-la-ak-tum*: YBC 7243: 46

nāḫu (= **ì-šaḫ**) 'lard'
ma-ḫi-ir na-ḫi-im: IM 54464 obv. 3
ì-giš *ù* **ì-saḫ**: IM 54464 obv. 5
ì-giš *ù* [**ì-ša**]**ḫ** *ša-ma-am* **kù-babbar ì-giš** u **ì-š[aḫ** *k*]*a-ma-r*[*a-am*]
ma-ḫi-ir **ì-giš** *ù* **ì-šaḫ** *ka-ma-ra-a*[*m*]: IM 52916, IM 52685, IM 52304 Section F: 49'–50'
i-na ma-ḫi-ir 1 **bán** 5 **sìla ì-šaḫ** 1 **bán ì-giš**: IM 54464 obv. 2
[*ma-ḫi-ir*] **ì-giš** [*ma-ḫi-i*]*r* **ì-sag-dug-ga** *ma-ḫi*[*-ir* **ì-šaḫ**]: IM 52916, IM 52685, IM 52304 Section F: 45'

nakāsu 'to fell trees, to hew, to cut dates from a tree, etc.' and general meanings of 'to cut'
^{giš}**erin** *ki ma-ṣi li-ik-ki-sú-nim*: VAT 8522 obv. i 5–6
i-na ap-pi-šu i-na-ki-sú-nim: VAT 8522 obv. i 8a, 8

nakbasu 'step(of a staircase)' (sometimes shortened to *na* in BM 85210)
na-ak-ba-súm: BM 85210 obv. i 1, 8, 13, 14, 17, 20
na<-ak-ba-súm>: BM 85210 obv. i 4

nakiru 'enemy, hostile'
uru^{ki} *na-ki-ir* ^{d}*Marduk a-ṣa-ba-at*: BM 85194 rev. ii 7, 22
BM 85210 obv. ii 1 (<**uru**^{ki}>)
uru^{ki} *na-ki-ir* ^{d}*Marduk lu-úṣ-ba-at*: BM 85210 obv. i 3

nalbantu (= **giš-ù-šub**) meaning uncertain, something having to do with brickmaking based on the root $\sqrt{}$*labānu*.
giš-ù-šub: BM 85196 obv. ii 23
BM 85210 rev. i 23, 24
n[*a-a*]*l-ba-tam i-na li-bu na-al-ba-tim e-pé-ša-am*: IM 52916, IM 52685, IM 52304 Section D: rev. 14'
i-na li-bu na-al-ba-tim ki-pa-ta-am
i-na li-bu ki-pa-tim na-al-ba-ta!!*-am*: IM 52916, IM 52685, IM 52304 Section D: rev. 20'–21'
coefficients:
15 **igi-gub** *mišlim* (referring to *nalbantu*) in BM 85210 rev. i 31–32 (= *TMB*: 51: no. 99: 9–10).

nalbantu is not to be confused with *nalbanum* 'brick mold', which has coefficients relating to the size of bricks. The dimensions of *nalbantu* have nothing to do with the size of bricks (cf. Salonen 1972: 88–99). As for the geometrical term, truncated pyramid, translators of mathematical texts should be cautioned that a given shape may be associated with more than one item. For analysis of this word, see Nemet-Nejat (1988: 297–98).

Thureau-Dangin (1935a: 16) suggested that *nalbantu* may be a platform for molding or drying bricks.

nalbanum 'brick mold'

na-al-ba-nu-um: Haddad 104 #9 iv 15
na-al-ba-ni: *TMS*, Text 3: 45
na-al-ba-an: YBC 5022: 3, 4, 5, 6, 9

coefficients:

5 *šà na-al-ba-ni*: *TMS*, Text 3: 45
7,12 *na-al-ba-an* **saḫar**: YBC 5022: 3
5,24 *na-al-ba-an* **sig$_4$-áb**: YBC 5022: 4
2,42 *na-al-ba-an* **sig$_4$-al-ùr-ra**: YBC 5022: 6
1,12 *na-al-ba-an* 1 **kùš sig$_4$**: YBC 5022: 9
4,48 *na-al-ba-an* ½ **kùš sig$_4$**: YBC 5022: 11
9 *na-al-ba-an* ⅓ **kùš sig$_4$**: YBC 5022: 13
8,20 *na-al-ba-an* **im-dugud**: YBC 5022: 14

Powell (1982: 119–21) defined *nalbanum* 'brickage' as the coefficient of the the number of brick-**sar** of a given type in a volume-**sar**. *nalbanum* and *nazbalum* are useful for calculating the weight of a single brick.

nalbanum may refer either to the mold, to the process of molding or to **giš-ù-šub** a structure mentioned in the mathematical texts (*MCT*: 132).

namḫartu (= **šu-ti-a**) 'goods, staples, etc. received'

šu-ti-a *Da-di-im*: M 6861: 14

namkarum 'irrigation canal'

^{íd}*nam-ka-rum*: BM 85196 rev. i 36

napāḫu (= **mú**) 'to become visible, rise, glow (said of stars, moon, sun, etc.)'

coefficients:

12 **mú** *ša* ^{d}**XXX**: CBS 10996 obv. iii 10'
Cf. 12 **šú** *ša* ^{d}**XXX**: CBS 10996 obv. iii 11'

napālu 'to convert'

40 **u$_4$**-*mu a-na* **ge$_6$** *na-pa-lu*
40 **ge$_6$** *a-na* **u$_4$**-*mu na-pa-lu*
40 *na-pal-tum*: CBS 10996 obv. iii 12'–14'

nappaltu 'difference in increasing or decreasing sequence'
coefficients:

40 *na-pal-tum*: CBS 10996 obv. iii 12'–14'

naptanu (= **kin-sig**) 'meal, time of the evening meal, evening'

gam (sic) **kin-sig é-kur**: BM 47860 rev. 4

naqāru (= **gul-gul**) 'to demolish (as preparation for rebuilding)'

(ḫé-)gul-gul: YBC 4669 rev. iii 8, 11, 15

naqû 'to sacrifice'
niqâ naqû: may refer explicitly to the slaughtering of animals

be-el ni-qí-[*im ni-qí-am úl i-n*]*a-qa-qí*: M 6829: 40' // *ARM* 24: no. 260

nāru (= **íd**) 'canal' > (?) 'strip'

[*na*]*-ra-am na-ar ta-ra-ḫi-i e-pé-ša-am*
[*n*]*a-ar ša-ka-na-ka-tim na-ra-am šu-zu-ba-am*: IM 52916, IM 52685, IM 52304 Section D: rev. 15'–16'
íd: Erm 15073 rev. i 1, 10
Strssbg. 364 obv. 1, 3, 5
VAT 8523 obv. i 1, 6, 9, 18, 22, 27; obv. ii 6, 16, 17, 20, 25, 29; rev. 1, 3, 15, 22, 25
íd-meš: Strssbg. 364 obv. 1, 4, 7
Strssbg. 367 obv. 2
id-an-na 'upper strip': YBC 4608 obv. [2], 3, 19
íd-ki-ta 'lower strip': YBC 4608 obv. 2, 3, 12, 21
íd-a-dug: AOT 305: 1
ìd: YBC 4696 obv. i 5, 9; obv. ii [8], [11]; rev. ii [2]; rev. iii 6, 13 [Stol (1988: 180–81) equated **ìd** with *tamirtu*, 'irrigated field, area of land under cultivation'; *tamirtu* was part of an irrigation district and could be irrigated at the disgression of farmers.]
ìd-an-ta: YBC 4696 obv. i 6; obv. ii 21; obv. iii 8, 12; rev. i [2], 10; rev. ii 5, 14
ìd-ki-ta: YBC 4696 obv. i 7; rev. ii 6, [21]; rev. iii 1

nasāḫu 'to remove mud from a canal, rubble from ruined buildings'
i-na **u$_4$ ½-kam** *is-sú-uḫ*: BM 85196 rev. ii 7
saḫar-ḫi-a *i-sú-uḫ*: BM 85196 rev. ii 5, 6
4 **gín saḫar** *is-sú-uḫ*: YBC 7164 obv. 18

In YBC 7164 #7 the verb *nasāḫu* is used to describe the excavation work of *šīlūtu*, and the veb *ḫerû* is used to describe the excavation work of both *šīlūtu* and **dusu**.

našpaku (= **é-uš-gíd-da, ì-dub**) 'granary, storehouse, silo (for barley, dates, oil)'
na-aš-pa-kum: IM 49949: 9
IM 52301 rev. 21
na-aš-pa-ki: Haddad 104 #4 iii 1
na-aš-pa-ki-im: Haddad 104 #1 i 8; #2 i 17; #3 i 29, 38, ii 10, 18, 29; #4 ii 34, iii 5; #5 iii 12, 18, 29
TMS, Text 3: 50
VAT 8522 obv. i 6a
na-aš-pa-ak: Haddad 104 #1 i 1; #3 i 31, 32, 36, ii 11, 15, 16
TMS, Text 3: 49, 51
TMS, Text 14 rev. 4
na-aš-pa-ak-šu: Haddad 104 #3 i 32, ii 5, 24
é-uš-gíd-da: IM 49949: 9 (sic)
YBC 7243: 23
ì-dub: AO 6484 obv. 23
BM 47860: obv. 1–13, [14], [15]

coefficients:
1,16 **é-uš-gíd-da in-nu-da**: YBC 7243: 23
6,0 *na-aš-pa-ki*: Haddad 104 #4 iii 1
6,0 *na-aš-pa-ki-im*: Haddad 104 #1 i 8; #2 i 17; #3 i 29, 38, ii 10, 18, 29; #4 ii 34, iii 5; #5 iii 12, 18, 29
6,0 *ša na-aš-pa-kum*: IM 52301 rev. 21
6,0 *šà na-aš-pa-ak šà-al-šu-di-im*: *TMS*, Text 3: 49
6,0 **igi-gub é-uš-gíd-da**(sic) *na-aš-pa-kum*: IM 49949: 9 [*CAD* s.v. *našpaku* reads 6 **igi-gub é-ùr**(!)-**ra** *na-aš-pa-ku-um*.]
6,0 **ì-dub**: AO 6484 obv. 23 [Cf. AO 6484 obv. 26; rev. 3, 6, 9 which uses the coefficient 6,0 without identifying it.]
6,0 **gán ì-dub** *me-*⌜*e*⌝ **igi-gub**-*ú*: BM 47860 obv. 8
6,0 **gán ì-dub** x DIŠ **igi-gu[b**-*ú*]: BM 47860 obv. 9
6,12 **gán ì-dub** x ŠE DU$_8$ NU **igi-g[ub**-*ú*]: BM 47860 obv. 10
6,12 **gán ì-dub** [x] x DIŠ [**igi-gub**-*ú*]: BM 47860 obv. 11
6,24 **gán ì-[dub ... igi-gub**-*ú*]: BM 47860 obv. 12

6,24 **g[án ì-dub ... igi-gub-*ú*]**: BM 47860 obv. 13
6,40 **gán ì-dub še**-im **igi-gub-*ú***: BM 47860 obv. 7
6,40 *sà na-aš-pa-ki-im*: *TMS*, Text 3: 50 [See Powell (1984a: 45) for a discussion of the factor 6,40.]
Cf. 6,40 *i-gi-gu-ub-bi-im gu-ub-bi-im*: IM 52301 rev. 20
6,45 *a-na* 6,40 **igi-gub** *na-aš-pa-ki-im*: VAT 8522 obv. i 6a
Cf. 6,45 **[gán ... igi-gub-*ú*]**: BM 47860 obv. 14
7,12 *šà na-aš-pa-ak* **gur$_7$**: *TMS*, Text 3: 51
Cf. 7,30 **[gán ...igi-gub-*ú*]**: BM 47860 obv. 15
a-na 8 *na-aš-pa-ak* **gur$_7$** *i-[ší-ma]*: *TMS*, Text 14 rev. 4
4,20,6,40 **gán** *am-am-ma* x **[ì]-dub igi-gub-*ú***: BM 47860 obv. 1
20,6,16 **gán** *ur-ru-uḫ* **sag-3 ki-min igi-gub-*ú***: BM 47860 obv. 2
30,3,46 **gán** *ki-ir-ṣi* **ki-min igi-gub-*ú***: BM 47860 obv. 3
16,40,2,30 **gán** *ka-re-e* **ki-min igi-gub-*ú***: BM 47860 obv. 4
4,10,3,7,30 **gán** *ka-ṣal* **ki-min ki-min igi-gub-*ú***: BM 47860 obv. 5
12,30,6,22,30 **gán gaba-ri ki-min igi-gub-*ú***: BM 47860 obv. 6
refers to storage places:
10 šà 1,2 ù 1: *TMS*, Text 3: 52 [Kilmer 1985]

našû (= **íl**) 'to carry'
in relation to bricks:
i-na-aš-ši: BM 85194 rev. iii 8
i-na-aš-ši-i: BM 85194 rev. iii 7
iš-ši: IM 54538 rev. 4
i-ši-a-am: AO 8862 iv 14
iš-ši-a-am: AO 8862 iii 39a; iv 1
íl(*-ma*): YBC 4673 obv. 7, 10
íl-íl(-ma): A 3553: 29, 35–36
YBC 4673 obv. i 17, 21, 25; obv. ii 3, 12, 15, 28
coefficients:
in relation to bricks:
4,30 **igi-gub sig$_4$ íl-í[l]**: A 3553: 29
in relation to volume of earth:
igi 6 saḫar ú-kin íl-íl du$_8$: YBC 4673 rev. iii 10–11
10(?),1,20 **igi-gub saḫar íl-íl sig$_4$ dù**
10(?),1,6,40(?) **igi-gub saḫar íl-íl im-dù-a**: A 3553: 35–36

nazālu 'to pour out, to drain'
me-e-ša az-zu-ul-ma: YBC 4186 obv. 2

Neugebauer and Sachs (*MCT*: 91 n. 230) analyzed *nazālu* as 'to pour, distill liquid' or in technical terminology 'to pour liquid mass over or onto various materials'.

nazbalu '(standard) load (carried in brick-making processes)'
coefficients:

45 **igi-gub** *n[a-az-ba-l]i-im*: *TMS*, Text 3: 36
na-az-bal **sig₄** 45 *i-gi-gu-bu*: IM 52916, IM 52685, IM 52304 Section E: rev. 22'
4,30 *na-az[-ba-al* x x]*-lu-um*: YBC 7243: 41
4,30 *na-az-ba-al* **sig₄**: YBC 5022: 2
4,30 **igi-gub sig₄** *na-az-ba-al* **sig₄**: IM 49949: 5
3,0 **igi-gub** *na-az-ba-al*(!) *ar-ḫu-um*: IM 49949: 15
6,40 *na-az-ba-al* **saḫar**: YBC 7243: 32
1,40 *na-az-ba-al* **saḫar**: YBC 5022: 8
na-az-ba-al **saḫar** 1,40 *i-gi-gu-bu*: IM 52916, IM 52685, IM 52304 Section E: rev. 23'
1,40 **igi-gub saḫar** na-az-ba-al **saḫar**: IM 49949: 4
na-az-ba-al **in-nu-da** 3,20 *i-gi-gu-bu*: IM 52916, IM 52685, IM 52304 Section E: rev. 24'
1,20 *ù* 15 *na-az-ba-al* **a**: YBC 5022: 44
30 **igi-gub** *na-za-ba-al a-gu-ri-im*: IM 49949: 14

na-az-ba-al-ša:

45 *na-az-ba-al-ša*: YBC 5022: 10 (referring to 1-**kùš** brick)
7,38 *na-az-ba-al-ša*: YBC 5022: 15 (referring to *kubtu*)
1,41,15 *na-az-ba-al-ša*: YBC 5022: 7 (referring to kiln-burnt brick)
3,22,30 *na-az-ba-al-ša*: YBC 5022: 5 (referring to ½-brick)
4,23,36 *na-az-ba-al-ša*: YBC 5022: 12 (referring to ½-brick)

na-az-ba-lu-um:

1,15 *na-az-ba-lu-um*: YBC 7243: 49

Lewy (1949b: 151–48) explained *nazbalum* 'carriage' as the number of bricks that a single worker can carry in a month (of 30 days) over the distance of 30 **nindan** (see YBC 4673 #2–3). Powell (1982: 121–22) added that *nazbalum* also includes the following concepts: (1) the number of bricks to be transported over a variable distance in a single day and (2) as a factor in calculating the wages of a worker, when the *nazbalum* is defined as the equivalent of 30 **bán** of barley.

In the coefficient lists *nazbalum* refers to the following items:

(1) *nazbalum* of bricks (e.g., YBC 5022: 2 and IM 49949: 5)

(2) *nazbalum* of (excavated) earth which could be used to make bricks (YBC 5022: 8; IM 52916, IM 52685, IM 52304 Section E: rev. 23'; IM 49949: 4; YBC 7243: 32)
(3) *na-az-ba-al* **a** '*nazbalum* of water' (YBC 5022: 44)
(4) *na-za-ba-al* **in-nu-da** '*nazbalum* of straw' (IM 52916, IM 52685, IM 52304 Section E rev. 24).

Neugebauer and Sachs (*MCT*: 132–33) were undecided as to whether *nazbalum* refers to a receptacle for transporting bricks and similar materials or, perhaps less likely, the transportation itself.

Both *nalbanum* and *nazbalum* are useful for calculating the weight of a single brick.

nēmedu 'cult platform'
kul-lat né-me-di: BM 47860 rev. 6

nēmelu '(commercial) profit'
ne-me-la: *TMS*, Text 13: 3, 5
ne-me-li: VAT 6469 obv. i 1;obv. ii 7
VAT 6546 obv. ii 2
ne-mé-l[*a-am*...]*-am* **kù-babbar** *ša-qa-la-am*: IM 52916, IM 52685, IM 52304 Section F: 48'

neqrebum 'start of building work' (Al-Rawi and Roaf 1984)
ne-eq-re-bu-um: Haddad 104 #9 iv 15
al-li ne-eq-re-bi-im: Haddad 104 #9 iv 20
coefficients:
5 *al-li ne-eq-re-bi-im*: Haddad 104 #9 iv 20

nēšu (= **ur-maḫ**) 'lion'
[...] 9 **gín kù-babbar ur-maḫ**: M 15192 + M 15236 obv. 6 // *ARM* 24: no. 146

nīd qabli (= **šub-murub$_4$**) (a mode of tuning)
coefficients:
sa ^{d}*é-a-***dù** *ù* **sa** *qud-mu-ú* 4,1 **sa šub-múrub**: CBS 10996 obv. i 17'

See Güterbock (1970: 45–52).

nikkassu (= **níg-šid**) 'account, accounting, result (of a mathematical operation or calculation)'
níg-šid: K 11097: 5, 10 (**níg-šid**-*šú*), 16

nikirtu(?) (meaning unknown)
coefficients:

15 *ni-di*(or KI)-*ir-tum*: YBC 7243: 25 [This coefficient appears between bitumen in the kiln and water of the field.]

niqû 'offering, sacrifice'

be-el ni-qí-[*im ni-qí-am úl i-n*]*a-qa-qí*: M 6829: 40' // *ARM* 24: no. 260
ma-ḫa-ar d[*Il-tim be-el ni-qí-im iz-za-az-ma*]: M 6829: [41'–42'] // *ARM* 24: no. 260

niṣirtu (= **sal-šeš**) 'arcana, secret'

sal-šeš lú**ḫal**-*t*[*i pirišti šamê u erṣeti*]: K 11097: 1

nīš gabarî 'rise of the duplicate' (name of the first interval and of a musical mode or octave type)
coefficients:

sa *qud-mu-ú ù* **sa** 5-*šú* 1,5 **sa** *nīš gaba-ri*: CBS 10996 obv. i 11'

See Güterbock (1970: 45–52).

niš(š)u 'carrying' (with reference to bricks)

ni-iš-ši (or *iš-ši*): IM 54538 rev. 4

NU-KAR-[x]-DÈ(?) (reading uncertain)
coefficients:

30 NU-KAR-[x]-DÈ(?): A 3553: 48a

nu-UD-*ki-tum* (reading uncertain)
coefficients:

24,5,20 *nu*-UD-*ki-tum*: CBS 10996 obv. iii 19'

nuḫurtu (= **nu-luḫ-ḫa-sar**) 'medicinal plant'
coefficients:

13,45 **igi-gub nu**(!)**-luḫ-ḫa-sar**: A 3553: 15

nurmû (= giš**nu-úr-ma**) 'pomegranate'
coefficients:

10,9,30 giš**nu-úr-ma**
10,7,30 giš**nu-úr-ma**
10,6,30 giš**nu-úr-ma** *ti-na-nu*
10,3,30 giš**nu-úr-ma** *ti-na-nu*: CBS 10996 obv. iii 15'–18'

nuzzuru meaning unknown

[*a-na n*]*u-zu-ri-im ù za-ba-lim*: BM 85196 rev. i 10

pagû (= **sa-li**) 'p.-lyre'
coefficients:

4[5] **igi-gub gán sa-li**: A 3553: 7

Kilmer (1990b: 85) analyzed the term **gán sa-li** as an 'unknown figure based on the *pagû*-instrument' and listed it among terms "based upon realia, but with no specific construction, illustration or dimensions available."

palû '(support) beam'

^{giš}*pa-lu-um*: BM 85196 obv. ii 7

pānu (= **gi-gur**) 'basket', perhaps used to collect barley or taxes in BM 85196 #14 or 'a **gur** measuring vessel' made from bronze (Salonen 1966: 285–86)

pa-nu-um: BM 85196 rev. i 9, 13
pa-na-am: BM 85196 rev. i 15
gi-gur: BM 85196 rev. i 9

pars/šiktu (= ^{giš}**ba-rí-ga**)

pa-ar-si-ka-tum: Haddad 104 #6 iii 31
pa-ar-sik-tam: Haddad 104 #6 iii 33
pa-ar-si-ka-tim: Haddad 104 #6 iii 34
pa-ar-si-ka!-*at*!: Haddad 104 #6 iii 32
^{giš}**ba-rí-ga**: YBC 4669 obv. i 1; rev. iii 1, 2, 3
CBS 10996 obv. ii 13' (^{giš}{**ba**}**-ba-rí-ga**), 14'–16', 22'–27' (^{giš}**<ba>-rí-ga**)
^{giš}**rí-ba-ga**: BM 85194 rev. i 44
^{giš}**ba<-rí>-ga-meš**: YBC 4669 rev. iii 5

coefficients:

40,5 **giš-{ba}-ba-rí-ga** 1 **bariga** 2 **bán**
40 ^{giš}**ba-rí-ga** 1 **bariga**
30 ^{giš}**ba-rí-ga** 2 **bán** (should be 3 **bán**)
20 ^{giš}**ba-rí-ga** 3 **bán** // 20 ^{giš}**bán** (should be 2 **bán**): CBS 10996 obv. ii 13'–16'
[...] **dal** ^{giš}**<ba>-rí-ga** 1 **gur**
[... d]**al** ^{giš}**<ba>-rí-ga** 2 (**bariga**) 2 **bán**
[...**dal** $^{gi]š}$**<ba>-rí-ga** 1 (**bariga**) 2 **bán**
[... **dal** ^{giš}**<ba>-rí-g**]**a** 1 (**bariga**)

[...**dal** ^{giš}**<ba>-rí-ga**] 3 **bán**
[...**dal** ^{giš}**<ba>-rí-ga**] 2 **bán**: CBS 10996 obv. ii 22'–27'

pāru 'cow'
[*šum*]*-ma pa-ru-um* 1 **ninda** *im-ta-*[*ḫar*...]: BM 85194 rev. iii 13
[*i-n*]*a* ½ *pa-ri-im*: BM 85194 rev. iii 14

This problem is fragmentary and the meaning cow in this context is highly improbable.

parzillu (= **an-bar**) 'iron'
pa-ar-zi-il-li: *TMS*, Text 3: 62
an-bar: YBC 4698 obv. i 12, 15
coefficients:
2,12 *šà pa-ar-zi-il-li*: *TMS*, Text 3: 62

pašāru 'to sell(?)' (parallel to *šâmu* 'to buy')
ganba *a-ša-am ù ki-ia ap-šu-ur*: MLC 1842 obv. 6–7
ki ma-ṣi a-ša-am u ki ma-ṣi ap-šu-ur: *TMS*, Text 13: 4–5
11 **sìla-ta-àm** *ta-šà-am* 7 **sìla** *ta-ap-šu-úr*: *TMS*, Text 13: 12
ta-ap-šu-ru: VAT 6469 obv. i 5
VAT 6546 obv. ii 7
ip-šu-ur: Haddad 104 #6 iii 36, 37, 38
i-pa-aš-ša-ra: Haddad 104 #6 iii 33
See *i-na* 7 **qa**$^{\text{ta-àm}}$ *ša ta-pa*!*-aš-*[*šà-ru šamnim*]: *TMS*, Text 13: 15 [von Soden (1964: 49) suggested this reading.]

pattu 'branch canal'
pa(?)*-ta-am e-pé-ša-am*: IM 52916, IM 52685, IM 52304 Section D: rev. 17'
a-meš *pat-ti*: CBS 10996 obv. iii 23'–24'
me-e pa-ta-a-tim: YBC 7164 rev. 17, 19
coefficients:
1,18 **a-meš** *pat-ti*
48 **a-meš** *pat-ti*: CBS 10996 obv. iii 23'–24'
See 48 *me-e* **a-šà**: YBC 7243: 26 [See *mû* regarding this coefficient.]
20 **pa**$_5$ *ši-lu-tum*: YBC 7243: 21

PEŠ in PEŠ **sig$_4$-al-ùr-ra** (meaning unknown)
coefficients:
5,24 **igi-gub** PEŠ **sig$_4$-al-ùr-ra**: A 3553: 25 [This phrase occurs in Tello lists of offerings of foodstuffs. The value 5,24 is applied to the number of burnt bricks (of ⅔ **kùš** a square side) used in laying 1 **sar** area in the problem text, YBC 4607 rev. 13–15 (Kilmer 1960: 289–90 n. 2).]

PI-*ya* [J. Friberg suggested to me that *pi-ia* should be read **pi**-*ia* = **ba-rí-ga**, that is, a standard measure or basket, thereby rendering this a *maḫīru* problem. Friberg's explanation does not eliminate the difficulties of translating *ku*-ZI-*im* and the use of *iš-ka-ra-am* in this context.]
PI-*ya*: YBC 9856 obv. 2, 3, 4, 6, 7

PI-*lu-ú-tum ša* G[I...] (meaning unknown)
coefficients:
24 PI-*lu-ú-tum ša* G[I...]: YBC 5022: 62

pirištu 'exclusive knowledge'
sal-šeš [lú]**ḫal**-*t*[*i pirišti šamê u erṣeti*]: K 11097: 1

pisannu (= **pisan**) 'container'
pí-sa-nu-um: BM 85194 rev. iii 35 (*pí*<*-sa-nu-um*>), 42
pisan: YBC 5022: 57
coefficients:
30 *ša* **pisan**: YBC 5022: 57

pitiqtu (= **im-dù-a**) 'wall of rammed earth'
pí-ti-iq-tum: IM 52301 rev. 21
IM 53961 obv. 2
IM 54011 obv. 2
pí-ti-iq-ti: *TMS*, Text 3: 43, 44
pí-[*t*]*i-iq-tim*: IM 49949 rev. 2
im-dù-a: A 3553: 8, 36
YBC 4673 obv. iii 9, 15; rev. i 1, 8
YBC 5022: 17
YBC 7243: 13
coefficients:
15 *šà pí-ti-iq-ti*: *TMS*, Text 3: 44
3,45 *pí-ti-iq-tum*: IM 52301 rev. 21
3,45 *šà pí-ti-iq-ti*: *TMS*, Text 3: 43
3,45 **igi-gub gú**(?)**-gal**(?) *pí*-[*t*]*i-iq-tim*: IM 49949: 13

3,45 **igi-gub im-dù-a**: A 3553: 8
3,45 *ša* **im-dù-a**: YBC 5022: 17
3,45 **im-dù-a**: YBC 7243: 13
referring to *pitiqtu*:
3,45 *i-gi-gu-bi-ka*: IM 53961 rev. 2
IM 54011 obv. 11
10(?),1,6,40(?) **igi-gub saḫar íl-íl im-dù-a**: A 3553: 36

Erm 15073 #8 may involve construction of an earth wall. The coefficient 3,45 is used repeatedly in rev. iii 6, 8, 9, 11, 14, 17, 18, 20, 21, 23)

For an analysis of the coefficient 0;3,45 **sar**, see Powell (1988: 167).

pitnu (= **sa**, **sa-a**) 'string of a musical instrument'
coefficients:
⌜2,4⌝ **s[a]** x x-*tum*
4,3 **sa** *kit-mu*
3,6 **sa** *ti*-x *i-šar-tum*
See: 7,4 *sa-ti-tum*
See: 4,6 *sa-muš-šum*
sa *qud-mu-ú ù* **sa** 5-*šú* 1,5 **sa** *niš gaba-ri*
sa 3 *uḫ-ri ù* **sa** 5-*šú* 7,5 **sa** *še-e-rì*
sa *ša*-**ge$_6$** *ù* **sa** 4 *uḫ-ri* 2,6 **sa** *i-šar-tum*
sa *qud-mu-ú ù* **sa** 4 *uḫ-ri* 1,6 **sa** *šal-šá-tum*
sa 3(!)-*šú* **sig** *ù* **sa** 3-*šú uḫ-ri* 3,4 **sa** *em-bu-bu*
sa *šá*-**ge$_6$** *ù* **sa** 3-*šú uḫ-ri* 2,4 **sa** 4-*tu*
sa [d]*é-a*-**dù** *ù* **sa** *qud-mu-ú* 4,1 **sa šub-múrub**
sa *qud-mu-ú ù* **sa** 3-*šú* **sig** 1,3 **sa giš-šub(!)-ba**
sa MIN 3-*šú ù* **sa** *šá*-**ge$_6$** 5,2 **sa murub$_4$**-*tu*
sa *šá*-**ge$_6$** *ù* **sa** [d]*é-a*-**dù** 2,4 **sa** ⌜*titur*⌝ **murub$_4$**-*tú*
sa 4 *uḫ-ri ù* **sa** 3-*šú* **sig** 6,3 **s[a...]**
sa 3-*šú* **sig** *ù* **sa** 5-*šú* ⌜3⌝, [5 **sa**...]
sa 3-*šú uḫ-ri ù* [**sa** 7,x **sa**...]
sa [d]*é-a*-**dù** [*ù* **sa** 4, x **sa**...]: CBS 10996 obv. i 6'–24'

pušikku [= **síg-ga-zum-ak(-a)**] 'carded wool'
síg-ga-zum-ak(-a): YBC 4673 rev. ii 14, 24, 25 (**síg-ga-zum-**<**ak**>), 29

qablītu (= **murub$_4$**-*tu*) 'interval between the second and fifth strings of a harp'
coefficients:
sa MIN 3-*šú ù* **sa** *šá*-**ge$_6$** 5,2 **sa murub$_4$**-*tu*: CBS 10996 obv. i 19'

qanû (= **gi**) 'reed'

qa-na-am [...]

qa-na-am [...]

qa-na-am 25 [...]

qa-na-am elqēma [*mindassu*] *ul idēma*: IM 52916, IM 52685, IM 52304 Section I (e): 2–5

qa-na-am el-qé-a-ma [*mi-in-da-su*] *ú-ul i-di*: IM 53965 obv. 2–3 [von Soden (1952: 53) emended the text to *qa-na-am el-qé-e*!-*ma* [*mi-in-da-su*] *ú-ul i-de-e*!-*ma*.]

Cf. *qa-na-am* [...] *za-qa-pa-am* 'to plant...a reed': IM 52916, IM 52685, IM 52304 p. 155: Section F (b) 2

qa-na-a: Haddad 104 #2 i 10

Cf. *i-na qa-na*: IM 54010 obv. 2

(x) *qa-né-e*: Haddad 104 #7 iii 39; #8 iv 10

(x) *qa-ni*: Haddad 104 #7 iii 41

qá-ni-ia: AO 6770 rev. 16, 17

qa-ni-ka: IM 54010 obv. 6 (15 *a-na ba-a qa-ni-ka i-ši-ma*), 8

gi: AO 6770 rev. 14

BM 34568 obv. ii 17, 18, 21

Strssbg. 362 rev. 11

Strssbg. 366 rev. 5

Strssbg. 368 obv. 1, 5, 7; rev. 10

VAT 7532 obv. 1, 7, 8, 12; rev. 8

VAT 7535 obv. 1, 8, 9, [12]; rev. 1, [7], 8, 11, 24, 25, 32

coefficients:

Cf. 7,12 **igi-gub** GI BUL IR: A 3553: 40

4,58 **igi-gub** GI-NÍGIN-NA: A 3553: 38

qanû ḫaṣbu (= **gi-kud-a**) 'cut reed'

gi-kud(-a) : A 3553: 45

VAT 7532 obv. 1

VAT 7535 obv. 1; rev. 1, 25

YBC 5022: 64

coefficients:

1 **gi-kud-a**: YBC 5022: 64

[1] **igi-gub gi-kud-⌜a⌝**: A 3553: 45

Kilmer (1960: 285–86) suggested this coefficient concerns a trapezoidal figure; however, in the cuneiform problem texts a cut reed is used to measure both rectangles and trapezoids.

qaštu (= ^{giš}**ban**) 'bow'

gán ^{giš}**ban**: BM 15285 Texts L, M

Kilmer (1990b: 85) distinguished between **gán** ^{giš}**ban** 'composite bow shape' and **gán ban-ak** = *eqel panakki* 'extended bow shape', both decorations found in extant works of art.

qātu (= **šu**) 'hand'

kiri$_6$ *a-na qa-ti zi-ki-ri iš-ša-ka-*[*a*]*n*: *TMS*, Text 3: 70
š[u ì]r...: VAT 6505 colophon

qišīru (or *kišīru*) meaning unknown, *CAD* s.v. *qišīru* suggests a possible by-form of *kušīru* 'profit'

qí-ši-ri: IM 31210 §3: 4
qí-ši-ri-ia: IM 31210 §3: 3
qí-ši-ri-ka: IM 31210 §3: 6
qí-šir-šu: IM 31210 §3: 2, 6
2 *at-ḫu-u qí-ši-ir-šu ši-na qí-ši-ri-ia se-be* x-*ra-ti ša iš-te-en qí-ši-ri*: IM 31210 §3: 2-4
aš-šum qí-ši-ir-šu ši-na qí-ši-ri-ka se-be x-*ra-tu*: IM 31210 §3: 6–7

qištu (= ^{giš}**tir**) 'forest'

$^{[g]iš}$**gi** *u* ^{giš}**tir** *ak-ba-ru*: *TMS*, Text 25 obv. ii 1 [von Soden (1964: 50) suggested the reading [*ša i-na a*]*pim u qištim ak-sú*!*-ru*.]

qitmu 'black dye'
coefficients:

2,13,20 *qí-it-mu-um*: YBC 7243: 27

In *AHw* s.v. *qitmu*, it is the black color which is used as a paste in *MCT*; *qitmu* also refers to the black coloring of gold.

qû (= **sìla**) a measuring vessel of standard capacity

qi-e: Haddad 104 #4 ii 40
qí-im: Haddad 104 #4 ii 32, 36, 42, iii 8
qí-ia: Haddad 104 #4 ii 32

qudmû 'fore', referring to first string of an instrument
coefficients:

sa *qud-mu-ú ù* **sa** 5-*šú* 1,5 **sa** *niš gaba-ri*: CBS 10996 obv. i 11'
sa *qud-mu-ú ù* **sa** 4 *uḫ-ri* 1,6 **sa** *šal-šá-tum*: CBS 10996 obv. i 14'

sa ^{d}*é-a-***dù** *ù* **sa** *qud-mu-ú* 4, 1 **sa šub-múrub**: CBS 10996 obv. i 17',
sa *qud-mu-ú ù* **sa** 3-*šú* **sig** 1,3 **sa giš-šub(!)-ba**: CBS 10996 obv. i 18'

Kilmer (1960: 299 n. 4) suggested that *satītum* may have the same meaning as **sa-di** = *qudmû* 'fore string'.

rabû (= **gal**) 'eldest son'
ra-bu-um: IM 31210 §7: 3
ra-bi-im: IM 31210 §2: 13
zi-it-ti ra-bi ṣe-eḫ-ri: IM 31210 §7: 7
gal: *TMS*, Text 10: 10, 13, 17(?), 18, [32], [35]
TMS, Text 24: 53
ḫa-la gal: *TMS*, Text 10: 4, 5, 9, [25], 26, 28, 31
ḫa-la gal-*i*: *LBAT*: no. 1648 obv.(?) 7', 11'

rabû (= **šú**) 'to set, disappear (of the moon)'
coefficients:
12 **šú** *ša* ^{d}**XXX**: CBS 10996 obv. iii 11'
Cf. 12 **mú** *ša* ^{d}**XXX**: CBS 10996 obv. iii 10'

raḫīṣu 'beaten-down threshing floor'
coefficients:
[...] **igi-gub-e** *ra-ḫi-ṣi*: CBS 10996 obv. iii 1'

(^{giš})*ra-pa-*[...]
coefficients:
^{giš}*ra-pa-*[...*i-gi*]*-gu-bu-šu*
^{giš}*ra-pa-*[...*i-gi*]*-gu-bu-šu*: IM 52916, IM 52685, IM 52304 Sections F and G: rev. 35'–36'

rāṭum (= **šita**$_3$) Neugebauer and Sachs (*MCT*: 134) suggested that the meaning of *rāṭum* lies in the realm of ditches, pipes and tubes.
coefficients:
[x,4]0 **igi-gub šita**$_3$ *ra-ṭe*$_4$*-em*: IM 49949: 10
1,12 *ra-ṭù-um ša* **urudu**: YBC 5022: 22
1,12 *ra*(?)-*ṭù*(?)-*um ša* **urudu**: YBC 7243: 36
1,30 *ra-ṭù-um ša* **kù-babbar**: YBC 5022: 23
1,20 *r*[*a-ṭ*]*ù-um* <*ša*> **guškin**: YBC 7243: 34
1,48 *ra-ṭù-um ša* **guškin**: YBC 5022: 24
1,48 [*ra-ṭù-um*] *ša* **[guš]kin**: YBC 7243: 37

Kilmer (1990b: 85) classified *rāṭum* among terms "based upon realia, but with not specific construction, illustration or dimensions available."

rāzu perhaps, to be identified with *ra'izu* 'pitchfork' (*MCT*: 141)
ra-a-zu: MM 86.11.404 obv. 5

rēdû (= **uku-ús**) 'soldier'
uku-ús *sí-ik-ka-as-su ku-ul-li-im-šu*: VAT 7531 obv. 5, 10, 12–13, 19–20; rev. 5–6
VAT 7621 obv. 4, 6

rēšu (= **sag**) 'head'
reš<-ka li->ki-il: VAT 6598 obv. 6 (*<reš->ka <li->ki-il*), 8 (*re*[*š<-ka li->k*]*i-il*), 16 (*<reš-ka> li<-ki-il>*), 18 (*reš<-ka li->ki-il*); rev. i 12 (*<reš-ka li->ki-il*),
re-eš-ka li-ki-il(*-ma*): Db_2-146 obv. 19
Haddad 104 #3 i 39–40, ii 19
IM 31210 §1: 22, §3: 9, §6: 10, §7: 13, 17-18, 20
IM 31248 rev. 3, 12
IM 52301 obv. 5–6, 23–24; rev. 2
IM 53965 rev. 1
Strssbg. 362 obv. 11
TMS, Text 11: 13
TMS, Text 19 rev. 3
TMS, Text 23: [8]
TMS, Text 24: 9, 43, 49
TMS, Text 25 obv. ii 12
VAT 6469 obv. ii 8 (*re-eš-k*[*a li-ki-il*])
VAT 6546 obv. ii 4
VAT 8389 obv. i 17; obv. ii 3; rev. i 15
VAT 8391 obv. ii 4, 12; rev. i 17, 22; rev. ii 24
VAT 8512 obv. 12, 26
VAT 8520 obv. 14; rev. 12, 14
VAT 8523 obv. i 26; obv. ii 11, 24
VAT 8528 rev. i 22
YBC 4186 obv. 9
YBC 4608 obv. 9, 12
ša re-eš-ka u-ki-il-lu: Haddad 104 #3 ii 2, 22; #5 iii 22
šà ri-iš-ka ú-ki-lu: *TMS*, Text 19 rev. 5 (*<-ka>*), 12
TMS, Text 23: 11
TMS, Text 24: 48

ša re-eš-ka ú-ka-al-lu: IM 31248 rev. 8
YBC 4186 obv. 11
YBC 4608 obv. 14
ša re-eš-ka ú-ka-lu: IM 31210 §1: 24–25, §7: 16
IM 52301 rev. 6–7
VAT 6598 obv. 7 (*ša re-eš<-ka ú->ka-lu*), 9 (<*ša*> *reš<-ka ú-ka-lu*>)
VAT 8389 obv. i 21; obv. ii 7; rev. i 22
VAT 8391 obv. ii 9, 16; rev. i 26, 29; rev. ii 28
VAT 8512 obv. 15; lower edge [2]
VAT 8520 obv. 18, 24, 27; rev. 18, 23, 25
VAT 8523 obv. ii [2], 15, 27
VAT 8528 rev. ii 6
ša re-eš-ka ú-ku-lu: IM 31210 §1: 27, §7: 25
See *ša tu-ka-lu*: VAT 6598 obv. 10, 17
20,6,16 **gán** *ur-ru-uḫ* **sag**-3 **ki-min igi-gub**-*ú*: BM 47860 obv. 2
coefficients:
20,6,16 **gán** *ur-ru-uḫ* **sag**-3 **ki-min igi-gub**-*ú*: BM 47860 obv. 2 [See *urḫu*]

rē'û (= **sipa**) 'shepherd'
sipa: *UET* 5: no. 121: 12, 13, 16

ruqqu (see discussion below)
ru-uq-qú: YBC 5022: 25, 26, 27, 28
ru-qú-um: YBC 7243: 51
ru-uq-qá-am im(?)-*ḫa-ṣú*(?): YBC 4669 rev. ii 7
coefficients:
2,8 *ru-uq-qú ša* **sud-á[g]**: YBC 5022: 28
2,15 *ru-uq-qú ša* **guškin**: YBC 5022: 25
2,24 *ru-uq-qú ša* **an-na**: YBC 5022: 27
4 *ru-uq-qú ša* **kù-babbar**: YBC 5022: 26
26,40 *ru-qú-um ša* **kù-babbar**: YBC 7243: 51

ruqqu may be a standardized sheet of metal (see Lewy 1949b: 161–62). Neugebauer and Sachs (*MCT*: 138–39) showed that the mention of area, not volume, seems to denote a flat object and the relation of silver to area seems to indicate standardized dimensions. In YBC 4669 #B6 a relation can be established between 1 **gín** silver and an area of 3 **šu-si** square for a *ruqqu*. 3 **šu-si** square can be explained in units of the area-**sar** as (0;0,30 **ninda**)^2 = 0;0,0,15 **sar**. The reciprocal of 0;0,0,15 namely 4,0,0, may be the 4 listed as a coefficient of a *ruqqu* of silver in YBC 5022: 26; however, 26,40 is also listed as the coefficient for a *ruqqu* of silver in YBC 7243: 51.

Oppenheim (1947b: 128) suggested the meaning 'grate' for *ruqqu*, based on comparison with the following terms: *raqqu* and *riqqu*, containers with a strainer or grating (censer), *riqqîtu*, reticulum of the ruminants, *raqqatu*, a netlike fabric, *raqqu*, the turtle sheet with its net markings, *raqqu* and *raqqatu* 'marsh', and *bīt ruqqi* 'cage'. Kilmer (1990b: 85) proposed the translation. 'kettle/hemisphere', while listing *ruqqu* among terms "based upon realia, but with not specific construction, illustration or dimensions available."

SA-[GIŠ?]-KIN (broken)
coefficients:
5,20 SA-[GIŠ?]-KIN: YBC 7243: 48

sag-ga-kum (= **kak**) meaning unknown
coefficients:
7,30 **igi-gub kak** *sag-ga-kum*: IM 49949: 2
Cf. 7,30 [**gán....igi-gub**-*ú*]: BM 47860 obv. 15

SAG-KI-GUD 'trapezoid'(?); meaning uncertain
SAG-KI-GUD: BM 15285 Text M

sammû (= **zà-mí**) 'lyre'
gán zà-mí: YBC 5022: 48
gán ^{giš}**zà-mí**: BM 15285 Text J
geštu zà-mí: A 3553: 6
YBC 5022: 50
YBC 7243: 4
coefficients:
33,20 *ša* **gán zà-mí**: YBC 5022: 48
26,40 **gán zà-mí**: YBC 7243: 4
26,40 *ša* **geštu zà-mí**: YBC 5022: 50
26,40 **igi-gub geštu zà-mí**: A 3553: 6

See: 26,40 **igi-gub** *šà a-bu-sà-am-mi-ki*: *TMS*, Text 3: 22
26,40 **igi-g[ub** *šà a-bu-sà-mi-ki*]: *TMS*, Text 20: 2
26,40 **igi-gub**: *TMS*, Text 20: 24 (referring to *apsamikku*)
26,40 **igi-gub** *a-bu-sà-mi-ki*: *TMS*, Text 21: 9
a-ab-sà-mi-kum 26,40 (sic) *i*-[*gi-gu-bu-šu*]: IM 52916, IM 52685, IM 52304 Section C: rev. 9'

ṣi-li-ip-ti a-ab-sà-mi-ki 1,20 i-gi-g[u-bu-šu]: IM 52916, IM 52685, IM 52304 Section C: rev. 11'
1,20 **bar**-*ta šà a-bu-sà-am-mi-ki*: *TMS*, Text 3: 23
1,20 **dal** *šà a-bu-sà-mi-(ik)-ki*: *TMS*, Text 21: 6, 25
33,20 *pi-ir-ku šà a-bu-sà-am-mi-ki*: *TMS*, Text 3: 24
48 *ta-al-*<*li*> *a-ab-sà-mi-ki i-gi-g[u-bu-šu]*: IM 52916, IM 52685, IM 52304 Section C: rev. 10'

Kilmer (1960: 286 n. 1; 1985) showed that both **geštú zà-mí** (= *ḫasīs sammî*) 'ear of the lyre' and **gán zà-mí** (= *eqel sammî*) 'plane of the lyre' denote the same shape as **áb-zà-mi** (= *apsamikku*) 'concave square or concave-sided tetragon', the figure illustrated in BM 15285. Lawergren and Gurney (1987: 37–52) identified *sammû* as a harp and *ḫasīs sammî* as 'sound hole'. Vaiman (1963: 81) connected the two meanings of **áb-zà-mi**, 'cow of the lyre' and 'concave square or concave-sided tetragon', suggesting that 'cow' refers to a square 'membrane (of cowhide)', which was pulled taut over a sound-hole and assumed the shape of a concave square. Kilmer (1985) demonstrated that the concave square may refer to a stylistic rendering of the bovine nose on the cow(-head) found on reconstructed lyres from Ur.

Oppenheim (1947: 127) argued that the figure is a field with one curved and two straight line borders; in other words, the outline of the field corresponds to a harp-like instrument. **geštú** 'ear' of the **zà-mí** is its handle.

samuššu 'second string of an instrument'
coefficients:
4,6 *sa-muš-šum*: CBS 10996 obv.i 10'

Kilmer (1960: 299 n. 4 and 5, 300 n. 2) suggested that *samuššu* might be identical to *šamūšu* (= *ša*-**ge**$_6$), also the second string of an instrument but noted that the consistent use of **sa** a determinative could contravene reading *samuššum*.

sanāqu 'to reach, border, check'
[a-n]a **bàd** *la sà-na-qám*: BM 85194 rev. ii 8
a-na i-si-iq-tim ú-ul i-sa-an-ni-iq: BM 85194 obv. ii 28-29 (<*-iq*>), 39, 43
aš-li ú-sà-an-ni-iq: BM 85196 obv. ii 31
sà-ni-iq: YBC 4675 obv. 5, 7; rev. 6, 16
YBC 9852 rev. 1
'to reduce':
1 **sukud** *sà-a[n]-qum*: VAT 8523 rev. 10
1 **sukud** *sà-a[n]-qí*: VAT 8523 rev. 13

sapānu 'to level, devastate, destroy, smooth'; 'to sow linseed, to sow (a field) with linseed' (*CAD* s.v. *sapānu*)

10 *sa-pa-ni(?)*: MM 86.11.404 obv. 5 (meaning unclear)

Oppenheim (1947: 128) suggested that *sapinu* might be connected with the tool *supinnu*, which is used in the manufacture of bricks, etc.

saparru (= **giš**-KAK+LIŠ) 'chariot, cart'
coefficients:

3 **bán še gún [gi]š**-KA[K+LI]Š-*ra*: CBS 10996 obv. iii 29'

sattītu 'first string (of a stringed instrument)'
coefficients:

7,4 *sa-ti-tum*: CBS 10996 obv. i 9'

Kilmer (1960: 299 n. 4) suggested that *sattītum* many have the same meaning as **sa-di** = *qud-mu-ú* but noted that the constant use of **sa** as a determinative argues against the reading *sa-ti-tum*.

sekēru 'to dam up (a canal, a watercourse)'

...*ša* **íd** *s[é]-ke-ri-im*: VAT 8523 obv. i 1
3,45 **éš-kàr** 1 **lú** *ša* **íd** *sé-ke-ri-im*: VAT 8523 rev. 14–15
is-sí-ik-ki-ir: VAT 8523 rev. 2, 19
is-sí-ki-ir: VAT 8523 rev. 20
is-sí-ki-ru: VAT 8523 rev. 22, 25

coefficients:

3,45 **éš-kàr** 1 **lú** *ša* **íd** *sé-ke-ri-im*: VAT 8523 rev. 14–15

sêru 'to plaster, cover with a clay slip'

si-ri-im: Haddad 104 #7 iv 1, 2; #8 iv 6, 11, 12
iš-ka-ar si-ri-im: Haddad 104 #7 iii 39
si-ri-ia: Haddad 104 #8 iv 7
i-se-er: Haddad 104 #7 iii [41], iv 5; #8 iv 14
e-se-ru: Haddad 104 #8 iv 7
i-se-ru: Haddad 104 #8 iv 9
te-se-er: Haddad 104 #8 iv 11

sikkatu 'foundation cone, wall cone driven in on a property as indication of ownership'

[giš]*sí-kà-tum*: BM 85196 obv. i 33

uku-ús *sí-ik-ka-as-su ku-ul-li-im-šu*: VAT 7531 obv. 5, 10, 12–13, 19–20; rev. 5–6
VAT 7621 obv. 4, 6

Powell (1984a: 36 n. 17) discussed ${}^{\text{giš}}$*sikkatu* 'peg, nail' in reference to the practice of recording a sale on a clay peg and embedding the peg in a house wall. This action may signify the ceremonial aspect of driving in a peg to hold the measuring line. For analysis of this term, see Ellis (1968: 46–93).

Thureau-Dangin (1935a: 11) suggested that ${}^{\text{giš}}$*sikkatu* is a funnel used to serve a pile of grain of given volume.

simmiltu [= **(giš-)kun$_4$**] 'slab, threshhold, doorsill'
giš-kun$_4$: BM 85210 obv. i 2, 8, 10, 13, 17, 18

CAD s.v. *askuppu* noted that Akkadian texts usually use **i+lu**. The Sumerian word includes not only the threshhold (of an outer door) with its step but also the steps of a ladder or staircase; however, Akkadian differentiates between *askuppu* 'threshhold' and *simmiltu* 'ladder, staircase'.

Thureau-Dangin (1936e: 58) transcribed **giš-i-lu** as *simmittum*.

simtu (equivalent of *šimtu*) mark, brand
[*si-im-ta-am ša* ${}^{\text{d}}$*Il-tim* **udu-ḫi-a** *ši-na-ša im-ma-ḫa-ṣá*]: M 6829: 42'ff. (restored by // text *ARM* 24: no. 260)

siparru (= **zabar**) 'bronze'
zabar: A 3553: 19
BM 34568 rev. i i 12, 14, 15, 17
TMS, Text 3: 56
coefficients:
1,20 **igi-gub zabar**: A 3553: 19
1,20 *šà* **zabar**: *TMS*, Text 3: 56

suḫḫu (= **suḫ$_4$**) 'reed matting'
coefficients:
3,45 *ša* **é-kišib suḫ$_4$**: YBC 5022: 32

suluppū (= **zú-lum**) '(ripened and plucked) dates'
ma-ḫi-ir **še-ì-giš** *ma-ḫi-ir* **zú-lum** *ma-ḫi-ir* **síg**: IM 52916, IM 52685, IM 52304 Section F: 44'

(^{giš})*sussulu* 'storage box' or trough made of wood, rarely of precious metals. It is listed among household objects used by cooks (for holding bread and meat); it is also used by fishermen to transport live fish .

^{giš}*sú-sú-lum*: BM 85196 obv. ii 17

sūtu (= ^{giš}**bán**) *sūtu* vessel

su-tim: Haddad 104 #5 iii 10, 14, 26
ṣí-ba-at sú-tim: VAT 8528 rev. i 23, 25; rev. ii 1, 2
^{giš}**ba-an**: YBC 4669 obv. i 5, 9
^{giš}**bán**: CBS 10996 obv. ii 16'
^{giš}**bán** *še-a-am ma-li-a-at*: BM 85194 rev. ii 1
^{giš}**bán lugal**: CBS 10996 obv. ii 17', 29'
^{giš}**bán-meš**: *LBAT*: no. 1648 rev.(?) 2', 3', 4', 7'

coefficients:

40,5 **giš-{ba}-ba-rí-ga** 1 **bariga** 2 **bán**
40 ^{giš}**ba-rí-ga** 1 **bariga**
30 ^{giš}**ba-rí-ga** 2 **bán** (should be 3 **bán**)
20 ^{giš}**ba-rí-ga** 3 **bán** // 20 ^{giš}**bán** (should be 2 **bán**)
12 ^{giš}**bán lugal**: CBS 10996 obv. ii 13'–17'
[...] **dal** ^{giš}**<ba>-rí-ga** 1 **gur**
[... **d]al** ^{giš}**<ba>-rí-ga** 2 **(bariga)** 2 **bán**
[...**dal** $^{\text{gi]š}}$**<ba>-rí-ga** 1 **(bariga)** 2 **bán**
[... **dal** ^{giš}**<ba>-rí-g]a** 1 **(bariga)**
[...**dal** ^{giš}**<ba>-rí-ga]** 3 **bán**
[...**dal** ^{giš}**<ba>-rí-ga]** 2 **bán**
[... ^{giš}**bán]** lugal: CBS 10996 obv. ii 22'–29'

ṣābu (= **erim**) 'men, workers'

ṣá-ba-am: IM 54538 obv. 4 [See rev. 2 *awīlūka*, which is used as a synonymn for *ṣābu*.]

erim: AO 8862 iv 21 tabulation
BM 85194 obv. ii 24
Strssbg. 362 obv. 7, 15; rev. 17, 18
TMS, Text 25 obv. i 5, 6
VAT 7528 obv. i 22; obv. iii 15(?)
VAT 7531 obv. 14
YBC 5037 obv. 6, 16

erim-ḫi-a: AO 8862 iii 34; iv 9, 18, 20, 21
TMS, Text 25 obv. i 4
VAT 7528 obv. i 10, 15, 16, 17, 21, 23; obv. iii 5, 13, 16, 23, [24]; rev. ii 3

VAT 7531 obv. 9, [18]
YBC 4657 obv. [3]; rev 9, 11, 13, 15, 16, 18, 20, 22, 25, 26
YBC 4662 obv. 42; rev. [4], 8, 10, 12, 18, 22, <25>, 29, <31>, 34
YBC 4663 obv. 2
YBC 4666 obv. 3, 4, 14, 16, 18; rev. 8, 9, 17, 19, 21
YBC 4669 rev. i 2, 5, 6
YBC 4673 obv. ii 7, 9; rev. iii 8, 13, 16, 20
YBC 5037 obv. 2
YBC 7164 obv. 2, 3, 7, 9, 11, 24, 28; rev. 2
YBC 8588 obv. 10, 11, 12; rev. 1, 7

erim-meš: BM 85194 obv. i 10, 32; obv. ii 2, 24
Erm 15073 rev. i 8, 14, 20; iii 7, 9, 12, 15
VAT 7531 obv. 11
VAT 7621 obv. 5

erim-meš-*šu-nu*: VAT 7621 obv. 7

For the apparently interchangeable use of **lúḫun-gá** and **erim**, see Weitemeyer (1962: 37).

ṣâtu 'explanatory word list' (commentary based on traditional interpretations, lit. excerpted words; see *CAD* s.v. *ṣâtu*)
ina ṣa-a-tim: K 11097: 10

ṣeḫru (= **tur**) 'small young, second in rank, child'
zi-it-ti ra-bi ṣe-eḫ-ri: IM 31210 §7: 7
tur: *TMS*, Text 10: 2, 11, 16(?), 20, 33
TMS, Text 24: 35, 53
ḫa-la tur(*-ma*): *TMS*, Text 10: 5, 6, 9, [24], 27, 28, [31]

ṣēnu (= **u$_8$-udu-ḫi-a**) 'sheep, (flock of) sheep and goats'
u$_8$-udu-ḫi-a *ù* **sila$_4$-gub**: YBC 4669 rev. ii 12, 14, 24
u$_8$-udu-ḫi-a ugu u$_8$-udu-ḫi-a: YBC 4669 rev. ii 20

ṣēru (= **líl**) 'steppeland, back country, open country'
líl: VAT 8389 obv. ii 14, 15, 16, 19, [20], [23], [24], [25], [26]; rev. i 3, 8, 12, 23, 24, 26, 27; rev. ii 1, 4, 5, 8, 9, 10
VAT 8391 obv. i 3, 8, 9, 11, 12, 15, 16, 17, 19, 20, 23, 28, 30; obv. ii 21, 22, 23, 24, 27, 31, 33; rev. i [1], [2], 5, 10, [14], 31, 32; rev. ii [1], [2], [4], [5], 8, 9, 12, 13, 14, 18, [21], 29

líl-*ia*: VAT 8389 obv. i 4
líl-meš: VAT 8389 obv. i 10, 11
líl-*ú-a*: VAT 8389 obv. i 5; rev. i 5
VAT 8391 obv. i 25; rev. i 7

ṣibtu (= **máš**) 'interest'
ṣí-ib-ta-am: VAT 8528 rev. i 15
ṣí-ib-tam: VAT 8521 rev. 13
ṣí-ib-tim: VAT 8521 rev. 20
a-na ṣi-ba-at i-di-in-ma: AO 6770 obv. 9
ṣí-ba-at sú-tim: VAT 8528 rev. i 23, 25; rev. ii 1, 2
máš(-bi): *TMS*, Text 22: 3, 9, 15, 19
VAT 8521 obv. 2, 3, 5, 9, 10, 11, 14, 15, 17, 21, 22, 24, 27; rev. 1, [3], 7, 8, 9, 10, 12, 15, 19, 21
VAT 8528 obv. i [3], 4, 8, 9, 10, 12, 14, 16, 26; obv. ii [-3], [-2], 0, 1, 2, [3], 5, 6, 9, 11, 14, 16, 20; rev. i 1, 3, 6, 8, 11, 19, 20, 24
YBC 4669 rev. iii 19
YBC 4698 obv. i 2, 3, 5, 6

See Steinkeller (1981: 113-145) showed the evolution from **máš** 'goat' to **máš** 'interest'.

ṣītu 'expenditure'
coefficients:
10 **igi-gub** *ma-li ṣi-wa-tum*: A 3553: left edge (b)

ṣubātu (= **túg**) 'cloth'
túg*ṣú-ba-tum*: BM 85194 rev. i 47

CAD s.v. *ṣubātu* was a garment was made of wool and used by men and women. It was an untailored, undecorated, blanket-like wrap which served as the main garb. *ṣubātu* was described in commercial texts according to quality; it was measured by counting rather than weighing. *ṣubātu* was a square piece of material of standard size and weight

ṣupru (= **umbin**) a technical building term, perhaps, a mistake for *išdu* (= **suḫuš**)
iš-tu **umbin saḫar-ḫi-a**: BM 85210 obv. ii 3

ša x-lu(?)-ú-um (broken)
coefficients:
57,36 *ša x-lu(?)-ú-um*: YBC 5022: 65

ša-[x x]-*tum* (broken)
coefficients:
1,41,15 *ša*-[x x]-*tum*: YBC 7243: 45

šaggilu 'requisition tax'
šag-gi-lu-ú: M 6861: 5

šaḫallûm (meaning unknown)
coefficients:
1,12 *ša-ḫa-al-lu-ú-um*: YBC 5022: 37

In *AHw* s.v. *šaḫallûm* von Soden questioned the meaning 'vessel', its origin as a Sumerian foreign word and a comparison with Ugaritic *suḫallu*.

Neugebauer and Sachs (*MCT*: 134) referred to *šaḫillu* in Meissner, *SAW*, 939, and vocalized the word as *šaḫallu*.

ša-KAK(-)UD-*šu-rum* (meaning unknown)
coefficients:
40 *ša*-KAK(-)UD-*šu-rum*: YBC 5022: 59 [Neugebauer and Sachs (*MCT*: 135) suggested the reading *ša*-KAK(-)UD-*šu-rum*, which Kilmer (1960: 301 n. 2) compared with *gušūru*.]

šakanakku 'governor'; pl.: *šakanakkū*
[*n*]*a-ar ša-ka-na-ka-tim na-ra-am šu-zu-ba-am*: IM 52916, IM 52685, IM 52304 Section D: rev. 16'

šalāmu 'to be healthy'
i-na a-mat ^{d}*Anu u An-tum liš-lim*: AO 6484 u.e.

šalāṭu 'to cut off' (used in canal work)
[*e-l*]*i-nu-um aš-lu-uṭ* 1,40 **ki-ta** *aš-lu-uṭ*: BM 85196 rev. i 20–21
sa ta-aš-lu-ṭú: BM 85196 rev. i 27, 28

šalšudu meaning uncertain
coefficients:
6,0 *šà na-aš-pa-ak šà-al-šu-di-im*: *TMS*, Text 3: 49

šaman rūšti (= **ì-sag**) 'best, fine oil'
[*ma-ḫi-ir*] **ì-giš** [*ma-ḫi-i*]*r* **ì-sag-dug-ga** *ma-ḫi*[*-ir* **ì-šaḫ**]: IM 52916, IM 52685, IM 52304 Section F: 45'

šamaššamu (= **še-giš-ì**, **še-ì-giš**) 'sesame'
še-giš-ì: CBS 10996 obv. iii 8'
gu-ru-un **še-giš-ì**: CBS 10996 obv. iii 4'
zi-qip **še-giš-ì**: CBS 10996 obv. iii 2', 3'
ma-ḫi-ir **še-ì-giš** *ma-ḫi-ir* **zú-lum** *ma-ḫi-ir* **síg**: IM 52916, IM 52685, IM 52304 Section F: 44'
coefficients:
⌜3⌝,20 *zi-qip* **še-giš-ì**
2,30 *zi-qip* **še-giš-ì**
1,20 *gu-ru-un* **še-giš-ì**
20 **ki-min**
1,20 **ki-min**
1 **ki-min**
3 (**bariga**),30 **še-giš-ì**: CBS 10996 obv. iii 2'–8'

šamnu (= **ì-giš**) 'oil'
[*mār*] **lúḫal nun-me** [*šamni*]: K 11097: 7
i-na 7 **qa**ta-àm *ša ta-pa*!*-aš-*[*šà-ru šamnim*]: *TMS*, Text 13: 15 [For this reading see von Soden 1964: 49.]
ì-giš: *TMS*, Text 13: 1, 2, 5, 7, 14, 17
ì-giš *ù* **ì-šaḫ**: IM 54464 obv. 5
i-na ma-ḫi-ir 1 **bán** 5 **sìla ì-šaḫ** 1 **bán ì-giš**: IM 54464 obv. 2
[*ma-ḫi-ir*] **ì-giš** [*ma-ḫi-i*]*r* **ì-sag-dug-ga** *ma-ḫi*[*-ir* **ì-šaḫ**: IM 52916, IM 52685, IM 52304 Section F: 45'
ì-giš *ù* [**ì-ša**]**ḫ** *ša-ma-am* **kù-babbar ì-giš** u **ì-š**[**aḫ** *k*]*a-ma-r*[*a-am*]
ma-ḫi-ir **ì-giš** *ù* **ì-šaḫ** *ka-ma-ra-a*[*m*]: IM 52916, IM 52685, IM 52304 Section F: 49'–50'

šâmu (= **šám**) 'to buy, to apply the price norm quoted for the actual purchase'
[*m*]*a-ḫ*[*i-ra-tim na*]*-sa-ḫa-am ša-ma-am ù k*[*a-ma-ra-*]*am*: IM 52916, IM 52685, IM 52304 Section F: 47'
ì-giš *ù* [**ì-ša**]**ḫ** *ša-ma-am* **kù-babbar ì-giš** u **ì-š**[**aḫ** *k*]*a-ma-r*[*a-am*]: IM 52916, IM 52685, IM 52304 Section F: 49'
a-ša-am: MLC 1842 obv. 1, 2, 6
TMS, Text 13: 4
ta-šà-am: *TMS*, Text 13: 12
šám: *TMS*, Text 13: 1

Cf. **ì šám** 'capacity equivalent' (*MCT*: 96-97). The transformation to capacity measure is obtained by multiplying the number of **sar** which constitute the volume (*1bh*) by the coefficient 5,0,0.

i šám saḫar-bi: YBC 4607 obv. 3, 5, 8, 9, 12, 15, 18; rev. 2

šamû 'sky, heaven'

sal-šeš [lú]**ḫal**-*t*[*i pirišti šamê u erṣeti*]: K 11097: 1

šamūšu (= *ša*-**ge6**, *šá*-**ge**$_6$) 'second string of an instrument'
coefficients:

sa *ša*-**ge**$_6$ *ù* **sa** 4 *uḫ-ri* 2,6 **sa** *i-šar-tum*: CBS 10996 obv. i 13'
sa *šá*-**ge**$_6$ *ù* **sa** 3-*šú uḫ-ri* 2,4 **sa** 4-*tu*: CBS 10996 obv. i 16'
sa MIN 3-*šú ù* **sa** *šá*-**ge**$_6$ 5,2 **sa murub**$_4$-*tu*
sa *šá*-**ge**$_6$ *ù* **sa** [d]*é-a*-**dù** 2,4 **sa** ⌜*titur*⌝ **murub**$_4$-*tú*: CBS 10996 obv. i 19'–20'

Kilmer (1960: 299 n. 5, 300 n. 2) suggested that *ša*-**ge**$_6$ may be a cryptographic writing for *šamūšum*.

šammūtu (meaning uncertain)

i-na ša-am-mu-ti-šu x x x [...]: IM 52916, IM 52685, IM 52304 Section I (e): 6

šammūtu occurs at the end of a series of agricultural activities concerning *še'um* 'seed barley' in I (d); it is followed by broken passages concerning *qanû* 'reed' in I (e).

šaqālu (= **lá**) 'to weigh'

ne-mé-l[*a-am*...]-*am* **kù-babbar** *ša-qa-la-am*: IM 52916, IM 52685, IM 52304 Section F: 48'
ì-lá: YBC 4652 obv. [2], [4], 7, 9, 11, 13, 15, 18; rev. 1, 3, 4, 6, 8, 12

šaqû 'to irrigate'
coefficients:

1,30 **a-šà a-meš** *ša-qu-ú* [...]
18 **a-šà a-meš** *ša-qu-ú* **ki-min**
2,46,40 **a-šà a-meš** *ša-qu-ú* **ki-min**: CBS 10996 obv. iii 26'–28'

šarāru (meaning uncertain)

ša-ri-ir: K 8705: 7
[*i-ša-r*]*a-ar-ma*: K 8705: 13

šarru (= **lugal**) 'king'
lugal šú: BM 47860 rev. 16
aga lugal: BM 15285 rev. 2

šāru 'wind'
a-na na-ap-ḫa-ar **uš**(!)*-ka iš-ši-ma* 4 *ša-ar er-bé-tim lu-*[*pu*]*-ut-ma*:
IM 52301 side 2–3 [See Bruins 1953c: 241–43.]

šassuku 'librarian'
dub [Id]**en**-*na-din*-**ibila** [lú]**gír-lá** [d]**amar-utu** [lú]*šà-suk-ku*: BM 47860 rev. 13–14

šattu (= **mu**) 'year'
ša-at-tam: VAT 8528 obv. i 8; rev. i 18
ša-at-tim: VAT 8528 rev. i 26; rev. ii 5
i-na ki ma-ṣi ša-na-tim: AO 6770 obv. 10
mu dù ù $\mathbf{u_4}$: AO 6770 obv. 17
mu-(x)-**kam**(-*ma*): AO 6770 obv. 11, 13, 14, 16
K 11097: 6
TMS, Text 22: 18, 20, 21, 22
VAT 8528 obv. i 11, 12, 14, 23, 24, 26; obv. ii [-1], [1], [2], 3, 7, 8, 9, 11, 12, 13, 16, 17, 18, 20, 21, 22; rev. i 3, 4, 5, 8, 12
YBC 4669 rev. iii 20

šemeru (= **ḫar**) 'ring'
ḫar kù-babbar: BM 85196 rev. ii 18, 20, 23

šēpu 'foot'
a-na še-pí-ia: BM 85194 rev. iii 24
a-na še-pí-šu: BM 85194 rev. iii 22

šēru (tuning interval of a stringed instrument)
coefficients:
sa 3 *uḫ-ri ù* **sa** 5-*šú* 7,5 **sa** *še-e-rì*: CBS 10996 obv. i 12'

See Güterbock (1970).

šer'u (= **ab-sín**) 'furrow'
ab-sín: BM 85194 rev. ii 34, 36

še'u (= **še**) 'barley'
še: AO 8862 iv 3 tabulation, 16 tabulation
BM 85194 rev. i 44; rev. ii 5, 34; rev. iii 10
BM 85196 obv. i 30, 31, 32; obv. ii 20, 21
CBS 10996 obv. iii 29', 34'(?)
CBM 12648 obv. i 5
Haddad 104 #3 i 31, 35, 37, ii 11, 14, 15, 18; #4 ii 35, iii 2, 6; #5 iii 10, 13, 17, 19, 25, 30; #6 iii 35
IM 53597 obv. 2
M 6861: 1, 2, 4, 8, 10, 13
TMS, Text 3: 5, 6, 16, 17, 18
TMS, Text 22: 1 (**sag še**), [5] (**sag [še]**); 19, 20, 24
VAT 7848 obv. 18
VAT 8389 obv. i 15
VAT 8528 rev. i 9
YBC 4657 rev. 24 (**še-ta-àm**), 25, 26
YBC 4666 obv. 2, 3, 7, 9; rev. 3, 6
YBC 4669 rev. i 19 (**sag še**), 20 (**še sag**)
YBC 4673 obv. i 8, 12, 13, 18; obv. ii 4
YBC 4698 obv. i 1, 2, 3, 4, 6; obv. ii 3 (**še-ta**); rev, i 13, 19; rev. ii 3, 5
YBC 5022: 30, 61
še-*um*: BM 85194 rev. ii 35
BM 85196 obv. i 33
Haddad 104 #3 i 31, 35; #4 iii 2; #6 iii 31
VAT 8389 obv. i 3, 9, 16, 20, 24; obv. ii 20, 24, [25]; rev. i 4, 9, 11, 13, 18, 19; rev. ii 5, 9, 12
VAT 8391 obv. i 12, [16]; obv. ii 2, 7, 13, 24, 28, 32; rev. i 14; rev. ii [2], 5, 9, 21, [23]
VAT 8528 rev. i 14
še-*am*: *TMS*, Text 14: rev. 6(?)
VAT 8389 obv. i 6, [8]; rev. i 6, 7, 12
VAT 8391 obv. i 6, 7, 26, 27; rev. i 1, 2, 8, 9; rev. ii 15, 16, [22], 24
VAT 8528 rev. i 13, 16, 17; rev. ii 7
še-a-am: AO 8862 iii 31; iv 3, 6, 15
BM 85194 rev. ii 1; rev. iii 8
Haddad 104 #1 i 9; #2 i 11, 18; #3 i 30, ii 30; #4 ii 40; #6 iii 32
YBC 4666 rev. 2, 5
YBC 10722 rev. 3
še-*e*: YBC 4666 obv. 4

YBC 4669 rev. i 18
VAT 8528 rev. i 21

Cf. *ši-e* x-*ra-am ša-ka-na-am*: IM 52916, IM 52685, IM 52304 Section I (d): 3

še-*im*: Haddad 104 #3 i 39
IM 52916, IM 52685, IM 52304 Section I (d) :5 ([*iš*]-*ka-ar za-ri-i ša* **še**-*im e-pé-š*[*a-a*]*m*); Section F: 43' (*ma-ḫi-ra-am e-pé-ša-am ma-ḫi-ir* **še**-*im ma-ḫi-ir* [...])
TMS, Text 22: 24 (*ri-iš* **še**-*im*)
VAT 8389 obv. i 3, 9, 19, 24; obv. ii 18, 22, 26; rev. i 4, 9, 11, 14, 17, 19; rev. ii 3, 7, 12
VAT 8391 obv. i 10, 14, 29; obv. ii 1, 6, 10, 13, 26, 30; rev. i 11, 13, 15, 20, 24; rev. ii [3], 7, 17, [20], 22, 26, 28
VAT 8528 rev. i 20

še-*ia*: YBC 4669 rev. i 22 (**sag še**-*ia*)

še-ú-ši-na: VAT 8389 obv. ii 16; rev. ii 1
VAT 8391 obv. i 5

še-a-ši-na: VAT 8391 obv. i 24; rev. ii 12

še-e-ši-na: VAT 8391 rev. i 6

še-meš: AO 6484 obv. 11, 18

še-gur: BM 85194 rev. iii 12
BM 85196 rev. i 9 ([-**gur**]), 10, 11, 12 (<-**gur**>), 13
MLC 1842 obv. 1, 2
YBC 4669 rev. iii 1
YBC 4698 obv. i 1, 4; obv. ii 1
VAT 8389 obv. i 1, 2; rev. i 1, 2
VAT 8391 obv. i 1, 2, 19, 20, 21, 22; rev. i 3, 4; rev. ii 10, 11
VAT 8528 rev. i 9

See: 30 *ša* GÁN(?)-GÁ ŠE(?) : IM 52301 rev. 22

coefficients:

3 **bán še gún [gi]š**-KA[K+LI]Š-*ra*: CBS 10996 obv. iii 29'
[...],3 *za-ba-lu* 1 **bariga š**[**e**(?)...]: CBS 10996 obv. iii 34'?

Cf. 6,12 **gán ì-dub** x ŠE DU_8 NU **igi-g[ub**-*ú*]: BM 47860 obv. 10
6,40 *ša* **é-kišib še**: YBC 5022: 30
6,40 **gán ì-dub še**-im **igi-gub**-*ú*: BM 47860 obv. 7
4,48 **é-kišib še**: YBC 5022: 61
16 **[igi]-gub** *šà* **gam** *šà* 2 **še** *i-na* **šà gam gar**: *TMS*, Text 3: 5
16,26,46,40 **igi-gub** *šà* **gam** *šà* 3 **še** *i-na* **šà gam gar**: *TMS*, Text 3: 6
13,20(?) **igi-gub** *šà* **a-šà še**: *TMS*, Text 3: 16
56,40 **ri** *šà* **a-šà še**: *TMS*, Text 3: 17
28,20 [*p*]*i-ir-ku šà* **a-šà še**: *TMS*, Text 3: 18

Kilmer (1990b: 85) listed **še** among "terms based upon realia, but with no specific construction, illustration, or dimensions available" and showed that this figure is constructed from 90° arcs. Vaiman (1963: 79–80) demonstrated that **a-šà še** in *TMS*, Text 3: 16–18 is a biconvex figure composed of two ¼-circle areas. However, he did not explain the 'circle with 2 (or 3) grains inscribed' in *TMS*, Text 3: 5-6.

For a discussion of the problem of translating **še** by *uṭṭetum*, see Sachs (1946: 208 n. 13).

šikin ubāni (= *šikin* **šu-si**) 'appearance of the finger'
šikin **šu-si**: K 8705: 6
K 11097: 5, 8, 9, 14

See Borger (1957: 191–92)

šikkatu 'harrowed'
si-ka-tum: CUA 34 rev. (c)

šīlūtum (work at the upper levels in the excavation of a canal) (*MCT*: 88)
20 **pa$_5$** *ši-lu-tum*: YBC 7243: 21
⅓ **ma-na (saḫar)** *ši-lu-tum*: YBC 7164 obv. 4, 6, 8, 10 (<⅓ **ma-na**>), 12, 15
20 *ši-i-lu-ta-am*: YBC 9874 rev. 11
igi-te-en u$_4$ *ši-lu-tam iḫ-re*: YBC 7164 obv. 16
igi-5-gal u$_4$ *ši-lu-tam* 4 **gín saḫar** *is-sú-uḫ*: YBC 7164 obv. 17–18
20 *ši-i-lu-ú-tim*: YBC 9874 rev. 6
coefficients:
20 **pa$_5$** *ši-lu-tum*: YBC 7243: 21
20 *ši-i-lu-ta-am*: YBC 9874 rev. 10
20 *ši-i-lu-ú-tim*: YBC 9874 rev. 6

The work of digging the canal is divided into different categories; the easier work is in the upper level and the slower work in the lower levels. Since the latter is always connected with the word **dusu**, meaning 'basket', (particularly in connection with construction projects), this may indicate that digging at the lower levels requires shovelling the excavated dirt into baskets which are raised to the top and emptied, whereas the work at the upper level merely involves digging out the earth and casting it from the shovel directly to the ground level. The work in the upper level is characterized by the abstract word *šīlūtum*, perhaps, connected with the verb *šalû* 'to cast, throw', via *šil'u* > *šīlu* (*MCT*: 88).

Lewy (1949b: 48–49) translated *šīlūtum* as 'the requirement, that which is asked for'. *šīlūtum* is not to be confused with *iškarum*, which fluctuates, while *šīlūtum* remains constant. She incorrectly interpreted these problems to mean that the amount of soil for which a worker is paid 1 **bán** is a constant. According to Lewy, a difficult task would receive the same compensation as an easy task, an explanation not in accord with the evidence.

šīmu (= **šám**) 'price'
  šám: *TMS*, Text 13: 1

šipātu (= **síg**) 'wool'
  síg: YBC 4673 rev. ii 17, 27
  síg(?)-ḫi-a(?): A 3553: 48 (b)
  síg nu-nu: YBC 4673 rev. ii 15, 30
  síg šu-ḫi-a: YBC 4673 rev. ii 13, 20, 28
  ma-ḫi-ir **še-ì-giš** *ma-ḫi-ir* **zú-lum** *ma-ḫi-ir* **síg**: IM 52916, IM 52685, IM 52304 Section F: 44'
coefficients:
  48 **igi-gub síg(?)-ḫi-a(?)**: A 3553: 48 (b)

For a discussion of different kinds of woolen textiles, see Hallo (1979).

šipru (= **kin**) 'work' (used in conjunction with volume)
  kin (*ù* **saḫar**): VAT 8523 obv. i 5, 11, 13, 14, 17, 20, 24, 25; obv. ii 5, 9, 10, 18, 22, 23, 31; rev. 5, 18, 21, 26
  kí-sa kin: BM 85196 obv. i 7

šubtu 'residential area, dwelling'
  šu-ub-tum: Böhl Coll. 1821 3

šūbultu 'shipping'
coefficients:
  1,12 *šà šu-bu-ul-ti*: *TMS*, Text 3: 34

šumu (= **mu**) 'name'
  4,22 **mu-bi nigin-meš**: *TMS*, Text 5 colophon
  mu-bi-im: Plimpton 322 column iv 1
  mi-nu šum-šu: *TMS*, Text 9: 16

šuqultu (= **ki-lá**) 'weight'
  šu-qú-ul-ta-ša: AO 6770 obv. 19

[k]i-lá: VAT 7848 obv. 17
ki-lá nu-na-tag: YBC 4652 obv. 1, 3, 6, 8, 10, 12, 14, 17; rev. 1, 4, 7, 11
ki-lá-bi: YBC 4698 rev. ii 14
YBC 7243: 33 **[ki-lá(?)-bi(?)]**
YBC 7284 rev. 2, 3
ki-l[á-*ša***]**: BM 34568 rev. i 12
ki(?)-lá-*šú*: VAT 7848 obv. 16
coefficients:
9 **ki-lá(?)-bi(?)** *ša* **urudu**: YBC 7243: 33

Powell (1971: 243) translated the term **ki-lá** = *šuqultu* as 'mass', which he listed among terms for weighing, but Oppenheim (1947: 126) suggested that **ki-lá** = *mašqaltu*. *CAD* s.v. *mašqaltu* does not cite the above passages.

šūru (= **gi-šu-kin**) 'faggot of reeds'
coefficients:
10 *šà šu-ri ka-sà-mi*: *TMS*, Text 3: 68
5 *šà šu-ri za-ba-li*: *TMS*, Text 3: 69
50 **gišmá-lá gi-šu-kin**: CBS 10996 obv. ii 7'

taddītum 'laying/layer (of bricks)'
1 **sar gagar** *ta-ad-di-ta-am* **en-nam ì-dib**: YBC 4607 rev. 5, 8, 11, 14, 17 (<*-am*>)
ta-ad-di-tum ša **sig**$_4$: YBC 5022: 38
coefficients:
14,24 *ta-ad-di-tum ša* **sig**$_4$: YBC 5022: 38

The weak link in translating *taddītum* as 'laying/layer of bricks' is the lack of the II/1 conjugation of *ndi* to which *taddītum* would correspond in the archaic infinitive (*MCT*: 94).

tallaktu 'going, step'
coefficients:
45 *ta-la-ak-tum*: YBC 7243: 46

Baumgartner (1925: 123–26) analyzed *tallaktu* among terms for construction or building and suggested that *tallaktu* may designate processional road (parallel to *mašdaḫu*), a pavement (synonym for *mutaliktu*, which has the same coefficient), or passageway.

tāmartu (= **igi-duḫ-a**) 'visibility (of the moon)'
coefficients:
 3,45 **igi-duḫ-a**: CBS 10996 obv. iii 9'

tamlâtu 'dam'
 ša 1-*et* 30 [*ta*]*m-la-tum*: BM 85210 colophon
 saḫar-ḫi-a *tam-lì-a-tim*: BM 85196 obv. i 13, 17

tappû 'peer'
 tap-pu-ú: *LBAT*: no. 1648 obv.(?) 9'

taraḫḫu 'side of a canal or excavation'; sometimes abbreviated to *ta*
 a-ta-ap ta-ra-aḫ-ḫi: BM 85196 rev. i 19
 [*na*]-*ra-am na-ar ta-ra-ḫi-i e-pé-ša-am*: IM 52916, IM 52685, IM 52304 Section D: rev. 15'
 [*a-*]*ta-a*[*p*] *ta-ra-aḫ-ḫi-im-mi*: YBC 9874 rev. 1
 te-er-di-is-sà ½ **kùš** *ta-ra-ḫi-sa ú-ša-mi-iṭ*: YBC 7164 obv. 20
 te/*de-er-di*(-*is*)-*sà* ½ **kùš** *ta ú-ša*(-*am*)-*mi-iṭ*: YBC 7164 obv. 21–22, 23–24, 27; rev. 1, 4, 7, 11, 14

Powell (1988: 165, 170) defined *taraḫḫu* as 'the earth dug out from the sides or bottom of a canal or irrigation ditch to create extra water capacity.'

Goetze (1948: 37) argued that *taraḫḫu*, used in parallelism with *kalakku*, is Sumerian in origin but the underlying word is not known. Lewy (1949b: 42 notes 1 and 3) analyzed *taraḫḫu*, 'lateral extension', as derived from ***da** + **raḫ** (**da** = *idu* 'side' and **ra** = *etēqu* 'to cross, to pass'). She suggested that *taraḫḫu* refers to the rectangular opening in the side of the main canal from which the extension branches off or to a cross-section of the extension.

tarbaṣu (= **tùr**) 'sheep-fold'
 ta-ar-ba-ṣú-ú: *ta-ar-ba-ṣú-ú e-li ta*<-*ar*>-*ba-ṣí*: YBC 7326 obv. 2
 ta-ar-ba-ṣí: YBC 7326 obv. 2
 (*ši-na*) *ta-ar-ba-ṣú-ú-a*(-*mi*): YBC 7326 obv. 1

tarkibu 'layer of bricks'
 tar-ki-bu-ta **sig$_4$/sig$_4$-áb**: AO 10822 rev. 5, 7

tašnintu 'rivalry, fight'
 taš-nin-ti: K 11097: 13
 taš-nin-tu$_4$ **um-me-a**: K 11097: 2

terdītu 'increase'

de-er-di-sà: YBC 7164 obv. 21; rev. 1 (<-*di*->)
te-er-di-sà: YBC 7164 obv. [23]; rev. 4, 7, 11, 14
te-er-di-is-sà: YBC 7164 obv. 20, 27

Powell (1988: 165, 170) explained *terdītu* as 'the additional volume of water gained by stripping off the sides and bottom of a canl to make it larger'.

Neugebauer and Sachs (*MCT*: 88) suggested translating *terdītu* as 'increase', based on the assumption that it is the archaic II/1 infinitive parallel to *ruddû* 'to add'. They were unable to explain the variant initial syllable *de-* instead of *te-*.

Goetze (1948: 36–37) suggested the translating *terdītu* as 'current', based on the verb √*redû* 'to drive'; he interpreted the passages in YBC 7164 to mean 'its (that is, the canal's) current swept away so and so much of the embankment'. Goetze supported his interpretation with a passage in *TCL* 7: no. 39:10-22 (Thureau-Dangin (1924a: 32): (10) *a-na pí-i nār Ug-dím-sa*$_4$ (11) *al-ka-ma* (12) *šum-ma mu-ú ma-ṭú-ú-ma* (13) *eqel bilti-šu-nu mu-ú la i-ka-aš-ša-du* (14) *i-na pí-i nār Ug-dím-sa*$_4$ (15) *me-e te-er-di-tam šu-uk-na-šu-nu-ši-im* (16) *šum-ma mu-ú ša i-na-an-na* (17) *i-na nār Ug-dím-sa*$_4$ *i-il-la* (18) *la i-ṣú-ú-ma ma-la eqel bilti-šu-nu ru-ub-bé-e-em* (19) *ma-ṣú-ú* (20) *mi-im-ma ši-ik-na-am* (21) *i-na pí-i nār Ug-dím-sa*$_4$ (22) *la ta-ša-ak-ka-na* 'Go to the 'mouth' of the canal Ugdimsa! If the waters are too low so that the waters cannot reach their assigned field, create for them a flow of water. (But) if the waters which just now are rising in the canal Ugdimsa are not deficient and sufficient to enrich their assigned field, then do not establish any stop in the mouth of the canal Ugdimsa'.

tibku 'layer of bricks'

ti-ib-ku-um: BM 85194 obv. iii 45, [46], 47

tibnu (= **in-nu**, **in-nu-da**) 'straw'
coefficients:

na-az-ba-al **in-nu-da** 3,20 *i-gi-gu-bu*: IM 52916, IM 52685, IM 52304 Section E: rev. 24'
4,30 *za-ba-lu* 6 **(gur) in-nu gún ki[-min]**
12,30 *za-ba-lu* 1 **(gur) in-nu gún [ki-min]**: CBS 10996 obv. iii 31'-32'
1,16 **é-uš-gíd-da in-nu-da**: YBC 7243: 23
3,36 *ša* **é-kišib in-nu-da**: YBC 5022: 31
3,36 **in-nu-da**: YBC 7243: 22
4,48 **igi-gub é-kišib in-nu-d[a]**: A 3553: 37

tilpanu (= ^{giš}**illar**) 'arc'
coefficients:
13,7,30 **igi-gub** *šà* **gán** ^{giš}**illar**
52,30 **ri** *šà* **gán** ^{giš}**illar**
30 *pi-ir-ku šà* **gán** ^{giš}**illar**: *TMS*, Text 3: 13–15

von Soden (1964: 47, 50) suggested the reading **giš-ru** with the meaning 'boomerang' for a figure which is a parallelogram inscribed in a ⅓ circle. Kilmer (1990b: 85) demonstrated that **gán-giš-šub** = *tilpānu* '(double) throw-stick, rhombus', is the geometric shapee formed from a configuration of two boomerangs facing each other (1985).

tinānu 'fig-like' (qualification of pomegranetes)
coefficients:
10,6,30 ^{giš}**nu-úr-ma** *ti-na-nu*
10,3,30 ^{giš}**nu-úr-ma** *ti-na-nu*: CBS 10996 obv. iii 17'–18'
3,20 *ti-nan-nu*: CBS 10996 obv. iii 20'

In Hh iii 45 *tinanû* describes a type of apple.

titur qablītu (musical interval)
coefficients:
sa *šá*-**ge**$_6$ *ù* **sa** ^{d}*é-a*-**dù** 2,4 **sa** ⌜*titur*⌝ **murub**$_4$-*tú*: CBS 10996 obv. i 20'

tubqu (= **ub**) 'corner'
BM 47860 rev. 11

tupšikku (= **dusu**) literally 'basket'; work at the lower levels in the excavation of a canal (*MCT*: 88)
tu-up-ši-ki: Haddad 104 #10 iv 31
^{giš}**dusu** *šu-up-ši-ki-im*: IM 49949: 3
8½(sic) **gín dusu**: YBC 7164 obv. 13
(**saḫar**) 10 **gín dusu**: YBC 7164 obv. 5, 7, 9, 10 (<**saḫar** 10 **gín**>), 13, 16
⅔ **u**$_4$ *ù* **igi**-5-**gál** ⅔ **u**$_4$ **dusu** *iḫ-re*: YBC 7164 obv. 18
igi-te-en u$_4$ **dusu** *iḫ-r*[*e*]: YBC 7164 obv. 17
coefficients:
2,13,20 *tu-up-ši-ki*: Haddad 104 #10 iv 31
2,13,20 **igi-gub** ^{giš}**dusu** *šu-up-ši-ki-im* (for *tupšikku*): IM 49949: 3
2,13,20 *ù* 1,15 *ša* GIŠ-DUB-ÍL: YBC 5022: 45

See *šīlūtu*.

ṭābiḫu (= ^{lú}**gír-lá**) 'butcher'
dub ^{Id}**en**-*na-din*-**ibila** ^{lú}**gír-lá** ^{d}**amar-utu** ^{lú}*šà-suk-ku*: BM 47860 rev. 13–14

ṭēnu (= ***àra****) 'ground, milled'*
zíz àra: BM 47860 rev. 11

ṭīṭu (= **im**) 'clay'
ṭi-da-am: Haddad 104 #7 iv 4; #8 iv 14
ṭi-dam: Haddad 104 #8 iv 8
ṭi-di: Haddad 104 #7 iii 42; #8 iv 7
ṭi-di-im: Haddad 104 #8 iv 10
im: YBC 5022: 46

coefficients:
2[...],30 *sa* **im**: YBC 5022: 46

ṭuppu (= **dub**) 'tablet'
dub: AO 6484 [colophon]
dub-(x)-**kam**(-*ma*): A 24194 rev. v 8
VAT 7528 colophon
YBC 4668 colophon
YBC 4673 colophon
YBC 4695 colophon
YBC 4697 colophon
YBC 4698 colophon
YBC 4708 colophon
YBC 4709 colophon
YBC 4710 colophon
YBC 4711 colophon
YBC 4712 colophon
YBC 4713 colophon
YBC 4714 colophon
YBC 4715 colophon
[du]b ḫa-la: K 11097: 1
ki pī **dub gaba-ri** *la-ba-ri-šú ša-ṭir-ma ba-ri*: BM 47860 rev. 12–13
dub PN: AO 6484 colophon
BM 47860 rev. 13–14

ṭupsarru (= ^{lú}**umbisag**, ^{lú}**dub-sar**) 'scribe'
^{lú}**umbisag** *Enuma* ^{d}*Anu* ^{d}**en-líl-lá**: AO 6484 colophon

ubānu (= **šu-si**) 'finger'
ši-kin **šu-si**: K 11097: 5, 8, 9, 14
K 8705: 6

ud-dà-sar 'kite'(?)
gán ud-dà-sar: BM 15285 Text K

uḫru 'behind', referring to strings of an instrument
coefficients:
sa 3 *uḫ-ri ù* **sa** 5-*šú* 7,5 **sa** *še-e-rì*
sa *ša*-**ge**$_6$ *ù* **sa** 4 *uḫ-ri* 2,6 **sa** *i-šar-tum*
sa *qud-mu-ú ù* **sa** 4 *uḫ-ri* 1,6 **sa** *šal-šá-tum*
sa 3(!)-*šú* **sig** *ù* **sa** 3-*šú uḫ-ri* 3,4 **sa** *em-bu-bu*
sa *šá*-**ge**$_6$ *ù* **sa** 3-*šú uḫ-ri* 2,4 **sa** 4-*tu*: CBS 10996 obv. i 12'–16'
sa 4 *uḫ-ri ù* **sa** 3-*šú* **sig** 6,3 **sa** [...]: CBS 10996 obv. i 21'
sa 3-*šú uḫ-ri ù* **s[a** 7,x **sa**...]: CBS 10996 obv. i 23'

Ú-KIN (= ?) 'work or volume output'
Ú-KIN: YBC 4673 rev. iii 1, 3, 6, 7, 10, 14

U + KUL-x (reading and meaning uncertain)
56 **igi-gub** U + KUL-x: A 3553: 9

ummānu (= **ugnim**) 'troop, army'
ugnim: Erm 15073 rev. i 3

ummânu (= **um-me-a**) 'scholar'
taš-nin-tu$_4$ **um-me-a**: K 11097: 2

ūmu (= **u**$_4$) 'day'
u$_4$-*mu*: CBS 10996 obv. iii 12'–13'
u$_4$-*ma* **ge**$_6$ mng. 'always': K 11097: 6
ú-ma-am: Strssbg. 366 obv. 1 [Thureau-Dangin (*TMB*: 205) translated *ú-ma-am* as 'today', in contrast to *ur-ra-am* 'tomorrow'.]
ú-ma ka-li-a(?)-*am*: IM 54538 obv. 4
ú-ma-ka-lu-tum: IM 54011 obv. 5
ú-ma-ka-lu-tu-un(?): IM 54538 rev. 3 [For this reading see von Soden 1952: 51.]
i-na ú-ma-ka-al: IM 54538 rev. 5
u$_4$-*mi*: Haddad 104 #8 iv 11
ki ma-ṣi **u**$_4$-*mi*: VAT 8528 obv. i 5

i-na ki-ma-ṣu-mi (for *i-na ki ma-ṣi* **u**$_4$*-mi*): BM 85194 rev. i 48
1 **u**$_4$*-mi-im*: Haddad 104 #10 iv 37
ka-la **u**$_4$*-mi-im*: Haddad 104 #7 iii 40, iv 5; #8 iv 8, 14
u$_4$*-mi-ia*: AO 8862 iv 19, 21
u$_4$*-mu-ú*: AO 8862 iv tabulation
u$_4$*-mu-ú-a*: AO 8862 iv 20
u$_4$*-ma-tim*: Strssbg. 366 obv. 3
u$_4$: AO 6770 obv. 17 (**mu dù** *ù* **u**$_4$)
AO 8862 iv 21 tabulation
BM 85196 rev. ii 5
M 8613 obv. 2 (2 **še** *i-na* **u**$_4$ *pa-né-e*)
UET 3: no. 1386: 10
VAT 7528 obv. i 10
YBC 4657 rev. 11
YBC 4662 rev. [4]
YBC 4666 obv. 16, 24; rev. 19
YBC 7164 obv. 9, 28
u$_4$-(x)(-**kam**): AO 8862 iv 21 tabulation
BM 85194 rev. i 47 (*i-na* **u**$_4$-1-**kam**), 50 (*i-na* **u**$_4$-1-**kam**)
BM 85196 rev. i 7; rev. ii 7
M 6861: 16
M 8613 obv. 3, 4, 5, 6, 7, 8; rev. 2', 3', 4'
TMS, Text 5 colophon
UET 3: no. 1386: 9 (*kal* 1-**še u**$_4$-60-**kam** 3⅔ <**sar**> 5 **gín**)
VAT 7528 obv. i 11, 17, 23
VAT 8528 rev. i 11 (*i-na* **u**$_4$-1-**kam**)
YBC 4657 rev. 12, 13, 15, 16, 18, 20, 22
YBC 4662 rev. 10, 12, 19, 25, 29, <31>
YBC 4666 obv. 16, 18, 24; rev. 19, 21
YBC 4673 obv. ii 8
YBC 7164 obv. 28; rev. 2
íb-tag$_4$ **u**$_4$: YBC 4669 rev. iii 16
igi-(x)-**gal** (⅔) u$_4$: BM 85196 rev. ii 4
YBC 4669 rev. iii 14
YBC 7164 obv. 17, 18
igi-te-en u$_4$: YBC 4669 rev. iii 10, 12
YBC 4673 obv. ii 14, 16
YBC 7164 obv. 16, 17
ud-da: VAT 7528 obv. i 15, 16, 17, 21, 22, 23
Note: **itu** 1 7½ **kam** (mng: 37½ days): YBC 7164 obv. 9, 11

coefficients:

40 **u$_4$**-*mu a-na* **ge$_6$** *na-pa-lu*
40 **ge$_6$** *a-na* **u$_4$**-*mu na-pa-lu*: CBS 10996 obv. iii 12'–13'

ÚR-BAD (meaning uncertain); 'rectangle(?)' (See Saggs 1960: 140–41 and Vaiman 1963: 85.)

ÚR-BAD: BM 15285 Text K, L

urḫu 'road, path'
coefficients:

20,6,16 **gán** *ur-ru-uḫ* **sag**-3 **ki-min igi-gub**-*ú*: BM 47860 obv. 2

The meaning of translations, 'volume of a path of three heads (or beginnings)' or, alternatively, 'volume of the three headed Urruḫ' (Livingstone 1986: 21), eludes the author. The phrase appears in a list of granary coefficients. Perhaps, the phrase is explained by the summary in rev. 4–8, which describes 'the circle of the evening meal of Ekur' and 'every one of the shrines... all of the cult platforms'. Note **sag**-(5, 6, 7) is used to describe polygons (see Kilmer 1990b: 86; Vaiman 1963: 82–83).

urru 'day'

ur-ra-am : Strssbg. 366 obv. 1

Thureau-Dangin (*TMB*: 205) translated *ur-ra-am* as 'tomorrow' in contrast in *ú-ma-am* 'today'.

uskaru [= UD-SAR > **u$_4$-sakar$_x$**$^{(sar)}$] 'crescent'

ús-ka-rum: IM 52916, IM 52685, IM 52304 Section C: rev. 13'
ús-ka$_4$-ri: *TMS*, Text 3: 7–9

gán-u$_4$-sakar$_x$: 'segment of a circle':

gán u$_4$-sakar$_x$: A 3553: 4
BM 15285 Texts K, O, P
BM 85210 rev. i 5, 9, 16, 17
MLC 1354 obv.1
YBC 5022: 21, 52, 54
YBC 7243: 8
gán-ud-a(?)**-sar**: YBC 7243: 7
ud-gán-sar: BM 85194 rev. iii 1

coefficients:

ús-ka-rum 2,30 *i-gi-gu-bu-šu ta-al-li ús-ka-rum*(!) [...*i-gi-gu-bu-šu*]:
IM 52916, IM 52685, IM 52304 Section C: rev. 13'

15 **igi-gub** *šà ús-ka$_4$-ri* (relates to the area of an *uskarum*): *TMS*, Text 3: 7
40 **ri** *šà ús-ka$_4$-ri* (relates to the diameter of an *uskarum*): *TMS*, Text 3: 8
40 *pi-ir-ku šà ús-ka$_4$-ri* (relates to the diameter of an *uskarum*): *TMS*, Text 3: 9
10,15,36 **gán u$_4$-sakar$_x$**: A 3553: 4
45(?) *ù* 10 *ša* **gán u$_4$-sakar$_x$**: YBC 5022: 21
5,24(?) **gán-ud-a(?)-sar**: YBC 7243: 7
4,20 *ša* **gán u$_4$-sakar$_x$**: ki-2: YBC 5022: 52
45 *ša* **gán u$_4$-sakar$_x$**: ki-3: YBC 5022: 54
14,36,10 **gán u$_4$-sakar$_x$**: *ki*(?) *in*-ŠUB: YBC 7243: 8

refers to **gán u$_4$-sakar$_x$**:
15 **igi-gub-ba**: BM 85210 rev. i 18 (?) (see *TMB* no. 98: 18)
MLC 1354 obv. 6

Goetze (1948: 34) argued that the reading is **gán-ud-sar** because a crescent-shaped area is indicated in the mathematical texts. For a discussion of **gán** with objects, see s.v. *eqlu*.

Kilmer (1985; 1990b: 86) showed that *uskaru* is a figure based on a half-circle. Vaiman (1963: 77–78) used formulas and illustrations to explain the 'half-moon'.

utūnu 'oven'
ú-tu-num: YBC 7997 obv. 1

uṭṭetu 'corn'
uṭ-ṭe$_4$-tim: BM 85196 obv. i 32
uṭ-ṭe$_4$-ti-ša: BM 85194 rev. iii 27, 29

zabalu (= **íl**) 'to transport'
za-ba-lu: CBS 10996 obv. iii 31'–34'
[*a-na n*]*u-zu-ri-im ù za-ba-lim*: BM 85196 rev. i 10
saḫar za-ba-li: *TMS*, Text 3: 42, 46
šu-ri za-ba-li: *TMS*, Text 3: 69
iz-bi-[*lam*]: BM 85196 rev. i 7
sig$_4$ *iz-bi-l*[*am*]: YBC 10722 rev. 2
sig$_4$ *iz-bi-la-am-ma*: AO 8862 iii 29–30; iv 5–6
i-za-bi-lam: BM 85196 rev. i 3
ta-az-bi-lam: Haddad 104 #10 iv 32
a-za-bi-il: IM 52916, IM 52685, IM 52304 Section E: rev. 25'–29'

saḫar *i-za-ab-bi-il*!: *TMS*, Text 3: 71
za-bi-lam-ma: Haddad 104 #10 iv 27
íl: YBC 5022: 45
íl-íl(*-ma*): YBC 4673 obv. ii 3, 28
(**ḫé-**)**íl-íl**: YBC 4669 rev. iii 9, 13, 17
saḫar íl-íl(*-ma*): A 3553: 35
YBC 4673 obv. ii 12, 15
sig$_4$ íl-íl: A 3553: 29
YBC 4673 obv. i 25
(x) **šu-si sig$_4$ íl**-*ma*: YBC 4673 obv. i 7, 10
(x) **šu-si sig$_4$ íl-íl**-*ma*: YBC 4673 obv. i 16–17, 20–21

coefficients:
5 *šà šu-ri za-ba-li*: *TMS*, Text 3: 69
10 *šà* **saḫar** *za-ba-li* **éš-kàr lú**-1: *TMS*, Text 3: 46
20 *šà* **saḫar** *za-ba-li*: *TMS*, Text 3: 42
[x],30 **saḫar** *i-za-ab-bi-il*!: *TMS*, Text 3: 71 [See von Soden 1964: 47 for this reading.]
a-na 40 **ninda** *a-za-bi-il* 2,13,20 *al-lu-um*
a-na 20 **ninda** *a-za-bi-il* 5 *al-lu-um*
a-na 15 **ninda** [*a-za-bi-il* x] *al-lu-um*
[*a-na*] 10 **ninda** *a-z*[*a-bi-il*] 6,40 *al-lu-um*
a-na 5 **ninda** *a-za-*[*bi-il*] 10 al-lu-um: IM 52916, IM 52685, IM 52304 Section E: rev. 25'–29'

See: 2,13,20(?) *ù* 1, 15 *ša* GIŠ-DUB-ÍL: YBC 5022: 45 YBC 5022: 45 [Neugebauer and Sachs (*MCT*: 135) read GIŠ-DUB-ÍL without translating the expression. Kilmer (1960: 289 n. 1) noted that the DUB looks like AB but ^{giš}**al** = *allu* is expected.]
4,30 *za-ba-lu* 6 **(gur) in-nu gún ki[-min]**
12,30 *za-ba-lu* 1 **(gur) in-nu gún [ki-min]**
3,45 *za-ba-lu* 2,40 6,40 [...]
[...],3 *za-ba-lu* 1 **bariga** š[e(?)...]: CBS 10996 obv. iii 31'–34'
4,30 **igi-gub sig$_4$ íl-í[l]**: A 3553: 29
10(?),1,20(?) **igi-gub saḫar íl-il sig$_4$ dù**: A 3553: 35

zakar šumi 'fame'
[*a-tar-tum za-k*]*ar šu-mi*: K 8705: 5

See *atartu*.

zaqāpu 'to plant'
qa-na-am [...] *za-qa-pa-am*: IM 52916, IM 52685, IM 52304 p. 155: Section F (b) 2

zaqpu 'planted'
zaq-pi: VAT 7848 obv. 20

zarû (agricultural activity)
[*iš*]-*ka-ar za-ri-i ša* **še**-*im e-pé-š*[*a-a*]*m*: IM 52916, IM 52685, IM 52304 Section I(d): 5

zarû [= [(giš)]**zarà**] 'wagon shaft'
coefficients:
12,30 **gán zarà**: YBC 7243: 5
12,30 **igi-gub gán** [giš]**zarà**: A 3553: 5
22,30 *ù* 12,30 *ša* **gán zarà**: YBC 5022: 49
3,30 *ša* **gán zarà ki**-2: YBC 5022: 51

Both meaning and reading are uncertain.
Kilmer (1960: 287 n. 3) argued that **giš** with **zarà** suggests [giš]**za-ra** = *zarû* 'shaft' of chariots, wagons, harrows and doors.

zâzu 'to divide (property), delimit shares'
ki-pa-tam a-na ši-na at-ḫi za-za-am e-pé-ša-am: IM 52916, IM 52685, IM 52304 Section D: rev. 19'
6 **šeš-meš** *i-zu-zu-uš*: YBC 4608 rev. 14
a-na te-še-et zu-uz: VAT 7621 obv. 2, 3
a-na 10 *zu-uz*$_5$: BM 85194 rev. ii 3

mitḫariš zâzu 'to divide into equal shares'
mi-it-ḫa-ri-iš l[*i-zu-uz*]: *TMS*, Text 26: obv. i 13
mi-it-ḫa-ri-iš zu-uz: VAT 7531 obv. 4, 18–19; rev. 4–5
VAT 7621 obv. 5

zēru (= **še-numun**) 'seed (of cereals and plants)'
še-numun: AO 6484 obv. 10, [13]
13,30 **še-numun**: BM 47860 rev. 10

zibbatu (= **kun**) 'storage basin of a canal'
kun: Erm 15073 rev. i 2, 14

zikru 'man'

kiri$_6$ *a-na qa-ti zi-ki-ri iš-ša-ka-*[*a*]*n*: *TMS*, Text 3: 70

ziqpu 'a container for sesame'
coefficients:

⌜3⌝,20 *zi-qip* **še-giš-ì**
2,30 *zi-qip* **še-giš-ì**: CBS 10996 obv. iii 2'–3'

ziqpu is a jar with a pointed bottom (*CAD* s.v. *ziqpu*); according to the Hh references *ziqpu* is a stone stand for the container.

zirīqu 'water-drawing device'
coefficients:

1,12 **a-meš** [giš]*zi-rí-qum*: CBS 10996 obv. iii 22'

zirīqu was later replaced by *zuruqqu*, as a simple apparatus for lifting water from a canal or well by the use of buckets; *zirīqu* appears beside makûtu 'poles' and *šummānu* 'beam for the rope' (*CAD* s.v. *zirīqu*).

Maekawa (1986b: 119–21) showed that the *ziriqum* apparatus was used essentially for irrigation of an orchard according to Ur III references.

zittu (= **ḫa-la**) 'share'

zi-it-tum: YBC 5022: 63
(x) *zi-it-ti*: IM 31210 §6: 2, §7: 5, 7
Strssbg. 362 obv. 9
zi-ti: IM 31210 §3: 3, §5: 6, 9, 11 (*zi-ti-šu*)
ḫa-la: A 3553: 44
K 11097: 1, 4
LBAT: no. 1648 obv.(?) 1', 2', 3', 4', 7', 8', 9', 11', 12'
Strssbg. 362 obv. 4 (**ḫa-la ki**-8)
TMS, Text 10: 15
TMS, Text 24: 41 (**ḫa-la** *si/sì-ra-ti*) [von Soden (1964: 50) suggested emending the text to **ḫa-la** *aš!-ši-ra-ti*.]
UET 5: no. 121: 3 (**igi-5-gal ḫa-la-kam**)
VAT 6597 obv. i 11; obv. ii 2, 5, 7, 10, 11, 12, 13; rev. 5, 6 (**ḫa** < **-la** >), 6, 8, 9
ḫa-la gal: *TMS*, Text 10: 4, 5, 9, [25], 26, 28, 31
ḫa-la tur(*-ma*): *TMS*, Text 10: 5, 6, 9, [24], 27, 28, [31]
ḫa-la-ba: A 3553: 12

coefficients:

1 *zi-it-tum*: YBC 5022: 63

[1] **igi-gub ḫa-la**: A 3553: 44
15 **igi-gub ḫa-la-ba**: A 3553: 12

Nemet-Nejat (1982: 425) noted that *zittu* (= **ḫa-la**) denotes both the object to be divided (real estate, assets, income, etc.) and the result of that division (the part, the share, etc.), as a consequence of inheritance or jointly owned property. The hapax legomenom *zuzātu* 'shares' may be the syllabic writing for **ḫa-la-meš** or **ḫa-la-ḫi-a** (Nemet-Nejat 1982: 426).

Kilmer (1960: 301 n. 4) argued that the sequence **ḫa-la** and **gi-kud-a** in A 3553: 44–45 and YBC 5022: 63–64 imply that *zittu* refers to the same type of problem as the 'cut reed', a suggestion is not borne by the cuneiform mathematical problem texts.

Broken (coefficient lists):
3,35 [...]-*tum*: YBC 7243: 42
48 1,40 1,20 *ú-k/lu*-[x x]-*um*: YBC 7243: 43

PERSONAL NAMES

[Id]*Anu-aba-utter*

[*ṭuppi* [I]][d]*Anu-aba-utēr* [lú]**umbisag** *Enūma* [d]*Anu* [d]**en-líl-lá ibila** *š*[*á* [I]][d]*Anu-bel-š*[*ú-n*]*u* [*apil* [Id]*Sîn-leqe-unninni*]: AO 6484 colophon

[Id]*Anu-bēlšunu*

[*ṭuppi* [I]][d]*Anu-aba-utēr* [lú]**umbisag** *Enūma* [d]*Anu* [d]**en-líl-lá ibila** *š*[*á* [I]][d]*Anu-bel-š*[*ú-n*]*u* [*apil* [Id]*Sîn-leqe-unninni*]: AO 6484 colophon

[Id]*Bēl-nādin-apli*

dub [Id]**en**-*na-din*-**ibila** [lú]**gír-lá** [d]**amar-utu** [lú]*šà-suk-ku*: BM 47860 rev. 13–14

Dāda-[d]*Šušinak*

im-gíd-da <*da*>*-da* [d]***múš-eren*** *ù* **XXX**-*iš-me-an-ni*: *TMS*, Text 5 colophon

Dādum

šu-ti-a *Da-di-im*: M 6861: 14

Darius

[I]*da-a-ri-ja-uš*$_{11}$ **lugal šú**: BM 47860 rev. 16

Ipquša

Ip-qú-ša: YBC 7273 rev.

Ili-[...]-[...]

Ì-lí-[...]-[...]: YBC 7273 rev.

Sîn-išmeanni

im-gíd-da <da>-da ^{d}**múš-eren** *ù* **XXX**-*iš-me-an-ni*: *TMS*, Text 5 colophon

^{Id}*Sîn-leqe-unninni*

[*ṭuppi* ^{I}]^{d}*Anu-aba-utēr* ^{lú}**umbisag** *Enūma* ^{d}*Anu* ^{d}**en-líl-lá ibila** *š*[*á* ^{I}]^{d}*Anu-bel-š*[*ú-n*]*u* [*apil* ^{Id}*Sîn-leqe-unninni*]: AO 6484 colophon

^{I}*Warad*-[...]

š[u i]r-[...]: VAT 6505 colophon

DIVINE NAMES

Antum

i-na a-mat ^{d}*Anu u An-tum liš-lim*: AO 6484 u.e.

^{d}*Anu*

i-na a-mat ^{d}*Anu u An-tum liš-lim*: AO 6484 u.e.

Cf. ^{d}**Ea-banû** 'Ea-creator', which refers to the fourth string of an instrument

sa ^{d}*é-a*-**dù** *ù* **sa** *qud-mu-ú* 4,1 **sa šub-múrub**: CBS 10996 obv. i 17'
sa *šá*-**ge$_{6}$** *ù* **sa** ^{d}*é-a*-**dù** 2,4 **sa** ⌜*titur*⌝ **murub$_{4}$**-*tú*: CBS 10996 obv. i 20'
sa ^{d}*é-a*-**dù** [*ù* **sa** 4,x **sa**...]: CBS 10996 obv. i 24'

^{d}*Gula*

a-šà *gu-la*: *UET* 5: no. 864: 16

^{d}**Igi-kur**

itu ^{d}**igi-kur**: M 6861: 15

^{d}*Iltum*

a-di **udu-ḫi-a** ^{d}*I*[*l-tim la ú-wa-š*]*e-ru*: M 6829: 39' // *ARM* 24: no. 260
[*i-nu-ma* **lú** *ik-ri-bi-šu* **udu-ḫi-a** *ma-da-tim a-na* ^{d}*Il-tim i-ka-ar-ra-ab*...]: M 6829: [1''–3''] (restored by // text *ARM* 24 #260: 1–3)

ma-ḫa-ar ^{d}[*Il-tim*]: M 6829: 41' // *ARM* 24: no. 260)
[*si-im-ta-am sa* ^{d}*Il-tim*]: M 6829: [42'ff.] (restored by // text *ARM* 24: no. 260)

^{d}*Marduk*
uru^{ki} *na-ki-ir* ^{d}*Marduk a-ṣa-ba-at*: BM 85194 rev. ii 7, 22
BM 85210 obv. ii 1 (<**uru**^{ki}>)
uru^{ki} na-ki-ir ^{d}*Marduk lu-úṣ-ba-at*: BM 85210 obv. i 3
dub ^{Id}**en**-*na-din*-**ibila** ^{lú}**gír-lá** ^{d}**amar-utu** ^{lú}*šà-suk-ku*: BM 47860 rev. 13–14
pa-liḫ ^{d}**amar-utu:** BM 47860 rev. 15

^{d}*Nabû*
BM 85210 right edge

^{d}*Nisaba*
AO 8862 u.e.
BM 85210 right edge
YBC 5022: 66

^{d}*Sîn*
12 **mú** *sa* ^{d}**XXX**
12 **šú** *sa* ^{d}**XXX**: CBS 10996 obv. iii 10'–11'

SUMERIAN CROSS-REFERENCE TO AKKADIAN GLOSSARY

a(-meš) = *mû*
á = *idū*
ab-sín = *šer'u*
ad = *abu*
ad-KID = *atkuppu*
aga = *agû*
ága = *middatu*
a-gar$_{5}$ = *abāru*
al = *allu*
an-bar = *parzillu*
an-na = *annaku*
ANŠE (meaning unknown)
àra = *ṭênu*
a-šà = *eqlu*

bà = *amūtu*

bàd = *dūru*

dù = *banû*

dù = *kalāma*

dub = *ṭuppu*

dug = *karpatu*

dug-ga = *ṭābu*

DUG-GU-ZI (meaning uncertain)

dumu = *māru*

dusu = *tupšikku*

e = *iku*

é = *bītu*

é-dù-a = *bītu epšu*

é-kišib = *bīt kunukki*

é-sig$_4$ = *igāru*

é-uš-gíd-da = *našpaku*

erim = *ṣābu*

éš-kàr = *iškaru*

esir = *ittû*

esír = *ittû*

esír(-è-a) = *kupru*

esír-ḫi-a = *kupru*

gaba-ri = *gabarû*, *māḫiru*

GÁ-DUB ŠE (meaning unknown)

gal = *rabû*

gán = *eqlu*

ganba = *maḫīru*

GÁN(?)-GÁ ŠE(?) (broken)

GÁN GI[Š...] (broken)

GÁN-MAN (meaning unknown)

ge$_6$ = *mūšu*, *mušītu*

geštú zà-mí = *ḫasīs sammî*

gi = *qanû*

GI BUL IR (meaning unknown)

gi-gur = *pānu*

gi-kud-a = *qanû ḫaṣbu*

gi-níg-sa-ḫi-a = *kiššû*

GI-NÍGIN-NA (meaning unknown)

gi-sa = *kiššû*

gi-sa-ḫi-a = *kiššû*

gi-šu-kin = *šūru*

gìr-gub(-ba) = *kilzappu*
gir$_4$ = *kīru*
gir$_4$-ad-KID = *kīr atkuppi*
giš = *iṣu*
giš-a-bal = *dilūtu*
gis-á-lá = *dilūtu*
giš-al = *allu*
giš-ba-rí-ga = *pars/šiktu*
giš-ban = *qaštu*
giš-bán = *sūtu*
giš-dib-dib = *dibdibbu*
GIŠ-DUB-ÍL (meaning unknown)
giš-dusu = *tupšikku*
giš-erin = *erinnu*
giš-gi = *apu*
giš-gu-za = *kussû*
giš-ig = *daltu*
giš-illar = *tilpānu*
giš-KAK+LIŠ = *saparru*
giš-kár = (no known Akkadian equivalent)
(giš-)kun$_4$ = *simmiltu*
GIŠ-LAGAB (meaning uncertain)
giš-má = *eleppu*
giš-má-gur$_8$ = *makurru*
giš-má-lá = *malallû*
giš-mar-gíd-da = *ereqqu*
giš-ná = *eršu*
giš-ninda = *akalu*
giš-nu-úr-ma = *nurmû*
giš-šub-ba = *isqu*
giš-tir = *qištu*
giš-ù-šub = *nalbantu*
giš-zà-mí = *sammû*
(giš-)zarà = *zarû*
gu$_4$ = *alpu*
gul-gul = *naqāru* (I/3)
gún = *biltu*
gur$_7$ = *karû*
guškin = *ḫurāṣu*
ḫa-la= *zittu*
ḫar = *ḫasû*

ḫar = *semeru*
ḪI-*rum* (meaning uncertain)
ì-dub = *našpaku*
ì-giš = *šamnu*
ì-sag = *šaman rūšti*
ì-šaḫ = *nāḫu*
ì-udu = *lipû*
ibila = *aplu*
íd = *nāru*
ìd = *nāru*
igi = *maḫrû*
igi-duḫ-a = *tāmartu*
igi $\textbf{gu}_4$ = *īn alpi*
íl = *našû, zabālu*
im = *ṭīṭu, ṭiddu*
im-babbar = *gaṣṣu*
im-dù-a = *pitiqtu*
im-dugud = *kubtu*
im-gíd-da = *imgiddû*
im-KI(?)-**a** (meaning unknown; reading uncertain)
im-lá = *imlû*
im-lá = *kirbānu*
im-šu = (no known Akkadian equivalent)
in-nu = *tibnu*
in-nu-da = *tibnu*
itu = *arḫu*
iz-zi, var. **i-zi** = *igāru*
ká = *bābu*
kak = *sag-ga-kum*
kàr = *iškaru*
ki = *ašru*
ki-gub = *manzazu*
ki-lá = *šuqultu*
ki-lá = *kalakku*
ki-lá-bi (see **im-lá**) meaning unknown
kí-sa = *kisû*
kin = *šipru*
kin-sig = *naptanu*
$\textbf{kir}_4$**-dib/dab** = *kartappu*
$\textbf{kiri}_6$ = *kirû*
kù = *kaspu*

kù-babbar = *kaspu*
kud = *ḫaṣābu*
kun = *zibbatu*
lá = *šaqālu*
làl = *dišpu*
lìl = *ṣēru*
lugal = *šarru*
lú = *amēlu*
lú-dub-sar = *ṭupšarru*
lú-gír-lá = *ṭābiḫu*
lú-ḫal = *bārû, bārûtu*
lú-ḫun-gá = *agru*
lú-umbisag = *ṭupšarru*
mar = *marru*
máš = *ṣibtu*
MÌN (meaning uncertain)
mu = *šattu*
mu = *šumu*
mú = *napāḫu*
murub$_4$-*tu* = *qablītu*
na$_4$ = *abnu*
na$_4$-kur-ra = *aban šadî*
níg-šid = *nikkassu*
nin = *aḫatu*
NU-KAR-[x]-DÈ(?) (reading uncertain)
nu-luḫ-ḫa-sar = *nuḫurtu*
nun-me = *apkallu*
pa$_5$ = *pattu*
pa$_5$-sig = *atappu*
PEŠ in PEŠ **sig$_4$-al-ùr-ra** (meaning unknown)
PI-*ya* (meaning unknown)
PI-*lu-ú-tum ša* G[I...] (meaning unknown)
pisan = *pisannu*
sa(-a) = *pitnu*
sa-li = *pagû*
sag = *rēšu*
SA-[GIŠ?]-KIN (broken)
SAG-KI-GUD (meaning uncertain)
saḫar = *eperu*
sal-šeš = *niṣirtu*
síg = *šipātu*

síg-ga-zum-ak(-a) = *pušikku*
sig$_4$ = *libittu*
sig$_4$.áb = *arḫu*
sig$_4$-al-ùr-ra = *agurru*
sig$_4$-anše = *amaru*
sìla = *qû*
sila$_4$-gub = *lillidu*
sipa = *rē'û*
sud-ág = *elmešu*
suḫ$_4$ = *suḫḫu*
suḫuš = *išdu*
sum = *nadānu*
ša-**ge$_6$** = *šamūšu*
šá-**ge$_6$** = *šamūšu*
ša-KAK(-)UD-*šu-rum* (meaning unknown) Akkadian word?
šám = šâmu
šám = *šīmu*
še = *še'u*
še-ba = *ipru*
še-giš-ì = *šamaššamū*
še-ì-giš = *šamaššamū*
še-lú-sar = *kisibirru*
še-numun = *zēru*
šeš = *aḫu*
šita$_3$ = *rāṭu*
šitim = *itinnu*
šu = *qātu*
šú = *rabû*
šub-murub$_4$ = *nīd qabli*
šuk, šuku = *kurummatu*
šu-si = *ubānu*
šu-ti-a = *namḫartu*
tag = *maḫāṣu*
túg = *ṣubātu*
túl = *būrtu*
tur = *ṣeḫru*
tùr = *tarbaṣu*
u$_4$ = *ūmu*
u$_4$-sakar$_x$ = *uskaru*
ub =*tubqu*
ud-dà-sar (meaning unknown)

ugnim = *ummānu*
u_8-udu-ḫi-a = *ṣēnu*
udu = *immeru*
udu-níta = *immeru*
Ú-KIN (meaning unknown)
U + KUL-X (reading and meaning uncertain)
uku-ús = *rēdû*
umbin = *ṣupru*
um-me-a = *ummânu*
úr = *išdu*
ÚR-BAD (meaning uncertain)
ur-maḫ = *nēšu*
ur_5-ra = *ḫubullu*
uru = *ālu*
uru-ki = *ālu*
urudu = *erû*
zabar = *siparru*
zà-mí = *sammû*
ZA-ZUM (Akkadian equivalent uncertain, possibly *sassum*)
zarà = *zarû*
zíz =*kunāšu*
zú-lum = *suluppū*

Appendix 2

Bibliography for Cuneiform Mathematical Texts

The bibliography of each tablet is organized as follows:
(1) *MCT*: tablets are listed by museum number
(2) *MKT* I–III: tablets are listed by museum number
(3) *TMS*: tablets are listed by text number
(4) The Coefficient Lists: tablets are listed by museum number; the tablets listed in this category are not recorded in this bibliography either in basic text volumes in which they appear or under the headings of Miscellaneous, Tell Harmal, etc..
(5) Tell Harmal: tablets are listed by museum number
(6) Mari: tablets are listed by museum number
(7) Miscellaneous: tablets are listed in alphabetical order by museum number with the following exceptions:

- (a) *LBAT* (Sachs 1955)
- (b) *UET* 3 (Legrain 1947)
- (c) *UET* 5 (Figulla and Martin 1953)
- (d) *UMB* 16/2 (Steele 1951)

(8) Old Akkadian texts are listed in alphabetical order by museum number with the following exceptions:

- (a) *BIN* 8 (Hackman 1958)
- (b) *DP* (Allotte de la Füye 1920)
- (c) Limet, *Liège* (Limet 1973)
- (d) *MAD* 5 (Gelb 1970)
- (e) *TMH* 5 (Pohl 1935)
- (f) *TSŠ* (Jestin 1937)

Each tablet or bibliographic entry includes photo, copy and bibliography; the first item(s) in the bibliography is (are) the text edition (see Nemet-Nejat 1989). In the category "Coefficient Lists" each tablet includes photo, copy and text edition; however, the bibliography is for the coefficient lists in their entirety because of the overlapping nature of the information contained in the lists.

MCT

A 24194

Photo: *MCT* pl. 41.
Copy: *MCT* pl. 15.

Bibliography: *MCT*: 107–19 • Friberg 1979b; 1981b • Gandz 1937 • *MKT* I: 455 (d) • *MKT* III: 45–48 (YBC 4711) • See A 24195.

A 24195

Photo: *MCT* pl. 42.
Copy: *MCT* pl. 16.
Bibliography: *MCT*: 119–29 • Goetsch 1968–69 • *MKT* III: 34–39 (YBC 4695) • See A 24194.

K 8705

Copy: *MCT* pl. 19.
Bibliography: *MCT*: 140 • Borger 1957 (duplicate of K 8865) • Jeyes 1989 • Reiner 1985 • Weidner 1917.

K 11097

Copy: *MCT* pl. 19.
Bibliography: *MCT*: 139–40 • Borger 1957 • Jeyes 1989 • // K 6055 (unpublished; *HKL* 2: 205) • Reiner 1985 • Weidner 1917.

MLC 1354

Photo: *MCT* pl. 47.
Copy: *MCT* pl. 21.
Bibliography: *MCT*: 56–57 • Goetze 1948 • Lewy 1947 • Seidenberg 1984–85.

MLC 1842

Photo: *MCT* pl. 48.
Copy: *MCT* pl. 22.
Bibliography: *MCT*: 106–07 • Allotte de la Füye 1928 • Curtis and Hallo 1959 • Friberg 1982: 57 [Thureau-Dangin (1937)] • Lewy 1949b: 40–67 • *MKT* I: 513; *MKT* III: 42–45 (YBC 4698) • Nemet-Nejat 1988 • Powell 1979.

MLC 1950

Photo: *MCT* pl. 27.
Copy: *MCT* pl. 2.
Bibliography: *MCT*: 48–49, 42 n. 125b • Becker 1966 • Bruins 1962; 1966 • Goetsch 1968-69 • Gundlach and von Soden 1963 • Neugebauer 1934: 34, 176 • Seidenberg 1984–85.

MM 86.11.404

Photo: *MCT* pl. 43.
Copy: *MCT* pl. 20.
Bibliography: *MCT*: 140–41 • Thureau-Dangin 1924b.

NBC 7934

Copy: *MCT* pl. 2.
Bibliography: *MCT*: 55–56 • Friberg 1982: 22 [Scheil (1915)] • Nemet-Nejat 1988.

NBC 8082

Copy: *MCT*: 10.
Bibliography: *MCT*: 10 • Neugebauer and Sachs 1984.

NCBT 1913

Copy: *MCT*: 10.
Bibliography: *MCT*: 10.

Plimpton 322

Photo: *MCT* pl. 22.
Bibliography: *MCT*: 38–41 • Anagnostakis and Goldstein 1973 • Bruins 1949; 1953b; 1955b; 1962; 1967 • Buck 1980 • Friberg 1981a • Gandz 1929; 1948 • Goetsch 1968-69 • Huber 1957 • Neugebauer 1969: 36–44, 50–51 • Pottage

1974 • Price 1964–65 • Seidenberg 1962b • Schmidt 1980 • Vaiman 1955 • Vygodskii 1959 • van der Waerden 1961: 77–80; 1980.

VAT 7848

Photo: *MCT* pl. 45.
Copy: *MCT* pl. 20.
Bibliography: *MCT*: 141–45 • Bruins 1953b • Geršman 1959 • Nemet-Nejat 1988 • Powell 1984a • Style similar to AO 6484.

YBC 4186

Photo: *MCT* pl. 36.
Copy: *MCT* pl. 11.
Bibliography: *MCT*: 91 • Jacobsen 1981 • Nemet-Nejat 1988 • Powell 1988.

YBC 4607

Photo: *MCT* pl. 37.
Copy: *MCT* pl. 12.
Bibliography: *MCT*: 91–94 • Bruno et al. 1968–69 • Dunham 1982 • Legrain 1913: no. 15; 1935 • Nemet-Nejat 1988 • Powell 1984a • Scheil 1915 • Thureau-Dangin 1903: nos. 137, 413; 1909b; 1932b; 1937b • Torraca et al. 1972.

YBC 4608

Photo: *MCT* pl. 28.
Copy: *MCT* pl. 13.
Bibliography: *MCT*: 49–53 • *MKT* III: 83 s.v. "Reihen" and "Verteilungsaufgaben" • Gandz 1938 • Goetsch 1968–69.

YBC 4612

Photo: *MCT* pl. 40.
Copy: *MCT* pl. 14.

Bibliography: *MCT*: 103–05 • Goetsch 1968-69 • Nemet-Nejat 1988.

YBC 4652

Photo: *MCT* pl. 39.
Copy: *MCT* pl. 13.
Bibliography: *MCT*: 100–03 • Goetsch 1968-69 • Keiser and Nies 1920: pls. 8, 73 [various stone weights and their inscriptions] • Nemet-Nejat 1988 • Oppenheim 1947b • Petruso 1985 • Powell 1971; 1979 • Thureau-Dangin 1921; 1930a: 54.

YBC 4657

Photo: *MCT* pl. 31.
Copy: *MCT* pl. 6.
Bibliography: *MCT*: 66–76 • Goetsch 1968-69 • Nemet-Nejat 1988 • Schwenzer 1915: 102.

YBC 4662

Photo: *MCT* pl. 33.
Copy: *MCT* pl. 8.
Bibliography: *MCT*: 71–76 • Goetsch 1968-69 • Seidenberg 1984-85 • See YBC 4657.

YBC 4663

Photo: *MCT* pl. 32.
Copy: *MCT* pl. 7.
Bibliography: *MCT*: 69–76 • Goetsch 1968-69 • See YBC 4657.

YBC 4666

Photo: *MCT* pl. 34.
Copy: *MCT* pl. 9.

Bibliography: *MCT*: 76–81 • Lewy 1949b: 40–67 • *MKT* III: 82 s.v. **kú ì-kú** 'Böschungen' • Nemet-Nejat 1988 • Powell 1988.

YBC 4675

Photo: *MCT* pl. 26.
Copy: *MCT* pl. 1.
Bibliography: *MCT*: 44–48, 51–52: no. 1 • *MKT* II: 31 s.v. *RI* • Bruins 1955a; 1962 • Høyrup 1984: 105.24–105.33 • Lewy 1938 • Neugebauer 1933–34 • Pottage 1974 • Powell 1984a.

YBC 5037

Photo: *MCT* pl. 30.
Copy: *MCT* pl. 5.
Bibliography: *MCT*: 59–66 • *MKT* I: 389–402 (YBC 4708); cf. *MKT* I: 187–229 • *TMB*: xviii–xix.

YBC 6295

Photo: *MCT* pl. 46.
Copy: *MCT* pl. 22.
Bibliography: *MCT*: 42, 49 n. 135d • Bruins 1954: 56 (IM 54472); 1962; 1971 • Friberg 1982: 64–65 • Goetsch 1968–69 • Høyrup 1984: 105.11–105.12 • Lewy 1949b: 40–67 • Neugebauer 1934: 199–202 • Sachs 1952.

YBC 6492

Photo: *MCT* pl. 48.
Copy: *MCT* pl. 22.
Bibliography: *MCT*: 105–06 • Nemet-Nejat 1988 • Seidenberg 1962b.

YBC 6967

Photo: *MCT* pl. 27.
Copy: *MCT* pl. 17.

Bibliography: *MCT*: 129–30 • Becker 1966: 29–37 • Goetsch 1968–69 • Høyrup 1984: 10–12, 126; 1985a: 58–62; 1990: 262–66 • Nemet-Nejat 1988.

YBC 7164

Photo: *MCT* pl. 35.
Copy: *MCT* pl. 10.
Bibliography: *MCT*: 81–90 • Goetsch 1968–69 • Goetze 1948 • Lewy 1949b: 40–67 • Nemet-Nejat 1988 • Powell 1988 • Salonen 1972: 88–99, pl. 9 • von Soden 1977

YBC 7273

Bibliography: *MCT*: 131 • Kraus 1966 • Nemet-Nejat 1988 • Postgate 1992: 159–61.

YBC 7284

Copy: *MCT*: 97.
Bibliography: *MCT*: 97–98 • Dunham 1982 • Lewy 1949: 137–70.

YBC 7289

Copy: *MCT*: 42.
Bibliography: *MCT*: 42–43 • Bruins 1962; 1971 • Friberg 1981a • Goetsch 1968–69 • Neugebauer 1934; 1969: 35 • Seidenberg 1962b; 1972 • van der Waerden 1961: 45.

YBC 7290

Copy: *MCT*: 44.
Bibliography: *MCT*: 44.

YBC 7302

Bibliography: *MCT*: 44 • Powell 1984a • Seidenberg 1962b.

YBC 7326

Photo: *MCT* pl. 43.
Copy: *MCT* pl. 17.
Bibliography: *MCT*: 130–31 • Goetsch 1968–69 • Kraus 1966 • Nemet-Nejat 1988 • Postgate 1992: 159–61 • von Soden 1937 • Young 1990.

YBC 7997

Photo: *MCT* pl. 49.
Copy: *MCT* pl. 23.
Bibliography: *MCT*: 98–99.

YBC 8588

Photo: *MCT* pl. 46.
Copy: *MCT* pl. 21.
Bibliography: *MCT*: 75–76 • Goetsch 1968–69.

YBC 8600

Photo: *MCT* pl. 47.
Copy: *MCT* pl. 21.
Bibliography: *MCT*: 57–59 • Bruins 1962 • Friberg 1982: 113–14 [Vaiman *DV* 2 (1976)] • Lewy 1947 • Nemet-Nejat 1988 • Powell 1984a • von Soden 1937; 1939 • Waschow 1937.

YBC 8633

Photo: *MCT* pl. 29.
Copy: *MCT* pl. 4.

Bibliography: *MCT*: 53–55 • Friberg 1982: 65 • Goetsch 1968-69 • Høyrup 1984: 105.6–105.14 • Lewy 1949b: 40–67.

YBC 9852
(Duplicate of YBC 4675 rev. 7–16)

Photo: *MCT* pl. 26.
Copy: *MCT* pl. 1.
Bibliography: *MCT*: 44–49 • Bruins 1962.

YBC 9856

Photo: *MCT* pl. 99.
Copy: *MCT* pl. 4.
Bibliography: *MCT*: 99–100 • Gandz 1938 • Goetsch 1968-69 • *MKT* III: 83 s.v. "Reihen" and "Verteilungsaufgaben." • Nemet-Nejat 1988.

YBC 9874

Photo: *MCT*: pl. 36.
Copy: *MCT* pl. 11.
Bibliography: *MCT*: 90–91 • Goetze 1948 • Lewy 1949b: 40–67 • Nemet-Nejat 1988.

YBC 10522

Photo: *MCT* pl. 44.
Bibliography: *MCT*: 131.

YBC 10722

Photo: *MCT* pl. 38.
Copy: *MCT* pl. 4.
Bibliography: *MCT*: 98 • Nemet-Nejat 1988.

YBC 11120

Bibliography: *MCT*: 44 • Powell 1984a • Seidenberg 1962b.

YBC 11126

Bibliography: *MCT*: 44 • Nemet-Nejat 1982.

No Museum Number

Copy: Scheil 1938
Bibliography: *MCT*: 6–10 • Goetsch 1968–69.

MKT I–III

AO 6484

Photo: *MKT* II pl. 1.
Copy: Thureau-Dangin 1922: no. 33: pls. lxi, lxii.
Bibliography: *MKT* I: 96–107 • *TMB*: 76–81 • *MCT*: 143–45 • Bruins 1953b; 1955b; 1962; 1967; 1971 • Friberg 1981a; 1982: 44 [Neugebauer *NGWG* I:43 (1933)] • Gandz 1937; 1947 • Geršman 1959 • Goetsch 1968–69 • Neugebauer 1933–1934 • Neugebauer and Waschow 1932 • Powell 1984a • Schott 1933 • Schuster 1931 • Thureau-Dangin 1932e; 1933d; 1936b • Vetter 1936 • van der Waerden 1961: 77–78 • Waschow and Neugebauer 1932.

AO 6770

Copy: Dossin 1934: no. 154.
Bibliography: *MKT* II: 37–42; *MKT* III: 62–65 • *TMB*: 71–73 • Bruins 1962; 1968; 1971; 1980 • Dunham 1980 • Gandz 1940; 1947; 1948 • Goetsch 1968–69 • Keiser and Nies 1920: pls. 8, 73 [various stone weights and their inscriptions] • Landsberger 1937: tablet 2 i 41–43 • Powell 1971; 1979 • Thureau-Dangin 1936d; 1936e • Torraca et al. 1972 • van der Waerden 1961: 73–74 • Waschow 1936–1937.

AO 8862

Photo: Thureau-Dangin 1932f.
Copy: *MKT* II pl. 35-38.
Bibliography: *MKT* I: 108-23 • *TMB*: 64-71 • Becker 1966: 32-33 • Bruins 1971 • Bruno et al. 1968-69 • Gandz 1937; 1947 • Goetsch 1968-69 • Høyrup 1984: 30-41; 1985a: 124-30; 1990: 309-20 • Neugebauer 1933c; 1933d • Sachs 1946 • Schott 1933 • Seidenberg 1978; 1984-85 • Thureau-Dangin 1931; 1932f; 1932g; 1937b • Torraca et al. 1972 • Vogel 1936 • van der Waerden 1961: 63-65.

AO 10642

Copy: de Genouillac 1925 pl. 6 (C 22).
Bibliography: *MKT* I: 123 • *TMB*: 204.

AO 10822

Copy: de Genouillac 1925 pl. 5 (D 63).
Bibliography: *MKT* I: 123-26 • *TMB*: 204-05 • Dunham 1980 • Scheil 1915 • von Soden 1937; 1939 • Torraca et al. 1972.

AO 17264

Photo: Thureau-Dangin 1934c
Copy: Thureau-Dangin 1934c
Bibliography: *MKT* I: 126-34 • *TMB*: 74-76 • Bruins 1962 • Caveing 1985 • Friberg 1982: 45 [Thureau-Dangin *RA* 31 (1934: 61-69)] • Gandz 1948 • Goetsch 1968-69 • Pottage 1974 • von Soden 1937 • Thureau-Dangin 1934c • Vetter 1936.

BM 13901

Copy: Thureau-Dangin 1936a.
Bibliography: *MKT* III: 1-14 • *TMB*: 1-10 • Bruins 1962; 1966; 1967 • Friberg 1981a; 1982: 60 [Thureau-Dangin *RA* 35 (1938: 71-77)] • Gandz 1937 • Goetsch 1968-69 • Gundlach and von Soden 1963 • Høyrup 1984: 12-15, 20-28, 49-50;

1985a: 62–82, 120–23, 143–44; 1989: 8–14; 1990: 266–80, 306–09 • Thureau-Dangin 1936a; 1936d; 1938b; 1939 • van der Waerden 1961: 68–70; 1980 • Young 1988; 1990.

BM 15285

Photo: *MKT* II pl. 3-4.
Join: Saggs 1960.
Copy: Gadd 1922.
Bibliography: *MKT* I: 137–42 • *TMB*: 53–57 • Bruins 1962 • Caratini 1957 • Friberg 1981b • Gadd 1922 • Goetze 1948 • Høyrup 1984: 21–23; 1985a: 82–83; 1990: 280–81 • Kilmer 1964; 1990b • Lawergren and Gurney 1987 • Leemans and Bruins 1951 • Neugebauer 1933-34; 1934: 168–70 • Saggs 1960 • Sayce 1876 • von Soden 1937; 1964 • Vaiman 1963.

BM 34568

Copy: *MKT* III pl. 1.
Bibliography: *MKT* III: 14–22 • *TMB*: 57-63 • Bruins 1953b; 1962; 1967; 1971 • Friberg 1981a • Gandz 1937; 1947 • Goetsch 1968–1969 • Høyrup 1984: 3–4; 1985a: 1-7, 152–53; 1990: 34–39 • Thureau-Dangin 1909a; 1937d; 1938c • van der Waerden 1961: 74–77, 79; 1980.

BM 85194

Photo: *MKT* II pl. 5-6.
Copy: King 1900, pl. 8-13.
Bibliography: *MKT* I: 142-93 • *TMB*: 21-39 • Bruins 1962 • Delitzsch 1885: 86-90 • Ellis 1970 • Eph'al 1984 • Friberg 1981b • Gadd 1954 • Gandz 1937; 1940; 1947; 1948 • Gelb 1982 • Geršman 1959 • Goetsch 1968-69 • Goetze 1948 • Groneberg 1990 • Høyrup 1984: 53-63; 1989: 35 • Kilmer 1964 • Landsberger 1957: tablets 2, 4 • Lewy 1944 • *MCT*: 96-97 • Michalowski 1977 • Neugebauer 1931d; 1933a; 1933b; 1933-34; 1934: 166-68, 170-71; 1945; 1947 • Neugebauer and Struve 1931 • Neugebauer and Waschow 1932 • Postgate 1978 • Powell 1972; 1982b; 1984a; 1984b; 1985; 1988 • Sachs 1944 • Sasson 1969 • Schuster 1931 • Seidenberg 1972 • Smith 1969 • von Soden 1937; 1939 • Thureau-Dangin 1909b; 1913; 1921; 1932a; 1932b; 1932c; 1932d; 1932e; 1932h; 1933a; 1933b; 1933c; 1933e; 1935a; 1935b; 1937a; 1937b; 1937c; 1937d;

1939 • Ungnad 1916; 1917–18 • Vaiman 1960 • Vetter 1936 • Vogel 1932–33; 1936 • van der Waerden 1961: 75–76; 1980 • Waschow 1932–1933; 1936–37 • Weidner 1916 • Zimmern 1916.

BM 85196

Photo: Thureau-Dangin 1935a: 4.
Copy: Thureau-Dangin 1935a: 2–3.
Bibliography: *MKT* II: 43–59 • *TMB*: 39–46 • Eph'al 1984 • Friberg 1981a 1982: 47 [Thureau-Dangin *RA* 32 (1935: 1–28)] • Gelb 1982 • Goetze 1948 • Groneberg 1990 • Lewy 1944 • *MCT*: 88–90 • Moorey 1985: 116–18 • Neugebauer 1934: 171 • Postgate 1978 • Powell 1978; 1988 • Sachs 1944 • Salonen 1966; 1972 • von Soden 1939; 1977 • Thureau-Dangin 1909a; 1932b; 1935a; 1937b; 1938c; 1939 • Ungnad 1917–18 • Vaiman 1960 • Vetter 1936 • van der Waerden 1961: 76–77; 1980 • Waschow 1936–37.

BM 85200 + VAT 6599

Photo: *MKT* II pl. 7–8.
Copy: *MKT* II pl. 39–40.
Bibliography: *MKT* I: 193–219 • *TMB*: 11–21 • Becker 1966: 34–37 • Bruins 1962; 1971 • Gandz 1937 • Goetsch 1968–69 • Neugebauer 1933–1934; 1934: 193–97 • von Soden 1939 • Thureau-Dangin 1936c; 1936e; 1937c • van der Waerden 1961: 71–72 • Waschow 1936–37.

BM 85210

Photo: *MKT* II pl. 9.
Copy: King 1900 pl. 14–15.
Bibliography: *MKT* I: 219–33 • *TMB*: 46–53 • Bruins 1950; 1962 • Delitzsch 1885: 86–90 • Ellis 1970 • Eph'al 1984 • Friberg 1982: 114 [Vaiman *DV* 2 (1976)] • Gandz 1947 • Goetsch 1968–1969 • Goetze 1948 • Høyrup 1984: 53–63 • Landsberger 1957: tablets 2, 4 • Lewy 1947 • Neugebauer 1933a; 1933b; 1945; 1947 • Neugebauer and Waschow 1932 • Reiner 1970 • Salonen 1972: 88–99, pl. 9 text 202 • Sasson 1969 • Seidenberg 1972 • Smith 1969 • von Soden 1977 • Thureau-Dangin 1932d; 1932e; 1932h; 1933a; 1937a • Vogel 1936 • van der Waerden 1980 • Waschow 1932–1933; 1936–37.

BM 96957
See VAT 6598.

CBM 12648

Copy: Hilprecht 1906 pl. 15: no. 25a.
Bibliography: *MKT* I: 234-35 • *TMB*: 214 • Friberg 1982: 13, 57 [Thureau-Dangin (1937)] • Waschow 1936-37.

Ist 0 4360

Copy: *MKT* II pl. 42.
de Genouillac 1925: no. A 567.
Bibliography: *MKT* I: 235-36 • *TMB*: 205.

Ist 0 4552

Copy: *MKT* II pl. 42.
de Genouillac 1924: no. B 168.
Bibliography: *MKT* I: 236 • *TMB*: 205.

Strssbg 362

Photo: *MKT* II pl. 10.
Copy: Frank 1928 pl. 4 (no. 6).
Bibliography: *MKT* I: 239-43 • *TMB*: 82-84 • Eph'al 1984 • Frank 1928: 19-20: no. 6 • Gandz 1937; 1940 • Goetsch 1968-69 • Neugebauer 1931a; 1933-34; 1934: 171-75, 185-86 • Powell 1979 • Schott 1933 • von Soden 1937 • Thureau-Dangin 1932c; 1936e; 1937d • Vogel 1936 • Waschow 1936-37 • Waschow and Neugebauer 1932.

Strssbg 363

Photo: *MKT* II pl. 11.
Copy: Frank 1928 pl. 5 (no. 7).

Bibliography: *MKT* I: 243-48 • *TMB*: 84-86 • Frank 1928: 20-21: no. 7 • Gandz 1937 • Goetsch 1968-1969 • Neugebauer 1931a; 1933-34 • Schuster 1931 • Thureau-Dangin 1939 • Vogel 1936.

Strssbg 364

Photo: *MKT* II pl. 11.
Copy: Frank 1928 pl. 6 (no. 8).
Bibliography: *MKT* I: 248-56 • *TMB*: 86-89 • Bruins 1962 • Frank 1928: 21-22: no. 8 • Gandz 1948 • Goetsch 1968-1969 • Neugebauer 1931d; 1933-34; 1934: 180-81 • Schuster 1931 • von Soden 1939 • Thureau-Dangin 1932d; 1940 • Vogel 1936 • Waschow 1932-1933; 1933.

Strssbg 366

Photo: *MKT* II pl. 12.
Copy: Frank 1928 pl. 7 (no. 9).
Bibliography: *MKT* I: 257-59; *MKT* III: 56 • *TMB*: 205 • Frank 1928: 22: no. 9 • Goetsch 1968-69 • Lewy 1947 • Neugebauer 1933-34 • Sachs 1947a • von Soden 1937 • Thureau-Dangin 1939 • Waschow 1936-37.

Strssbg 367

Photo: *MKT* II pl. 12.
Copy: Frank 1928 pl. 8 (no. 10).
Bibliography: *MKT* I: 259-63; *MKT* III: 60-61 • *TMB*: 90-91 • Bruins 1962; 1966 • Frank 1928: 22-23: no. 10 • Goetsch 1968-69 • Høyrup 1984: 73-75, 105.34-105.42 • Neugebauer 1931d; 1934: 175-80 • Thureau-Dangin 1940.

Strssbg 368

Photo: *MKT* II pl. 13.
Copy: Frank 1928 pl. 9 (no. 11).
Bibliography: *MKT* I: 311-14 • *TMB*: 91-92 • Frank 1928: 23: no. 11 • Friberg 1982: 49, 60 [Thureau-Dangin *RA* 35 (1938: 71-77)] • Gandz 1940 • Goetsch 1968-69 • Schuster 1931 • Thureau-Dangin 1933c; 1934d; 1936b; 1937c; 1938b • Vogel 1936.

VAT 672

Photo: *MKT* II pl. 13.
Copy: *MKT* II pl. 43.
Bibliography: *MKT* I: 267-68 • *TMB*: 205 • Goetsch 1968-69 • Thureau-Dangin 1937d.

VAT 6469

Photo: *MKT* II pl. 14.
Copy: *MKT* II pl. 43.
Bibliography: *MKT* I: 268-69; *MKT* III: 56 • *TMB*: 205 • Waschow 1936-1937.

VAT 6505

Photo: *MKT* II pl. 14.
Copy: *MKT* II pl. 43.
Bibliography: *MKT* I: 270-73 • *TMB*: 131-33 • Bruins 1962; 1967; 1971 • Goetsch 1968-69 • Sachs 1947a • Thureau-Dangin 1936b.

VAT 6546

Photo: *MKT* II pl. 14.
Copy: *MKT* II pl. 43.
Bibliography: *MKT* I: 268-69; *MKT* III: 56 • *TMB*: 205 • Waschow 1936-37.

VAT 6597

Photo: *MKT* II pl. 15.
Copy: *MKT* II pls. 43, 45.
Bibliography: *MKT* I: 274-77 • *TMB*: 205-06 • von Soden 1939.

VAT 6598
(Joined with BM 96957 by C.B.F. Walker, Letter to author, 23 May, 1986)

Photo: *MKT* II pls. 16, 17.
Copy: *MKT* II pl. 44.
Bibliography: *MKT* I: 277–87 • *TMB*: 127–31 • Bruins 1962; 1971 • Dunham 1980; 1982 • Friberg 1981a • Goetsch 1968–69 • Neugebauer 1931–32; 1933–34 • von Soden 1937; 1939 • Thureau-Dangin 1936d • Torraca et al. 1972 • Ungnad 1916 • van der Waerden 1954–66: 65–66 • Weidner 1916 • Zimmern 1916.

VAT 6599
Joined with BM 85200.

VAT 7528

Photo: *MKT* II pl. 18.
Copy: *MKT* II pl. 45.
Bibliography: *MKT* I: 508–13 • *TMB*: 206–07 • Goetsch 1968–69.

VAT 7530

Photo: *MKT* II pl. 19.
Copy: *MKT* II pl. 46.
Bibliography: *MKT* I: 287–89 • *TMB*: 100–01 • Goetsch 1968–69 • *MCT*: 18–19 • Powell 1978; 1979 • Sachs 1946 • von Soden 1937 • Thureau-Dangin 1936b • Waschow 1937.

VAT 7531

Photo: *MKT* II pl. 20.
Copy: *MKT* II pl. 46.
Bibliography: *MKT* I: 289–94 • *TMB*: 98–99 • Bruins 1962; 1966; 1967 • Goetsch 1968–69 • Gundlach and von Soden 1963 • von Soden 1937 • Thureau-Dangin 1924a: 8–10 (no. 31), 19–20 (no. 41), 27–28 (no. 77) • Vaiman 1961: 258–63 • Waschow 1936–37.

VAT 7532

Photo: *MKT* II pl. 21.
Copy: *MKT* II pl. 46.
Bibliography: *MKT* I: 294–303 • *TMB*: 96–98 • Friberg 1982: 49, 60 [Thureau-Dangin *RA* 35 (1938: 71–77)] • Goetsch 1968-69 • Høyrup 1984: 52–53; 1985a: 144–46; 1989: 15–22; 1990: 342–43 • Thureau-Dangin 1933c; 1936b; 1937c; 1938b.

VAT 7535

Photo: *MKT* II pl. 22.
Copy: *MKT* II pl. 47.
Bibliography: *MKT* I: 303–10 • *TMB*: 93–96 • Friberg 1982: 49, 60 [Thureau-Dangin *RA* 35 (1938: 71–77)] • Goetsch 1968-69 • von Soden 1937; 1939 • Thureau-Dangin 1936b; 1937c; 1938b • Vogel 1958.

VAT 7537

Photo: *MKT* II pl. 23.
Copy: *MKT* II pl. 48.
Bibliography: *MKT* I: 466–78 • *TMB*: 207 • Neugebauer 1933-34; 1934: 192–93.

VAT 7620

Photo: *MKT* II pl. 23.
Copy: *MKT* II pl. 48.
Bibliography: *MKT* I: 314–17; *MKT* II: 58 • *TMB*: 208.

VAT 7621

Photo: *MKT* II pl. 23.
Copy: *MKT* II pl. 48.
Bibliography: *MKT* I: 290–94 • *TMB*: 99–100 • von Soden 1937 • Waschow 1936-37 • See VAT 7531.

VAT 8389

Photo: *MKT* II pl. 24.
Copy: *MKT* II pl. 49.
Bibliography: *MKT* I: 317–35 • *TMB*: 103–07 • Bruins 1971 • Friberg 1982: 49, 60 [Thureau-Dangin *RA* 35 (1938: 71–77)] • Geršman 1959 • Goetsch 1968–69 • Høyrup 1984: 76–79, 126; 1985a: 99–105; 1990: 289–95 • Neugebauer 1934: 181–83 • Thureau-Dangin 1937c; 1938b; 1939; 1940 • Vogel 1960: 90–91 • van der Waerden 1961: 66–68; 1980.

VAT 8390

Photo: *MKT* II pl. 25.
Copy: *MKT* II pl. 50.
Bibliography: *MKT* I: 335–40; *MKT* III: 58 • *TMB*: 112–15 • Bruins 1971 • Gandz 1937 • Goetsch 1968–69 • Høyrup 1984: 15–19, 28; 1985a: 83–89; 1990: 281–85 • Neugebauer 1934: 189–90 • Young 1990.

VAT 8391

Photo: *MKT* II pl. 26.
Copy: *MKT* II pl. 51.
Bibliography: *MKT* I: 319–35 • *TMB*: 107–12 • Becker 1966: 31–32 • Bruins 1962; 1971 • Friberg 1982: 49, 60 [Thureau-Dangin *RA* 35 (1938: 71–77)] • Geršman 1959 • Goetsch 1968–69 • Høyrup 1984: 76–79, 127, 128; 1985a: 105–10; 1990: 295–99 • Neugebauer 1933–34; 1934: 181–83 • Thureau-Dangin 1938b; 1939; 1940.

VAT 8512

Photo: *MKT* II pl. 27.
Copy: *MKT* II pl. 52.
Bibliography: *MKT* I: 340–45 • *TMB*: 101–03 • Høyrup 1984: 126, 132–33 n. 22, 105.15–105.23 • Gandz 1937; 1948 • Goetsch 1968–1969 • Huber 1955 • Neugebauer 1934: 183–85 • von Soden 1939 • Thureau-Dangin 1940 • van der Waerden 1961: 72–73; 1980.

VAT 8520

Photo: *MKT* II pl. 28.
Copy: *MKT* II pl. 53.
Bibliography: *MKT* I: 346–51 • *TMB*: 115–18 • Gandz 1937 • Goetsch 1968–1969 • Høyrup 1984: 80–83, 126; 1985a: 135–37 • Neugebauer 1934: 186–88.

VAT 8521

Photo: *MKT* II pl. 29.
Copy: *MKT* II pl. 54.
Bibliography: *MKT* I: 351–67; *MKT* III: 59 • *TMB*: 121–23 • Landsberger 1937: tablet 2 i 41–43 • Neugebauer 1933–34; 1934: 197–99 • Thureau-Dangin 1936d • van der Waerden 1961: 44–45 • Waschow 1936–37.

VAT 8522

Photo: *MKT* II pl. 30.
Copy: *MKT* II pl. 55.
Bibliography: *MKT* I: 367–73; *MKT* III: 61 • *TMB*: 208 • Friberg 1982: 49, 113–14 • Kraus 1966 • Lewy 1947 • *MCT*: 59 n. 1521 • Neugebauer 1949 • Postgate 1992: 159–61 • Powell 1984a • von Soden 1937; 1939 • Thureau-Dangin 1936e • Waschow 1936–37.

VAT 8523

Photo: *MKT* II pl. 31.
Copy: *MKT* II pl. 55, 56.
Bibliography: *MKT* I: 373–80; *MKT* III: 61 • *TMB*: 124–27 • Goetsch 1968–69 • Powell 1988 • Thureau-Dangin 1936b • von Soden 1939 • Waschow 1936–37.

VAT 8528

Photo: *MKT* II pl. 32.
Copy: *MKT* II pl. 56, 57.

Bibliography: *MKT* I: 353–67; *MKT* III: 59–61 • *TMB*: 118–21 • Goetsch 1968–69 • Landsberger 1937: tablet 2 i 41–43 • Lewy 1947 • Neugebauer 1933-34; 1934: 197–99 • Thureau-Dangin 1936d; 1936e • Waschow 1936–37.

YBC 4668

Copy: *MKT* III pl. 2.
Bibliography: *MKT* I: 420–66; *MKT* III: 62 • *TMB*: 162–90 • Gandz 1937 • Goetsch 1968–69 • Neugebauer 1934: 188–89; 1935 • Thureau-Dangin 1936e; 1940 • Waschow 1936–37 • Young 1990.

YBC 4669

Copy: *MKT* III pl. 3.
Bibliography: *MKT* I: 514–16; *MKT* III: 26–29, 62 • *TMB*: 208–10 • *MCT*: 103, 138–39 • Dunham 1980 • Gelb 1982 • Groneberg 1990 • Kraus 1966 • Landsberger 1937: tablet 2 i 41–43 • Lewy 1944; 1949b: 137–70 • Moorey 1985: 46–50 • Michel 1989 • Postgate 1992: 159–61 • Powell 1984b • Postgate 1978 • Thureau-Dangin 1921; 1936e; 1937b; 1937d • Torraca et al. 1972 • Waschow 1936–37; 1937.

YBC 4673

Copy: *MKT* III pl. 3.
Bibliography: *MKT* I: 506–08; *MKT* III: 29–34 • *TMB*: 210–13 • *MCT*: 93–96 • Bruno et al. 1968–69 • Lewy 1949b: 137–70 • Scheil 1915 • Thureau-Dangin 1937b; 1937d • Torraca et al. 1972 • Waschow 1936–37.

YBC 4695

Copy: *MKT* III pl. 4.
Bibliography: *MKT* I: 501–02; *MKT* III: 34–39 • *TMB*: 214 • Goetsch 1968–69.

YBC 4696

Copy: *MKT* III pl. 4.
Bibliography: *MKT* I: 504–06; *MKT* II: 60–64; *MKT* III: 66 • *TMB*: 214 • Goetsch 1968–69 • Stol 1988: 176–81.

YBC 4697

Copy: *MKT* III pl. 5.
Bibliography: *MKT* I: 485–86; *MKT* III: 40–41 • *TMB*: 214 • Becker 1966: 34 • Gandz 1937 • van der Waerden 1961: 70–71.

YBC 4698

Copy: *MKT* III pl. 5.
Bibliography: *MKT* I: 513; *MKT* III: 42–45 • *TMB*: 214 • Bjorkman 1989 • Curtis and Hallo 1959 • Friberg 1982: 39 [Schneider (1931)], 57 • Goetze 1951 • Keiser and Nies 1920: pl. 8, 73 [various stone weights and their inscriptions] • Landsberger 1937: tablet 2 i 41–43 • Legrain 1947: no. 1377 • Powell 1979 • Salonen 1970: 80–82 • Scheil 1918 • Thureau-Dangin 1937d • Waschow 1937 • // MLC 1842.

YBC 4708

Copy: *MKT* II pl. 57.
Bibliography: *MKT* I: 389–402; *MKT* III: 61 • *TMB*: 194–203 • *MCT*: 94–95 • Goetsch 1968–69 • Neugebauer 1934–36 • Thureau-Dangin 1936b.

YBC 4709

Photo: *MKT* II: pl. 33.
Copy: *MKT* II: pl. 58.
Bibliography: *MKT* I: 412–20 • *TMB*: 155–62 • Gandz 1937 • Goetsch 1968–69 • Neugebauer 1933–34; 1934: 190–92.

YBC 4710

Copy: *MKT* II pl. 58.
Bibliography: *MKT* I: 402–12; *MKT* III: 61 • *TMB*: 148–55 • Goetsch 1968–69 • Waschow 1936–37.

YBC 4711

Bibliography: *MKT* I: 503–04; *MKT* II: 45–48 • *TMB*: 214 • Goetsch 1968–69.

YBC 4712

Photo: *MKT* II pl. 33.
Copy: *MKT* II pl. 59.
Bibliography: *MKT* I: 420–66; *MKT* II: 62 • *TMB*: 162–90 • Neugebauer 1935 • Thureau-Dangin 1936e.

YBC 4713

Copy: *MKT* II pl. 59.
Bibliography: *MKT* I: 420–66; *MKT* II: 62 • *TMB*: 162–90 • Neugebauer 1934: 188–89 • Waschow 1936–37.

YBC 4714

Photo: *MKT* II pl. 34.
Copy: *MKT* II pl. 60.
Bibliography: *MKT* I: 487–501 • *TMB*: 136–48 • Goetsch 1968–69.

YBC 4715

Copy: *MKT* II pl. 60.
Bibliography: *MKT* I: 478–85 • *TMB*: 190–94.

YBC 6504

Copy: *MKT* III pl. 6.
Bibliography: *MKT* III: 22–25 • *TMB*: 134–36 • Bruins 1966; 1971 • Gandz 1937 • Goetsch 1968–69 • Gundlach and von Soden 1963 • Høyrup 1984: 41–49; 1985a: 131–35; 1989: 26–31 • Young 1988; 1990.

TMS

Text 1

Photo: Bruins 1964a: 175.
Copy: *TMS* pl. 1.
Bibliography: *TMS*: 22–23 • Bruins 1953b; 1964a • Friberg 1982: 88 • Price 1964a • Seidenberg 1984-85 • von Soden 1964.

Text 2

Photo: obv.: Bruins 1971 pl. xviii.
Copy: *TMS* pl. 2, 3.
Bibliography: *TMS*: 23–24 • Becker 1961 • Bruins 1950; 1959; 1962 • Friberg 1982: 88, 90 [Becker (1961)] • von Soden 1964.

Text 5 (=Tablet Aa)

Photo: Bruins and Rutten 1954.
Copy: *TMS* pl. 7–10.
Bibliography: *TMS*: 35–49 • Bruins 1962; 1966; 1971 • Bruins and Rutten 1954 • Friberg 1981b • Goetsch 1968–69 • Lawergren and Gurney 1987 • von Soden 1964.

Text 6 (=Tablet Bb)

Copy: *TMS* pl. 11–13.
Bibliography: *TMS*: 49–51 • See Text 5.

Text 7 (=Tablet A-B)

Copy: *TMS* pl. 14, 15.
Bibliography: *TMS*: 52–57 • Bruins 1962; 1964a; 1971 • Goetsch 1968–69 • von Soden 1964.

Text 8 (=Tablet V)

Copy: *TMS* pl. 16.
Bibliography: *TMS*: 58–62 • Goetsch 1968–69 • von Soden 1964.

Text 9 (=Tablet H)

Copy: *TMS* pl. 17.
Bibliography: *TMS*: 63–69 • Bruins 1962; 1971 • Goetsch 1968–69 • Høyrup 1984: 91–98; 1985a: 137–42; 1990: 320–28 • von Soden 1964.

Text 10 (=Tablet G)

Copy: *TMS* pl. 18, 19.
Bibliography: *TMS*: 69–74 • Bruins 1962 • Goetsch 1968–69 • von Soden 1964.

Text 11 (=Tablet P)

Copy: *TMS* pl. 20.
Bibliography: *TMS*: 74–78 • von Soden 1964.

Text 12 (=Tablet M)

Copy: *TMS* pl. 21.
Bibliography: *TMS*: 78–81 • Bruins 1962 • von Soden 1964.

Text 13 (=Tablet L)

Copy: *TMS* pl. 22.
Bibliography: *TMS*: 82–83 • Bruins 1962; 1966 • Friberg 1982: 89 • Gundlach and von Soden 1963 • Høyrup 1984: 7–10, 84–86; 1985a: 149–50 • *MCT*: 107 • Powell 1979 • von Soden 1964.

Text 14 (=Tablet N)

Copy: *TMS* pl. 23.
Bibliography: *TMS*: 84–88 • von Soden 1964.

Text 15 (=Tablet E)

Copy: *TMS* pl. 24.
Bibliography: *TMS*: 88–91 • Friberg 1982: 89 • von Soden 1964.

Text 16 (=Tablet F)

Copy: *TMS* pl. 25.
Bibliography: *TMS*: 91–95 • Bruins 1962 • Friberg 1982: 89 • Goetsch 1968–69 • Høyrup 1984: 86–91; 1985a: 110–19; 1989: 23–25; 1990: 299–305 • von Soden 1964.

Text 17 (=Tablet O)

Copy: *TMS* pl. 26.
Bibliography: *TMS*: 95–97 • Bruins 1962 • Goetsch 1968–69 • von Soden 1964.

Text 18 (=Tablet S)

Copy: *TMS* pl. 27.
Bibliography: *TMS*: 97–101 • Bruins 1950; 1962 • Friberg 1982: 89 • Seidenberg 1984–85 • von Soden 1964.

Text 19 (=Tablet C-D)

Copy: *TMS* pls. 28, 29.
Bibliography: *TMS*: 101–05 • Bruins 1950; 1962; 1971 • Friberg 1982: 89 • Goetsch 1968–69 • Nemet-Nejat 1988 • von Soden 1964.

Text 20 (=Tablet Z)

Copy: *TMS* pl. 30.
Bibliography: *TMS*: 105–08 • *CAD* s.v. *apsamikku* • Goetsch 1968–69 • Kilmer 1985 • Lawergren and Gurney 1987 • Saggs 1960 • Vaiman 1959.

Text 21 (=Tablet R)

Copy: *TMS* pls. 31, 32.
Bibliography: *TMS*: 108–11 • *CAD* s.v. *messētu* • Kilmer 1964 • Lawergren and Gurney 1987 • Vaiman 1959 • See *TMS*, Text 20.

Text 22 (=Tablet Q)

Copy: *TMS* pl. 33.
Bibliography: *TMS*: 111–14 • Bruins 1962 • Friberg 1982: 90 • Goetsch 1968–69 • Nemet-Nejat 1988.

Text 23 (=Tablet T)

Copy: *TMS* pl. 34.
Bibliography: *TMS*: 114–17 • Friberg 1982: 90 • von Soden 1964.

Text 24 (=Tablet W)

Copy: *TMS* pls. 35, 36.
Bibliography: *TMS*: 117–22 • *AHw* s.v. *tāluku* • *CAD* s.v. *manâtu* • Friberg 1982: 90 • *MCT*: 130 s.v. *taktītu* • von Soden 1964.

Text 25 (=Tablet U)

Copy: *TMS* pls. 37, 38.
Bibliography: *TMS*: 122–24 • Friberg 1982: 90 • Nemet-Nejat 1988 • Powell 1988 • von Soden 1964.

Text 26 (=Tablet Y)

Copy: *TMS* pl. 39.
Bibliography: *TMS*: 124–29 • Friberg 1982: 90 • Nemet-Nejat 1988 • von Soden 1964.

COEFFICIENT LISTS

A 3553

Photo and Text Edition: Kilmer 1960: 275–78.

BM 47860

Copy and Text Edition: Livingstone 1986: 18–23.

CBS 10996

Photo and Text Edition: Kilmer 1960: 278–83 • Duchesne-Guillemin 1984.

IM 49949

Copy: Edzard 1971: no. 236.
Photo and Text Edition: Edzard 1970: no. 236.

IM 52301 rev.20–22

Copy and Text Edition: Baqir 1950b.

IM 52916, IM 52685, IM 52304

Copy and Text Edition: Goetze 1951

TMS, Text 3

Copy: *TMS* pls. 4, 5.
Text Edition: *TMS*: 25–34.

YBC 5022

Photo: *MCT* pl. 44.
Copy: *MCT* pl. 18.
Text Edition: *MCT*: 132–36.

YBC 7243

Photo: *MCT* pl. 49.
Copy: *MCT* pl. 23.
Text Edition: *MCT*: 136–37.

BIBLIOGRAPHY: Barnett 1969 • Bruins 1953b; 1959; 1962; 1950a; 1950b; 1964a • Crocker 1978 • Crocker and Kilmer 1984 • Delitzsch 1885: K 4378 col i: 10–11, 16–22; vi: 15–22 • Duchesne-Guillemin 1965; 1984 • Dunham 1980 • Ellis 1970 • Foxvog 1986 • Friberg 1982: 91-92, 113–14, 120–21 • Goetsch 1968–69 • Goetze 1948; 1951 • Groneberg 1990 • Gurney 1968 • Güterbock 1970 • Høyrup 1984: 105.14 • Kilmer 1985; 1960; 1965; 1971; 1974; 1984; 1990a; 1990b • Kilmer, Brown and Crocker 1986 • Kilmer and Civil 1986 • Kümmel 1970 • Lambert 1971 • Landsberger 1934; 1957: tablets 2, 4 • Laroche 1973 • Lawergren and Gurney 1987 • Lewy 1947; 1949b • Lieberman 1980 • Nemet-Nejat 1988 • Oppenheim 1947b • Postgate 1978 • Powell 1971: 85 n. 4; 1982 • Reiner 1970 • Salonen 1966; 1972: 88-99 • Seidenberg 1972; 1984-85 • von Soden 1964; 1977 • Torraca et al. 1972 • Vaiman 1959; 1963 • Vitale 1982 • Wulstan 1968; 1974.

TELL HARMAL TABLETS

Db_2-146

Photo: Baqir 1962 pl. 1.
Copy: Baqir 1962 pl. 2.
Bibliography: • Baqir 1962 • Friberg 1981a • von Soden 1967.

IM 31210

Bibliography: Bruins 1954: 57–61; 1971 • Friberg 1982: 76–77 • Nemet-Nejat 1988.

IM 31247

Bibliography: Bruins 1953c: 249–53; 1962.

IM 31248

Bibliography: Bruins 1953d: 256–57; 1962.

IM 43996
(teacher and student copy)

Photo: Bruins 1964b pl. 2.
Bibliography: Bruins 1953d: 255–56; 1955a; 1962 • Vaiman 1958.

IM 52301

Copy: Baqir 1950b
Bibliography: Baqir 1950b • Bruins 1951a; 1953c; 1962; 1966 • Friberg 1982: 76 [Bruins *Sumer* 9 (1953: 241–53)] • Gundlach and von Soden 1963 • Høyrup 1984: 105.43–105.55; 1985a: 146–49; 1990: 338–43 • von Soden 1952.

IM 52672

Bibliography: Bruins 1954: 61 • Bruins and Rutten 1954 • Goetze 1951: 129–136 • *TMS*, Text 4.

IM 52916, IM 52685, IM 52304

Copy: Goetze 1951.
Bibliography: Goetze 1951 • See Friberg 1981b // Goetze: 129–136: Sections A, B • Friberg 1982: 73 [Goetze (1951)] • Kilmer 1960 • von Soden 1977 • Vaiman 1959.

IM 53953

Copy: Baqir 1951.
Bibliography: Baqir 1951: 31–32: no. 2 • Bruins 1953c • *MCT*: 48, n. 135b • von Soden 1952.

IM 53957

Copy: Baqir 1951.
Bibliography: Baqir 1951: 37–38: no. 5 • Bruins 1953c; 1962 • *MCT*: 103 (YBC 4669 #4) • Nemet-Nejat 1988 • von Soden 1952.

IM 53961

Copy: Baqir 1951.
Bibliography: Baqir 1951: 35–36: no. 4 • Bruins 1953c • Kilmer 1960 • *MCT*: 165; *TMB*: 242 s.v. ***éš-gàr*** • Nemet-Nejat 1988 • von Soden 1952.

IM 53963

Bibliography: Bruins 1953c: 249.

IM 53965

Copy: Baqir 1951.
Bibliography: Baqir 1951: 39–40: no. 7 • Bruins 1953c; 1962 • Friberg 1982: 72 [Baqir (1951)] • Nemet-Nejat 1988 • von Soden 1952.

IM 54010

Copy: Baqir 1951.
Bibliography: Baqir 1951: 38–39: no. 6 • Bruins 1953c; 1962 • Friberg 1982: 72 [Baqir (1951)] • von Soden 1952.

IM 54011

Copy: Baqir 1951.
Bibliography: Baqir 1951: 45: nos. 10, 11 • Bruins 1953c • Friberg 1982: 72 [Baqir (1951)] • Nemet-Nejat 1988 • von Soden 1952.

IM 54464

Copy: Baqir 1951.
Bibliography: Baqir 1951: 43–44: no. 9 • Bruins 1953c; 1962 • Nemet-Nejat 1988 • Powell 1979 • von Soden 1952.

IM 54472

Bibliography: Bruins 1954: 56; 1962; 1971 • *MCT*: 42.

IM 54478

Copy: Baqir 1951.
Bibliography: Baqir 1951: 30–31: no. 1 • Bruins 1953c • Friberg 1982: 72 [Baqir (1951)] • von Soden 1952.

IM 54538

Copy: Baqir 1951.
Bibliography: Baqir 1951: 33–35: no. 3 • Nemet-Nejat 1988 • von Soden 1952 • See *TMB*: nos. 141–143.

IM 54559

Copy: Baqir 1951.
Bibliography: Baqir 1951: 41–42: no. 8 • von Soden 1952 • See IM 54478.

IM 55357

Copy: Baqir 1950a.
Bibliography: Baqir 1950a • Bruins 1951a; 1953b; 1955a; 1962; 1966 • Drenckhahn 1951; 1952 • Friberg 1981a; 1982: 70 [Baqir (1950)].

MAH 16055

Photo: Bruins 1962: 304.
Copy: Bruins 1955a: 47–48.
Bibliography: Bruins 1955 • Bruins 1962.

MARI TABLETS

M 6829

Photo: Soubeyran 1984: 31.
Copy: Soubeyran 1984: 38.
Bibliography: Soubeyran 1984: 35–41: no. 7 • // *ARM* 24: no. 260 • Nemet-Nejat 1988.

M 6861

Photo: Soubeyran 1984: 42.
Bibliography: Soubeyran 1984: 46–47: no. 9 • Nemet-Nejat 1988.

M 8613

Photo: Soubeyran 1984: 31.
Copy: Soubeyran 1984: 25.
Bibliography: Soubeyran 1984: 30–35: no. 6 • Nemet-Nejat 1988 • Sachs 1946.

M 15192 + M 15236

Photo: Soubeyran 1984: 42.
Copy: Soubeyran 1984: 38.
Bibliography: Soubeyran 1984: 41–45: no. 8 • // *ARM* 24: no. 146 • Moorey 1985: 116–19 • Nemet-Nejat 1988 • Thureau-Dangin 1909a.

MISCELLANEOUS

AO 9071

Bibliography: Bruins 1962: 286 [an unpublished text which is a long serial text on linear equations using the terms *UŠ* and *SAG*].

AOT 304

Copy: Thureau-Dangin 1903: no. 413.
Bibliography: Vaiman 1961: 246–47 • *MCT*: 94–96.

AOT 305

Copy: Thureau-Dangin 1903: no. 412.
Bibliography: Allotte de la Füye 1907: 75–77 • Friberg 1982: 15 [Allotte de la Füye (1907)] • Jacobsen 1981.

BM 80209

Copy: Pinches 1963: no. 39.
Bibliography: Friberg 1979b; 1981b • Goetze 1951 • Kilmer 1964 • See *TMS*, Texts 5, 6.

Böhl Collection, No. 1821

Photo: Leemans and Bruins 1951.
Bibliography: Kilmer 1964: 142–43 • Leemans and Bruins 1951 • Friberg 1979b; 1981b; 1982: 71 [Leemans and Bruins (1951)] • Nemet-Nejat 1988 • See BM 85194 i 37–50 and *UET* 5: no. 864.

CBS 11318

Copy: Neugebauer and Sachs 1984: 251.
Bibliography: Neugebauer and Sachs 1984: 247.

Const. 236

Bibliography: Allotte de la Füye 1907: 77–78.

CUA 34

Bibliography: Goetze 1957: 40: no. 33 • Codex Hammurabi §43 • *MCT*: 6–10

Erm 15073

Copy: Vaiman 1961.
Bibliography: Vaiman 1961: 232–44 • Vaiman 1960.

Erm 15188

Photo, Copy and Bibliography: Vaiman 1958.

Erm 015189

Photo, Copy and Bibliography: Vaiman 1955 • Bruins 1962.

Haddad 104

Copy and Photo: Al-Rawi and Roaf 1984.
Bibliography: Al-Rawi and Roaf 1984 • Dunham 1980 • Lewy 1947; 1949: 137–70 • Nemet-Nejat 1988 • Neugebauer 1949 • Oppenheim 1947b • Powell 1982c; 1984a; 1984b • Salonen 1966 • Thureau-Dangin 1921; 1936d.

LBAT: no. 1647

Copy: Sachs 1955: no. 1647.
Bibliography: See *TMS*, Text 14.

LBAT: no. 1648

Copy: Sachs 1955: no. 1648.
Bibliography: Gandz 1938 • Kraus 1969 • Nemet-Nejat 1988 • Powell 1984b Salonen 1966.

M 10

Copy: Sachs 1952: 152.
Bibliography: Sachs 1952: 152–53.

MLC 1731

Photo: Sachs 1946 pl. 6.
Copy: Sachs, 1946: 204.
Bibliography: Sachs 1946: 203–14 • Cocquerillat 1965 • Friberg 1979c • Goetze 1946 • Lewy 1951 • Peet 1923: 27–31 • Powell 1971; 1974; 1984a; 1986 • Sachs 1947a • Seidenberg 1962b • Thureau-Dangin 1939.

UET 3: no. 1386

Copy: Legrain 1947 pl. 151.
Bibliography: Vaiman 1961: 244–46.

UET 5: no. 121

Copy: Figulla and Martin 1953 pl. 26.
Bibliography: Vaiman 1961: 248-50 • Gandz 1938 • Kraus 1966; 1969 • Postgate 1992: 159-61.

UET 5: no. 858

Copy: Figulla and Martin 1953 pl. 138.
Bibliography: Vaiman 1961: 250-54.

UET 5: no. 859

Copy: Figulla and Martin 1953 pl. 138.
Bibliography: Vaiman 1961: 254-56.

UET 5: no. 864

Copy: Figulla and Martin 1953 pl. 139.
Bibliography: Kilmer 1964: 140-42 • Vaiman 1961: 256-58 • See Böhl Collection, No. 1821.

UM 29 15 192

Copy: Neugebauer and Sachs 1984: 251.
Bibliography: Neugebauer and Sachs 1984: 248.

UMB 16/2 pl. 7

Photo: Steele 1951 25 (pl. 7).
Bibliography: Friberg 1982: 71 [Steele (1951)].

VAT 8547

Photo: Sachs 1952: 154.
Bibliography: Sachs 1952: 153–56 • See YBC 6295.

2N-T 30

Bibliography: Neugebauer and Sachs 1984: 247.

2N-T 116

Bibliography: Neugebauer and Sachs 1984: 247.

2N-T 174

Bibliography: Neugebauer and Sachs 1984: 245.

2N-T 472

Bibliography: Neugebauer and Sachs 1984: 246–47.

OLD AKKADIAN TABLETS

A 681

Copy: Luckenbill 1930 pl. 36: no. 70.
Bibliography: Edzard 1968 • Friberg 1982: 29 [Deimel (1923)], 99 [Edzard (1968)] • Whiting 1984.

A 5443 and A 5446

Copy: Whiting 1984.
Bibliography: Whiting 1984 • Friberg 1987.

BIN 8: no. 24

Copy: Hackman 1958 pl. 11: no. 24.
Bibliography: Hackman 1958: no. 24 • Friberg 1982: 37 [Luckenbill (1930)], 81–82 • Luckenbill 1930: no. 116.

DP 604–612

Bibliography: Allotte de la Füye 1920 pl. 150–52: DP 604–612.

Limet, *Liège* nos. 36–39

Photo and Copy: Limet 1973 pl. 11–12: nos. 36–39 (PUL 27–29, 31)
Bibliography: Limet 1973: 64–67: nos. 36–39 (PUL 27–29, 31) • Powell 1976a • Whiting 1984.

MAD 5: 112

Copy: Gelb 1970: no. 11.
Bibliography: Gelb 1970: no. 112 • Friberg 1982: 101 [Gelb (1970)]; 1983; 1986a • Powell 1976a • Whiting 1984.

TM.75.G.1392

Bibliography: Friberg 1982: 134–35; 1983; 1986a • Pomponio 1981.

TMH 5: 65 (HS 815)

Photo and Copy: Pohl 1935: no. 65 (HS 815).
Collations: Westenholz 1975: no. 65.
Bibliography: Powell 1976a • Whiting 1984.

TSŠ 50, 77, 188, 671

Copy: Jestin 1937: nos. 50, 77, 188, 671.
Bibliography: Archi 1980 • Brandes 1980 • Friberg 1982: 138 [Høyrup (1982)]; 1983; 1986a; 1987 • Guitel 1963 • Høyrup 1982 • Lambert 1953; 1954 • Pomponio 1981 • Powell 1972; 1976a • Whiting 1984.

VAT 12593

Bibliography: Allotte de la Füye 1915b • Deimel 1922 • Neugebauer 1933–1934 • Whiting 1984.

YBC 1793

Copy: Keiser 1919: no. 293.
Bibliography: Powell 1976a.

BIBLIOGRAPHY

Aaboe, A.

1964 *Episodes from the Early History of Mathematics.* New York: Random House.

1965 Some Seleucid Mathematical Tables (Extended Reciprocals and Squares of Regular Numbers). *JCS* 19: 79–86.

1969 Two Atypical Multiplication Tables from Uruk. *JCS* 22: 88–91.

Abusch, T.

1981 Notes on a Pair of Matching Texts: A Shepherd's Bulla and An Owner's Receipt. In *Studies on the Civilization and Culture of Nuzi and the Hurrians*, edited by M. A. Morrison and D. I. Owen, pp. 1–9. Winona Lake: Eisenbrauns.

Adams, R. McC.

1965 *Land Behind Baghdad.* Chicago and London: The University of Chicago Press.

1966 *The Evolution of Urban Society.* Chicago: Aldine Publishing Co.

Adamson, P. B.

1985 Problems over Storing Food in the Ancient Near East. *WO* 16: 5–15.

Allotte de la Füye, F-M.

1907 La mesure des volumes dans les textes archaïques de la Chaldée. *RA* 6: 75–78.

1915a Un cadastre de Djokha, *RA* 12: 47–54.

1915b Mesures agraires et formules d'arpentage. *RA* 12: 117–146.

1920 *Documents présargoniques*, fascicule supplémentaire. Paris: Éditions Ernest Leroux.

1928 La sens du mot **kar** dans les comptes rendus de Larsa. *RA* 25: 25–30.

Al-Rawi, F., and Roaf, M.
1984 Ten Old Babylonian Mathematical Problems from Tell Haddad, Hamrin. *Sumer* 43: 175–218.

Anagnostakis, C., and Goldstein, B. R.
1973 On an Error in the Babylonian Table of Pythagorean Triangles. *Centaurus* 18: 64–66.

Archi, A.
1980 Un testo matematico d'età proto-siriana, *Studi eblaiti* 3: 63–64 + pl.

Archibald, R. C.
1936 Babylonian Mathematics. *Isis* 26: 63–81

Badawy, A.
1958 Architectural Provision against Heat in the Orient. *JNES* 17: 122–28.

Baqir, T.
1950a An Important Mathematical Problem Text from Tell Harmal. *Sumer* 6: 39–54 + 2 pl.

1950b Another Important Mathematical Text from Tell Harmal. *Sumer* 6: 130–48 + 5 pl.

1951 Some More Mathematical Texts. *Sumer* 7: 28–45 + 13 pl.

1962 Tell Dhiba'i: New Mathematical Texts. *Sumer* 18: 11–14. + 3 pl.

Barnett, R. D.
1969 New Facts about Musical Instruments from Ur. *Iraq* 31: 96–103.

Barton, G. A.
1918 *The Haverford Library Collection of Cuneiform Texts from the Temple Archives of Telloh*. Vol. 2. New Haven: Yale University Press.

Baumgartner, W.
1925 Untersuchungen zu den akkadischen Bauausdrücken. *ZA* 36: 123–38.

Becker, O.

1961 Die Bedeutung des Wertes 3⅛ für π in der babylonischen Mathematik. *Praxis der Mathematik* 3: 58–62.

1966 *Das Mathematische Denken der Antike*. Göttingen: Verlag Vandenhoeck & Ruprecht.

Beckman, G.

1983 Mesopotamians and Mesopotamian Learning at Ḫattuša. *JCS* 35: 97–114.

Berriman, A. E.

1955 A New Approach to the Study of Ancient Metrology. *RA* 49: 193–201.

Bjorkman, J.K.

1989 Second Millennium B.C. Prices of Iron and Copper. *N.A.B.U.* 1 (March): 12–13.

Blome, F.

1928 **du/GAB** in Angaben über die Grösse der Brote. *Or* 34–35: 129–35.

Borger, R.

1957 *niṣirti bārûti*, Geheimlehre der Haruspizin (Zu Neugebauer-Sachs, *MCT*, V und W, und einigen verwandten Texten). *BiOr* 14: 190–95.

1967–75 *Handbuch der Keilschriftliteratur*. 3 vols. Berlin: Walter de Gruyter & Co. [*HKL*]

Bottéro, J.

1987 The Culinary Tablets at Yale. *JAOS* 107: 11–19.

Brandes, M. A.

1980 Modelage et imprimerie aux debuts de l'écriture en Mésopotamie. *Akkadica* 18: 1–30.

Brinkman, J. A.

1963 New Evidence on Old Assyrian *ḫamuštum*. *Or* NS 32: 387–94.

1965 Note on Old Assyrian *ḫamuštum*. *JNES* 24: 118–20.

Bruins, E. M.

1949 On Plimpton 322: Pythagorean Numbers in Babylonian Mathematics. *Indagationes Mathematicae* 11: 191–94.

1950 Quelques textes mathématiques de la mission de Suse. *Indagationes Mathematicae* 12: 369–77.

1951a Comments on the Mathematical Tablets of Tell Harmal. *Sumer* 7: 179–85.

1951b Nouvelles découvertes sur les mathématiques babyloniennes. In *Les conférences du Palais de la Découverte*, pp. 5–29. Paris.

1953a La classification des nombres dans les mathématiques babyloniennes. *RA* 47: 185–88.

1953b A Contribution to the Interpretation of Babylonian Mathematics: Triangles with Regular Sides. *Indagationes Mathematicae* 15: 412–22.

1953c Revision of the Mathematical Texts from Tell Harmal. *Sumer* 9: 241–53.

1953d Three Geometrical Problems. *Sumer* 9: 255–59.

1954 Some Mathematical Texts. *Sumer* 10: 55–61.

1955a On the System of Babylonian Geometry. *Sumer* 11: 44–49.

1955b Pythagorean Triads in Babylonian Mathematics; the Errors on Plimpton 322. *Sumer* 11: 117–21.

1959 Regular Polygons in Babylonian and Greek Mathematics. *Janus* 48: 5–23.

1962 Interpretation of Cuneiform Mathematics. *Physis* 4: 277–341.

1964a Babylone et Héron versus Euclide. *RA* 58: 173–81.

1964b *Codex Constantinopolitanus Palatii Veteris*, I/3. Leiden: Brill.

1966 Fermat Problems in Babylonian Mathematics. *Janus* 53: 194–211.

1967 Reciprocals and Pythagorean Triangles. *Physis* 9: 373–92.

1968 Le premier problème de AO 6770. *RA*: 80–82.

1970 La construction de la grande table de valeurs réciproques AO 6456. *CRRA* 17: 99–115.

1971 Computation in the Old Babylonian Period. *Janus* 58: 222–67.

1980 On the History of Logarithms: Appendix. *Janus* 67: 241–60.

Bruins, E. M., and Rutten, M.
1954 La notation des fractions: un nouveau texte de série. *CRRA* 3: 129–43.

1961 *Textes mathématiques de Suse* (MDP 34). Paris: Librairie orientaliste Paul Geuthner. [*TMS*]

Brunés, T,
1967 *The Secrets of Ancient Geometry and Its Use*. 2 vols. Copenhagen: Rhodos International Science Publishers.

Bruno, A.; Chiari, G.; Tossarelli, C.; and Bultnick, G.
1968–69 Contributions to the Study of the Preservation of Mud Brick Structures. *Mesopotamia* 3–4: 252–70.

Buck, R. C.
1980 Sherlock Holmes in Babylon. *American Mathematical Monthly* 87: 335–45.

Caratini, R.
1957 Quadrature du cercle et quadrature des lunules en Mésopotamie. *RA* 51: 11–20.

Cardascia, G.
1955 *Les villes de Mésopotamie: leurs institutions économiques et sociales. La ville (2e partie) institutions économiques et sociales* (Recueils de la Sociéte Jean Bodin 7). Brussels.

Castellino, G. R.
1972 *Two Sulgi Hymns (BC)* (Studi Semitici 42). Rome: Istituto di studi del vicino oriente.

Caveing, M.
1985 La tablette babylonienne AO 17264 du Musée du Louvre et la problème des six frères. *Historia Mathematica* 12: 6–24.

Civil, M.
1969 *The Series lú = ša and Related Texts* (MSL 12). Rome: Pontifical Biblical Institute.

1973 Notes on Sumerian Lexicography II. *JCS* 25: 171–77.

Clay, A. T.
1915 *Miscellaneous Inscriptions in the Yale Babylonian Collection* (YOS 1). New Haven: Yale University Press.

Cocquerillat, D.
1965 Les calculs pratiques sur les fractions à l'époque séleucide. *BiOr* 22: 239–42.

Crocker, R. I.
1978 Remarks on the Tuning Text *UET* VII 74 (U. 7/80). *Or* 47: 99–104.

Crocker, R. I. and Kilmer, A. D.
1984 The Fragmentary Music Text from Nippur. *RA* 78: 81–85.

Curtis, J. B. and Hallo, W. W.
1959 Money and Merchants in Ur III. *HUCA* 30: 103–39.

Dahood, M.
1981 Ebla, Ugarit, and the Bible. In *The Archives of Ebla* by G. Pettinato, pp. 271–321. Garden City, New York: Doubleday & Co.

Damerow, P.
1988 Die Exponentialtafel BM 22706. Paper presented at 4. Workshop über Begriffsentwicklung in der Babylonischen Mathematik, May, at Berlin.

Damerow, P., and Englund, R.,
1989 The Proto-Elamite Texts from Tepe Yahya. Chapter 5 in *Excavations at Tepe Yahya: The Third Millennium* (American School of Prehistoric Research, Bulletin 30), ed. by C. C. Lamberg-Karlovsky and D. Potts. Harvard University Press.

1987 Die Zahlzeichen systeme der Archaischen Texte aus Uruk. Chapter 3 in *Zeichenliste der Archaischen Texte aus Uruk* (ATU 2), by M. Green and H. Nissen, in cooperation with P. Damerow and R. Englund (Berlin 1987).

Damerow, P.; Englund, R.; and Nissen, H.
1988a Die Entstehung der Schrift. *Spektrum der Wissenschaft* (February): 74–85.

1988b Die ersten Zahldarstellungen und die Entwicklung des Zahlbegriffs. *Spektrum der Wissenschaft* (March): 46–55.

David, A.
1928 La mesure de capacité "*matu*" sur les tablettes de Kirkuk; la mesure de superficie "*epinnu.*" *RA* 25: 99–100.

Deimel, A.
1909 Studien zu *CT* I, III, V, VII, IX und X. *ZA* 23: 107–44.

1922 *Die Inschriften von Fara*. Vol. 2, *Schultexte aus Fara, in Umschrift herausgegeben und bearbeitet* (WVDOG 40). Leipzig.

1924 *Die Inschriften von Fara*. Vol. 3, *Wirtschaftstexte aus Fara* (WVDOG 45). Leipzig.

Delaporte, L.
1911 Document mathématique de l'époque des rois d'Our. *RA* 8: 131–33.

Delitzsch, F.
1885 *Assyrische Lesestücke*. 3rd edition. Leipzig.

Diakonoff, I. M.
1974 Ancient Writing and Ancient Written Language: Pitfalls and Peculiarities in the Study of Sumerian. In *Jacobsen Festschrift* (AS 20), pp. 99–121. Chicago: The University of Chicago Press.

Dossin, G.
1927 *Autres textes sumériens et accadiens* (MDP 18). Paris.

1934 *Lettres de la première dynastie babylonienne* (TCL 18). Paris.

Drenckhahn, F.
1951 A Geometrical Contribution to the Study of the Mathematical Problem-Text from Tell Harmal (IM.55357) in the Iraq Museum, Baghdad. *Sumer* 7: 22–27.

1952 News and Correspondence. *Sumer* 8: 234–35.

Duchesne-Guillemin, M.
1965 Notes complémentaires sur la découverte de la gamme babylonienne. In *Studies in Honor of B. Landsberger* (AS 16), pp. 268–72. Chicago: The University of Chicago Press.

1984 *A Hurrian Musical Score from Ugarit: The Discovery of Mesopotamian Music* (SANE 2/2). Malibu: Undena Publications.

Dunham, S.
1980 *A Study of Ancient Mesopotamian Foundations*. Ph.D. diss., Columbia University.

1982 Bricks for the Temples of Šara and Ninurra. *RA* 76: 27–41.

1986 Sumerian Words for Foundation, Part 1: **temen**. *RA* 80: 31–64.

Edzard, D. O.
1968 Eine altsumerische Rechentafel (*OIP* 14,70). In *Festschrift von Soden (lišān mitḫurti)* (AOAT 1), pp. 101–04. Kevelaer & Neukirchen-Vluyn.

1970 *Altbabylonische Rechts- und Wirtschaftsurkunden aus Tell ed-Dēr im Iraq Museum, Baghdad* (ABAW N.F. 72). Munich: Verlag der Bayerischen Akademie der Wissenschaften.

1971 *Cuneiform Texts. Altbabylonische Rects- und Wirtschaftsurkunden aus Tell ed-Dēr bei Sippar* (TIM 7). Wiesbaden.

Ellis, R. S.
1968 *Foundation Deposits in Ancient Mesopotamia* (YNER 2). New Haven and London, 1968.

Ellis, M. de J.
1970 A Note on the "Chariot's Crescent." *JAOS* 90: 266–69.

Eph'al, I.
1984 The Assyrian Siege Ramp at Lachish: Military and Lexical Aspects. *Tel Aviv* 11: 60–70.

Falkenstein, A.
1936 *Archaische Texte aus Uruk* (Ausgrabungen der Deutschen Forschungsgeminschaft in Uruk-Warka 2). Berlin and Leipzig.

1937 Archaische Texte des Iraq-Museums in Bagdad. *OLZ* 40: 402–08.

1948 Der "Sohn des Tafelhauses." *WO* 1: 172–86.

1953 Die babylonische Schule. *Saeculum* 4: 125–37.

1965 Fluch über Akkade. *ZA* 23.

1966 Sumerische Bauausdrücke. *Orientalia*, NS 35: 229–46.

Figulla, H. H., and Martin, W. J.
1953 *Letters and Documents of the Old-Babylonian Period* (UET 5). London. [*UET* 5]

Finkelstein, J.J.
1968 An Old Babylonian Herding Contract and Genesis 31: 38f. *JAOS* 88: 30–36.

Foster, B.
1983 Self-Reference of an Akkadian Poet. *JAOS* 103: 123–30.

Foxvog, D. A.
1986 A Summary of Non-Sealed Labor Assignments from Umma. *Acta Sumerologica* 8: 59–75.

Frank, C.
1928 *Strassburger Keilschrifttexte*. Berlin and Leipzig: Water de Gruyter & Co.

Freudenthal, H.
1977 What is Algebra and What Has It Been in History. *Archive for History of Exact Sciences* 16: 189–201.

Friberg, J.
1978 The Third Millennium Roots of Babylonian Mathematics I: A Method for the Decipherment, through Mathematical and Metrological Analysis, of Proto-Sumerian and Proto-Elamite Semi-Pictographic Inscriptions. Göteborg, Sweden: Department of Mathematics CTH-GU.

1979a The Early Roots of Babylonian Mathematics II: Metrological Relations in a Group of Semi-Pictographic Tablets of the Jemdet Nasr Type, Probably from Uruk-Warka. Göteborg, Sweden: Department of Mathematics CTH-GU.

1979b Methods and Traditions of Babylonian Mathematics II: An Old-Babylonian Catalogue Text with Equations for the Square and the Circle. Göteborg, Sweden: Department of Mathematics CTH-GU.

1979c A Proposal for the Use of New Parallel Multiplication and Division Algorithms, Modelled after Egyptian and Babylonian Prototypes. Göteborg, Sweden: Department of Mathematics CTH-GU.

1981a Methods and Traditions of Babylonian Mathematics, Plimpton 322, Pythagorean Triples and the Babylonian Triangle Parameter Equations. *Historia Mathematica* 8: 277–318.

1981b Methods and Traditions of Babylonian Mathematics 2: An Old-Babylonian Catalogue Text with Equations for the Square and the Circle. *JCS* 33: 57–64.

1981c On the Babylonian Standard Tables of Reciprocals and Squares of "Six-Place" Regular Sexagesimal Numbers. *Historia Mathematica* 8: 465.

1982 *A Survey of Publications on Sumero-Akkadian Mathematics, Metrology and Related Matters (1854–1982)*. Göteborg, Sweden: Department of Mathematics CTH-GU.

1983 The Early Roots of Babylonian Mathematics III: Three Remarkable Texts from Ebla. Göteborg, Sweden: Department of Mathematics CTH-GU.

1984 Numbers and Measures in the Earliest Written Records. *Scientific American* 250: 110–17.

1986a The Early Roots of Babylonian Mathematics III: Three Remarkable Texts from Ancient Ebla. *Vicino Oriente* 6: 3–25.

1986b *Mathematics and Metrology in Ancient Near Eastern Texts*, I and II. Göteborg, Sweden: Department of Mathematics CTH-GU.

1987 Measures of Area in Ancient Mesopotamia: in terms of "fields" and "gardens," "asses," of "seed-barley," or of "reeds." Göteborg, Sweden: Department of Mathematics CTH-GU.

Gadd, C. J.

1922 Forms and Colours. *RA* 19: 149–58.

1954 Inscribed Prisms of Sargon II from Nimrud. *Iraq* 16 (1954), pp. 173-201 + pl.

1956 *Teachers and Students in the Oldest Schools*. London: School of Oriental and African Studies.

1957 Fragments of Assyrian Scholastic Literature. *BSOAS* 20: 255–65.

1967 Omens Expressed in Numbers. *JCS* 21: 52–63.

Gandz, S.

1929 The Origin of Angle-Geometry. *Isis* 12: 452–81.

1936a The Babylonian Tables of Reciprocals. *Isis* 26: 426–31.

1936b Mene Mene Tekel Upharsin, a Chapter in Babylonian Mathematics. *Isis* 26: 82–94.

1937 The Origin and Development of the Quadratic Equations in Babylonian, Greek and Early Arabic Algebra. *Osiris* 3: 405–557.

1938 The Algebra of Inheritance. *Osiris* 5: 319–91.

1940 Studies in Babylonian Mathematics II: Conflicting Interpretations of Babylonian Mathematics. *Isis* 31: 405–25.

1947 Studies in Babylonian Mathematics III: Isoperimetric Problems and the Origin of the Quadratic Equations. *Isis* 32: 103–15.

1948 Studies in Babylonian Mathematics I: Indeterminate Analysis in Babylonian Mathematics. *Osiris* 8: 12–40.

Gelb, I.J.

1967 Growth of a Herd of Cattle in Ten Years. *JCS* 21: 64–69.

1970 *Sargonic Texts in the Ashmolean Museum, Oxford* (MAD 5). Chicago:The University of Chicago Press.

1982 Measures of Dry and Liquid Capacity. *JAOS* 102: 585–90.

de Genouillac, H.

1924 *Premières recherches archéologiques à Kich (Fouilles françaises d'El-'Akhymer, mission d'Henri de Genouillac, 1911–1912)* 1. Paris: É. Champion.

1925 *Premières recherches archéologiques à Kich (Fouilles françaises d'El-'Akhymer, mission d'Henri de Genouillac, 1911–1912)* 2. Paris: É. Champion.

Gerdes, P.
1985 Three Alternate Methods of Obtaining the Ancient Egyptian Formula for the Area of a Circle. *Historia Mathematica* 2: 261–68.

Geršman, I.G.
1959 Nerazresennye voprosy vavilonskoi metrologii (Unresolved problems in Babylonian metrology). *VDI* 68: 100–12.

Goetsch, H.
1968–69 Die Algebra der Babylonier. *Archive for the History of Exact Science* 5: 79–160.

Goetze, A.
1945 The Akkadian Dialects of the Old-Babylonian Mathematical Texts. In *Mathematical Cuneiform Texts* (AOS 29), by O. Neugebauer and A. J. Sachs, pp. 146–51. New Haven, American Oriental Society and the American Schools of Oriental Research.

1946 Number Idioms in Old Babylonian. *JNES* 5: 185–202.

1948 Review of *Mathematical Cuneiform Texts* by O. Neugebauer and A. J. Sachs. *JCS* 2: 33–37.

1951 A Mathematical Compendium from Tell Harmal. *Sumer* 7: 126–55.

1957 Old Babylonian Documents from Sippar in the Collection of the Catholic University of America. *JCS* 11: 14–40.

Gordon, E.I.
1959 *Sumerian Proverbs* (Museum Monographs 19). Philadelphia, The University Museum.

1960 **é-dub-ba** Compositions. *BiOr* 17: 142–44.

Greengus, S.
1979 *Old Babylonian Tablets from Ischali and Vicinity* (PIHANS 44). Istanbul.

Groneberg, B.
1990 Bord de jarre inscrit du Palais de Mari. *N.A.B.U.* 2 (July): 46–47.

Grosz, K.
1986 Daughters Adopted as Sons at Nuzi and Emar. *CRAA* 33: 81-86.

Grizmek, B.
1972 Editor. *Grizmek's Animal Life Encyclopedia*, Vol. 3 of *Mammals*. New York: van Nostrand Reinhold Company.

Guitel, G.
1963 Signification mathématique d'une tablette sumérienne. *RA* 57: 145–50.

Gundlach, K-B. and von Soden, W.
1963 Einige altbabylonische Texte zur Lösung "quadratischer Gleichungen." *Abhandlung aus dem Mathematishcen Seminar der Universität Hamburg* 26: 248-63.

Gurney, O.R.
1968 An Old Babylonian Treatise on the Tuning of the Harp. *Iraq* 30: 229–33.

1973 *Middle Babylonian Legal Documents and Other Texts* (UET 7). London.

Gurney, O.R., and Finkelstein, J.J.
1957 *The Sultantepe Tablets*, I. London: The British Institute of Archaeology at Ankara.

Güterbock, H.
1970 Musical Notation in Ugarit. *RA* 64: 45–52.

Hackman, G. G.
1958 *Sumerian and Akkadian Administrative Texts from Predynastic Times to the End of the Akkad Dynasty* (BIN 8). New Haven: Yale University Press.

Hall, B. S.
1979 Der Meister sol auch kennen schreiben und lesen: Writings about Technology ca. 1400–ca. 1600 A.D. and Their Cultural

Implications in Early Technologies. In *Early Technologies* 3, ed. by D. Schmandt-Besserat, pp. 47–58. Malibu: UNDENA.

Hallo, W. W.
1963 Review of *Le travail du métal au pays de Sumer* by H. Limet. *BiOr* 20: 136–42.

1970 Antediluvian Cities. *JCS* 23: 57–67.

1976 Review of *La Collezione Schollmeyer* by G. Pettinato and H. Waetzoldt. *BiOr* 33: 38–40.

1977 New Moons and Sabbaths: A Case-Study in the Contrastive Approach. *HUCA* 48: 1–18.

1979 Obiter Dicta Ad SET. In *Studies in Honor of Tom B. Jones* (AOAT 203), pp. 1–14. Kevelaer: Verlag Butzon and Bercker Kevelaer.

Hallo, W. W., and Simpson, Wm. K.
1971 *The Ancient Near East*. New York: Harcourt Brace Jovanovich, Inc.

Hanson, A.W.
1952 Field Plans. *MCS* 2: 1-3, 21–26.

Heinrich, E. and Seidl, U.
1967 Grundrisszeichnungen aus dem Alten Orient. *MDOG* 98: 24–45.

Helbaek, H.
1960 Ecological Effects of Irrigation in Ancient Mesopotamia. *Iraq* 22: 186–96.

Hilprecht, H. V.
1906 *Mathematical, Metrological and Chronological Tablets from the Library at Nippur* (*BE* 20/1). Philadelphia: Department of Archaeology, The University of Pennsylvania.

Høyrup, J.
1982 Investigations of an Early Sumerian Division Problem c. 2500 B.C. *Historia Mathematica* 9: 19–36.

1984 *Babylonian Algebra from the View-Point of Geometrical Heuristics*. Roskilde, Denmark: Roskilde University Centre Institute of Educational Research, Media Studies and Theory of Science.

1985a *Algebra and Naive Geometry: An Investigation of Some Basic Aspects of Old Babylonian Mathematical Thought*. Roskilde, Denmark: Roskilde University Centre Institute of Educational Research, Media Studies and Theory of Science. Photocopy.

1985b Varieties of Mathematical Discourse in Pre-Modern Socio-cultural Contexts: Mesopotamia, Greece, and the Latin Middle Ages. *Science and Society* 49: 4–41.

1989 Zur Frühgeschichte algebraischer Denkweisen. *Mathematische Semesterberichte* 36: 1–46.

1990 Algebra and Naive Geometry: An Investigation of Some Basic Aspects of Old Babylonian Mathematical Thought I. *AfO* 17: 27–69, 262–354.

Huber, P.

1955 Zu einem mathematischen Keilschrifttext (VAT 8512). *Isis* 46: 104–06.

1957 Bemerkungen über mathematische Keilschrifttexte. *L'enseignement mathématique* 3: 19–27.

Hulin, P.

1963 A Table of Reciprocals with Sumerian Spellings. *JCS* 17: 72–76.

Hunger, H.

1973 *Spätbabylonische Texte aus Uruk* 1. Berlin.

Hunt, R. C.

1988 Hydraulic Management in Southern Mesopotamia in Sumerian Times. *Bulletin on Sumerian Agriculture* 4: 189–206.

Hussey, M. I.

1912 *Sumerian Tablets in the Harvard Semitic Museum* (HSS 1). Cambridge: Harvard University Press.

Jacobsen, Th.
1935 *Sennacherib's Aqueduct at Jerwan* (OIP 24). Chicago, The University of Chicago Press.

1981 *Salinity and Irrigation Agriculture in Antiquity* (BM 14). Malibu: UNDENA.

1982 Oral to Written. In *Societies and Languages of the Ancient Near East (Studies in Honour of I.M. Diakonoff*, pp. 129–37. Warminister, England: Arts & Phillips, Ltd.

Jeyes, U.
1989 *Old Babylonian Extispicy: Omen Texts in the British Museum* (Nederlands Historisch-Archaeologisch Instituut te Istanbul 64). Netherlands.

Jestin, R. R.
1937 *Tablettes sumériennes de Šuruppak conservées au Musée de Stamboul* (Mémoires de l'Institut français d'archéologie de Stamboul 3). Paris. [*TSŠ*]

Joannes, F.
1990 Cadastre et titres de propriété en Babylonie achéménide. *N.A.B.U.* 1 (March): 7–8.

Jordon, J.
1932 *UVB* 3 (APAW 1932/2). Berlin: Verlag der Akademie der Wissenschaften.

Keiser, C.E.
1919 *Selected Temple Accounts of the Ur Dynasty* (YOS 4). New Haven: Yale University Press.

Keiser, C. E. and Nies, J. B.
1920 *Historical, Religious and Economic Texts and Antiquities* (BIN 2). New Haven: Yale University Press.

Kilmer, A. D.
1960 Two New Lists of Key Numbers for Mathematical Operations. *Or* NS 29: 273–308 + 3 pl.

1964 The Use of Akkadian *dkš* in Old Babylonian Geometry Texts. In *Studies Presented to A. Leo Oppenheim*, pp. 140–46. Chicago.

1965 The Strings of Musical Instruments: Their Names, Numbers and Significance. In *Studies in Honor of Benno Landsberger* (AS 16), pp. 261–68. Chicago: The University of Chicago Press.

1971 The Discovery of an Ancient Mesopotamian Theory of Music. *PAPS* 115: 131–49.

1974 The Cult Song with Music from Ancient Ugarit. *RA* 68: 69–82.

1984 A Music Tablet from Sippar(?): BM 65217 + 66616. *RA* 78: 69–80 + 2 pl.

1985 Additions and Corrections to Akkadian Lists of Mathematical Coefficients (*igigubbû*) (*Orientalia* 29: 273–308). Photocopy.

1990a Review of *Musikgeschichte in Bildern: Mesopotamien* by A. Rashid. *JAOS* 110: 758-60.

1990b Sumerian and Akkadian Names for Designs and Geometric Shapes. In *Investigating Artistic Environments in the Ancient Near East*, edited by A. Gunter, pp. 83-91. Smithsonian Institution.

Kilmer, A. D.; Brown R. R.; and Crocker R. I.
1986 *Sounds from Silence: Recent Discoveries in Ancient Near Eastern Music*. Berkeley: Bīt Enki Publications.

Kilmer, A. D. and Civil, M.
1986 Old Babylonian Musical Instructions Relating to Hymnody. *JCS* 38: 94–98.

King, L. W.
1900 *Cuneiform Texts from Babylonian Tablets in the British Museum* 9. London.

Klein, J.
1986 On Writing Monumental Inscriptions in Ur III Scribal Curriculum. *RA* 80: 1–7.

Klíma, J.

1940 *Untersuchungen zum Altbabylonischen Erbrecht* (Monographie des Archiv Orientálni 7). Prague: Orientalisches Institut.

Kramer, S. N.

1949 Schooldays: A Sumerian Composition Relating to the Education of a Scribe. *JAOS* 69: 199–215.

1963 *The Sumerians: Their History, Culture and Character.* Chicago and London: The University of Chicago Press.

Kraus, F. R.

1966 *Staatliche Viehhaltung im Altbabylonischen Lande Larsa* (MKNAW Nieuwe Reeks 29/V). Amsterdam.

1969 Von Altmesopotamischen Erbrecht. In *Essays on Oriental Laws of Succession* (Studia et Documenta 9), pp. 1–57. Leiden, E.J. Brill.

Kümmel, H.

1970 Zur Stimmung der babylonische Harfe. *Or* NS 39: 252–63.

Kyrala, A.

1972 Speculations on Babylonian Telescopes, Planetary Distances and Sizes. *Sumer* 28: 21–28.

Labat, R.

1965a Cryptographie cunéiforme. *RA* 59: 183.

1965b Jeux numériques dans l'idéographie susienne. In *Studies in Honor of Benno Landsberger* (AS 16), pp. 257–60. Chicago: The University of Chicago Press.

Laessøe, J.

1951 The Irrigation System at Ulḫu, 8th Century B.C. *JCS* 5: 21–32.

1953a Literacy and Oral Tradition in Mesopotamia. In *StOr Ioanni Pedersen...dicata*, pp. 205–18. Einar Munsgaard.

1953b Reflexions on Modern and Ancient Oriental Waterworks. *JCS* 7: 5–26.

Lambert, M.

1953 La vie économique à Shuruppak. *Sumer* 9: 198–213.

1954 La vie économique à Shuruppak. *Sumer* 10: 150–90.

1965 La vie économique à Umma à l'époque d'Agade. *RA* 59: 61–126.

Lambert, W. G.

1957 Ancestors, Authors and Canonicity. *JCS* 11: 1–14, 112.

1971 The Converse Tablet: A Litany with Musical Instructions. In *Near Eastern Studies in Honor of Wm. Foxwell Albright*, ed. by H. Goedicke, pp. 335–53. Baltimore.

Landsberger, B.

1934 Aus der 3 Tafel **DIR** = *siiaku*. *ZA* 42: 155–63.

1937 *Die Serie ana-ittišu* (MSL 1). Rome: Pontifical Biblical Institute.

1954 Babylonian Scribal Craft and Its Terminology. *ICO* 23: 123–26.

1957 *The Series ḪAR-ra = ḫubullu, Tablets I–IV* (MSL 5). Rome: Pontifical Biblical Institute.

1960a Einige unerkannt gebliebene oder verkannte Nomina des Akkadischen 1. *ūṭu* = "halbe Elle," "Spanne." *Wiener Zeitschrift für die Kunde des Morgenlandes* 56: 109–12.

1960b Scribal Concepts of Education. In *City Invincible*, ed. by C. H. Kraeling and R. Mc. Adams, pp. 94–123. Chicago: The University of Chicago Press.

Landsberger, B.; Hallock, R. T.; Jacobsen, Th.; and Falkenstein, A.

1956 *Neo-babylonian Grammatical Texts* (MSL 4). Rome: Pontifical Biblical Institute.

Langdon, S.

1911 *Tablets from the Archives of Drehem, with a Complete Account of the Origin of the Sumerian Calendar, Translation, Commentary and 23 Plates*. Paris.

1918 Assyriological Notes: Mathematical Observations on the Scheil-Esagila Tablet. *RA* 15: 110–12.

1936 The Length of Daylight in Each Month. *RA* 33: 194–95.

Laroche, E.
1973 Notation musicale. *RA* 67: 124–29.

Lawergren, B., and Gurney, O. R.
1987 Sound Holes and Geometrical Figures: Clues to the Terminology of Ancient Mesopotamian Harps. *Iraq* 49: 37–52.

Leemans, M., and Bruins, E. M.
1951 Un texte vieux-babylonien concernant des cercles concentriques. *CRRA* 2: 31–35.

Legrain, L.
1913 Collection Louis Cugnin, textes cunéiformes. Catalogue, transcription et traduction. *RA* 10: 41–68.

1935 Quelques textes anciens. *RA* 32: 128–29.

1947 *Business Documents of the Third Dynasty of Ur* (UET 3). London. [*UET* 3]

Leichty, E.
1987 Omens from Doorknobs. *JCS*: 190–96.

Lewy, H.
1938 La mesure de l'*imeru* dans les textes de Nuzi. *RA* 35: 33–35.

1944 Assyro-Babylonian and Israelite Measures of Capacity and Rates of Seeding. *JAOS* 64: 65–73.

1945 A Propos of Babylonian Metrology. *BASOR* 98: 25–26.

1947 Marginal Notes on a Recent Volume of Babylonian Mathematical Texts. *JAOS* 67: 305–20.

1949a Origin and Development of the Sexagesimal System of Numeration. *JAOS* 69: 1–11.

1949b Studies in Assyro-Babylonian Mathematics and Metrology. *OR* NS 18: 40–67, 137–70.

1951 Studies in Assyro-Babylonian Mathematics and Metrology. *Or* NS 20: 1–12.

1964 The Assload, the Sack and Other Measures of Capacity. *RSO* 39: 181–97.

Lieberman, S.
1980 Of Clay Pebbles, Hollow Clay Balls, and Writing: A Sumerian View. *AJA* 84: 339–58.

Limet, H.
1960 *Le travail du métal au pays de Sumer au temps de la IIIe dynastie d'Ur*. Paris: Sociéte d'Édition Les Belles Lettres.

1973 *Étude de documents de la période d'Agadé appartenant à l'Université de Liège*. Paris. [Limet, *Liège*]

1982 Une science "bloquée": le cas de la Mésopotamie ancienne. *Akkadica* 16: 17–34.

Liverani, M.
1990 The Shape of Neo-Sumerian Fields. *Bulletin on Sumerian Agriculture* 5: 147–86.

Livingstone, A.
1986 *Mystical and Mythological Explanatory Works of Assyrian and Babylonian Scholars*. Oxford: Clarendon Press.

Luckenbill, D. D.
1930 *Inscriptions from Adab* (OIP 14). Chicago.

Maekawa, K.
1974 Agricultural Production in Ancient Sumer, Chiefly from Lagash Materials. *Zinbun* 13: 1–60.

1981 The Agricultural Texts of Ur III Lagash of the British Museum (I). *Acta Sumerologica* 3: 37–61.

1986a The Agricultural Texts of Ur III Lagash of the British Museum (III). *Acta Sumerologica* 8: 85–120.

1986b The Agricultural Texts of Ur III Lagash of the British Museum (IV). *Zinbun* 21: 91–157 + pls.

1987 Collective Labor Service in Girsu-Lagash: The Pre-Sargonic and Ur III Periods. In *Labor in the Ancient Near East* (AOS 68), ed. by M. A. Powell, pp. 49–71. New Haven, Conn.

Mahoney, M. S.
1970–71 Babylonian Algebra: Form vs. Content. *Studies in the History and Philosophy of Science* 1: 269–380.

Matouš, L.
1933 *Die lexikalischen Tafelserien der Babylonier und Assyrer in den Berliner Museen I: Gegenstandslisten (Serie ḪAR-ra = ḫubullu)*. Berlin.

May, K. O.
1973 *Bibliography and Research Manual of the History of Mathematics*. Toronto and Buffalo, University of Toronto Press.

Mayer, W.
1985 *Zur Unterteilung des Sekels im spätzeitlichen Babylonien*. *Or* NS 54: 203–15.

Meek, T. J.
1935 *Excavations at Nuzi III, Old Akkadian, Sumerian, and Cappadocian Texts from Nuzi* (HSS 10). Cambridge.

van der Meer, P. E.
1935 *Textes scolaires de Suse* (MDP 27). Paris.

1938 *Syllabaries A, B^1 and B with Miscellaneous Lexicographical Texts from the Herbert Weld Collection*. Oxford and London.

Meissner, B.
1937 Warenpreisen in Babylonien. *APAW*: 5–7.

Michalowski, P.

1977 Ur III Topographical Names. *OrAnt* 16: 287–96.

1983 Review of *Briefe aus dem Iraq Museum* (AbB 8) by L. Cagni. *JCS* 35: 221–28.

Michel, C.

1989 Un problème relatif à la métallurgie. *N.A.B.U.* 4 (December): 82–83.

Millard, A. R.

1988 The Bevelled-Rim Bowls: Their Purpose and Significance. *Iraq* 50: 49–57.

Moorey, P.R.S.

1985 *Materials and Manufacture in Ancient Mesopotamia: The Evidence of Archaeology and Art*, BAR International Series 237.

Moorey, P. R. S.; Curtis, J. E.; Hook, D. R.; and Hughes, M. J.

1988 New Analyses of Old Babylonian Metalwork from Tell Sifr. *Iraq* 50: 39–48.

Muhly, J.D.

1973 *Copper and Tin: the Distribution and the Nature of the Metals Trade in the Bronze Age* (Transactions of the Connecticut Academy of Arts and Sciences 43). Hamden, Conn.: Archon Books.

Nemet-Nejat, K. R.

1982 *Late Babylonian Field Plans in the British Museum* (Studia Pohl, Series Maior 11). Rome: Pontifical Biblical Institute.

1988 Cuneiform Mathematical Texts as Training for Scribal Professions. In *A Scientific Humanist: Studies in Memory of Abraham Sachs* (Occasional Publications of the Samuel Noah Kramer Fund 9), edited by E. Leichty, M. deJ. Ellis, and P. Gerardi, pp. 285–300. Philadelphia.

1989 A Survey of Publications on Cuneiform Mathematical Texts. In *Semitic Studies in Memory of Moshe Held* (Journal of the Ancient

Near Eastern Society 19), edited by E. Greenstein and D. Marcus, 119-33. New York.

Neugebauer, O.

1931a Beiträge zur Geschichte der babylonischen Arithmetik. *Quellen und Studien zur Geschichte der Mathematik* B 1: 120-30.

1931b Sexagesimalsystem und babylonische Bruchrechnung II. *Quellen und Studien zur Geschichte der Mathematik* B 1: 425-57.

1931c Sexagesimalsystem und babylonische Bruchrechnung III. *Quellen und Studien zur Geschichte der Mathematik* B 1: 458-63.

1931d Zur Geschichte der babylonischen Mathematik. *Quellen und Studien zur Geschichte der Mathematik* B 1: 67-80.

1931-32 Über die Approximation irrationaler Quadratwurzeln in der babylonischen Mathematik. *AfO* 7: 90-99.

1932-33 Zur Transkription mathematischer und astronomischer Keilschrifttexte. *AfO* 8: 221-23.

1933a Babylonische "Belagerungsrechnung." *Quellen und Studien zur Geschichte der Mathematik* B 2: 305-10.

1933b Das Pyramidenstumpf-Volumen in der vorgriechischen Mathematik. *Quellen und Studien zur Geschichte der Mathematik* B 2: 347-51.

1933c Studien zur Geschichte der antiken Algebra I. *Quellen und Studien zur Geschichte der Mathematik* B 2: 1-27.

1933d Verbesserungen zu "Studien zur Geschichte der antiken Algebra I." *Quellen und Studien zur Geschichte der Mathematik* B 2: 253-54.

1933-34 Zur Terminologie der mathematischen Keilschrifttexte. *AfO* 9: 199-204.

1934 *Vorlesungen über Geschichte der antiken mathematischen Wissenschaften*. Bd. 1, *Vorgriechische Mathematik*. Berlin: Springer.

1934–36 Serientexte in der babylonischen Mathematik. *Quellen und Studien zur Geschichte der Mathematik* B 3: 106–14.

1935 Der Verhältnisbegriff in der babylonischen Mathematik. *AnOr* 12: 235–58.

1935–37 *Mathematische Keilschrifttexte*. 3 vols. Berlin: Verlag von Julius Springer. [*MKT*]

1936 Zur geometrischen Algebra. *Quellen und Studien zur Geschichte der Mathematik* B 3: 245–59.

1941 On a Special Use of the Sign "Zero" in Cuneiform Astronomical Texts. *JAOS* 61: 213–15.

1945 The History of Ancient Astronomy Problems and Methods. *JNES* 4: 1–38.

1947 The Water Clock in Babylonian Astronomy. *Isis* 37: 37–43.

1949 Comments on Publications by Mrs. Hildegard Lewy on Mathematical Cuneiform Texts. *Or* NS 18: 422–26.

1963 The Survival of Babylonian Methods in the Exact Sciences of Antiquity and the Middle Ages. *PAPS* 107: 528–35.

1969 *The Exact Sciences in Antiquity*. New York: Dover Edition.

Neugebauer, O., and Sachs, A. J.

1945 *Mathematical Cuneiform Texts* (AOS 29). New Haven: American Oriental Society and the American Schools of Oriental Research. [*MCT*]

1984 Mathematical and Metrological Texts. *JCS* 36: 243–51.

Neugebauer, O. and Struve, V. V.
1931 Über die Geometrie des Kreises in Babylonien. *Quellen und Studien zur Geschichte der Mathematik* B 1: 81–92.

Neugebauer, O. and Waschow, H.
1932 Bemerkungen über Quadratwurzeln und Quadratwurzel-Approximationen in der babylonischen Mathematik. *Quellen und Studien zur Geschichte der Mathematik* B 2: 291–97.

Nikol'skii, M. V.
1908 Dokumenty hozyaistvennoi otcetnosti drevneisei epohi Haldei (Economical documents from the oldest period of the Chaldeans). *Drevnosti vostocnyya* 3/2.

1915 Dokumenty...2: Epoha dinasti Agade i epoha dinastii Ura "Documents..." (The epoch of the dynasty of Agade and the epoch of the dynasty of Ur). *Drevnosti Vostocnyya* 5.

Nougayrol, J.
1968 Textes suméro-accadiens des archives et bibliothèques privées d'Ugarit. Chapter 1 in *Ugaritica* 5 (Mission de Ras Shamra 16), by J. Nougayrol, E. Laroche, C. Virolleaud, C. F. A. Schaeffer. Paris.

Oates, D.
1985 Walled Cities in Northern Mesopotamia in the Mari Period. *MARI* 4: 585–94.

Oppenheim, A. L.
1938 Seqel, Mine and Talent in Nuzi. *OLZ* 41: 485–86.

1947a A Fiscal Practice of the Ancient Near East. *JNES* 6: 116–20.

1947b Review of *Mathematical Cuneiform Texts* by O. Neugebauer and A. J. Sachs. *JNES* 6: 126–28.

1949 Akk. *arad ekalli* 'Builder.' In *F. Hrozný Festschrift* (ArOr 17), pp. 227–35. Prague.

1958 On an Operational Device in Mesopotamian Bureaucracy. *JNES* 17: 121–28.

1964 *Ancient Mesopotamia*. Chicago and London: The University of Chicago Press.

1965 A Note on the Scribes in Mesopotamia. In *Studies in Honor of Benno Landsberger* (AS 16), pp. 253–56. Chicago: The University of Chicago Press.

Oppert, J.
1886a Les Mesures assyriennes de capacité et de superficie. *RA* 1: 124–27.

1886b La notation des mesures de capacité dans les documents juridiques cunéiformes. *ZA* 1: 87–90.

1889 Les signes numériques des mesures babyloniennes de capacité. *ZA* 4: 371–73.

Owen, D.
1972 Neo-Sumerian Texts from American Collections, I. *JCS* 24: 137–73.

Parrot. A.
1958 *Mission archéologique de Mari* 2. Paris.

Peet, T. E.
1923 *The Rhind Mathematical Papyrus*. London.

Pemberton, W., Postgate, J.N., and Smyth, R.F.
1988 Canals and Bunds, Ancient and Modern. *Bulletin on Sumerian Agriculture* 4: 207–21.

Petruso, K. M.
1985 Additive Progression in Prehistoric Mathematics: A Conjecture. *Historia Mathematica* 12: 101–06.

Petschow, H.
1974 *Mittelbabylonische Rechts- und Wirtschaftsurkunder der Hilprecht-Sammlung Jena* (ASAW 64). Berlin.

Pettinato, G.

1969 *Texte zur Landwirtschaft in der Ur III-Zeit: die runden Tafeln* (AnOr 45). Rome: Pontifical Biblical Institute.

1977 TSS 242, Fondazione della città *UNKEN*[ki]. *OrAnt* 16: 173–76.

1981a *The Archives of Ebla*. Garden City, New York: Doubleday & Co.

1981b Lista di numeri sumerici: texto N. 54. *Materiali epigrafici di Ebla* 3: 212–13.

1981c Il problema dello scriba di Kis: n. 73. *Materiali epigrafici di Ebla* 3: 269–70.

Pettinato, G., and Waetzoldt, H.

1975 *Saatgut und Furchenabstand beim Getreidenanbau* (StOr 46). Helsinki.

Picchioni, S. A.

1980 La direzione della scrittura cuneiforme e gli archivi di Tell Mardikh Ebla. *Or* NS 49: 225–51.

Pinches, Th. G.

1891 *The Cuneiform Inscription of Western Asia*. Vol. 4, *Selection from the Miscellaneous Inscriptions of Assyria*, prepared...by...Sir H.C. Rawlinson. 2nd ed. London.

1908 *The Amherst Tablets Part I: Texts of the Period extending to and including the Reign of Bûr-Sin*. London.

1909a *The Cuneiform Inscription of Western Asia*. Vol. 5, *A Selection from the Miscellaneous Inscriptions of Assyria*, prepared...by...Sir H.C. Rawlinson. London.

1909b Some Mathematical Tablets of the British Museum. In *Hilprecht Anniversary Volume*, pp. 73–78. Leipzig.

1963 *Cuneiform Texts from Babylonian Tablets in the British Museum 44, Miscellaneous Texts*. London: British Museum Publications.

Pohl, A.
1935 *Vorsargonische und sargonische Wirtschaftstexte* (Texte und Materialien der Frau Hilprecht Collection of Babylonian Antiquities im Eigentum der Universität Jena 5). Leipzig.

Pomponio, F.
1980 AO 7754 ed il sistema ponderale di Ebla. *OrAnt* 19: 171–86.

1981 Tabella di concordanze di misure per aridi. *Materiali epigrafici di Ebla* 3: 270–71.

Postgate, J. N.
1978 An Inscribed Jar from Tell Al Rimah. *Iraq* 40: 71–75.

1986 Middle Assyrian Tablets: the Instruments of Bureaucracy. *AfO* 13: 10–39.

1992 *Early Mesopotamia: Society and Economy at the Dawn of History*. London: Routledge.

Pottage, J.
1974 The Mensuration of Quadrilaterals and the Generation of Pythagorean Triads: A Mathematical, Heuristical and Historical Study with Special Reference to Brahmagupta's Rules. *Archive for the History of Exact Sciences*: 299–354.

Powell, M. A.
1971 *Sumerian Numeration and Metrology*. Ph.D. diss., University of Minnesota.

1972 Sumerian Area Measures and the Alleged Decimal Substratum. *ZA* 62: 165–221.

1973a A Note on the "*imērum*" Measure at Mari. *RA* 67: 77–78.

1973b On the Reading and Meaning of GANA$_2$. *JCS* 25: 178–84.

1973c Review of *CT* 50 by E. Sollberger. *ZA* 63: 99–106.

1974 Notes on Akkadian Numbers and Number Syntax. *JSS* 24: 13–18.

1975 Review of *Étude de documents de la période d'Agadé appartenant à l'Université de Liège* by H. Limet. *JCS* 27: 180–88.

1976a The Antecedents of Old Babylonian Place Notation and the Early History of Babylonian Mathematics. *Historia Mathematica* 3: 417–39.

1976b Two Notes on Metrological Mathematics in the Sargonic Period. *RA* 70: 97–102.

1978 A Contribution to the History of Money in Mesopotamia prior to the Invention of Coinage. In *Festschrift Lubor Matouš*, ed. by B. Hruška and G. Komoróczy, pp. 211–43. Budapest.

1979 Ancient Mesopotamian Weight Metrology: Methods, Problems and Perspectives. In *Studies in Honor of Tom B. Jones* (AOAT 203), pp. 71–109. Kevelaer: Verlag Butzon and Bercker.

1982a Aspects of Cuneiform Writing. *ArOr* 51: 80–82.

1982b Merodach-Baladan at Dur-Jakin. *JCS* 34: 59–61.

1982c Metrological Notes on the Esagila Tablet and Related Matters. *ZA* 72: 106–23.

1983 The Standard of Dilmun. In *Dilmun*, ed. by D. T. Potts, pp. 141–42. Berlin.

1984a Late Babylonian Surface Mensuration. *AfO* 31: 32–66.

1984b On the Absolute Value of the Assyrian *qa* and *emār*. *RA* 46: 57–61.

1985 Salt, Seed, and Yields in Sumerian Agriculture. A Critique of the Theory of Progressive Salinization. *ZA* 75: 7–38.

1986 *mun-du* as an Akkadian Plural Loan Word in Sumerian. *ZA* 76: 12–16.

1988 Evidence for Agriculture and Waterworks in Babylonian Mathematical Texts. *Bulletin on Sumerian Agriculture* 4: 161–72.

1990 Masse und Gewichte. In *RlA* 7, ed. by D.O. Edzard, pp. 481–517. Berlin: Walter de Gruyter.

Price, D. J. de Solla.
1964–65 The Babylonian "Pythagorean Triangle" Tablet. *Centaurus* 10: 1–13.

1979 Proto-Astrolabes, Proto-Clocks and Proto-Calculators: The Point of Origin of High Mechanical Technology in Early Technologies. In *Early Technologies* 3, ed. by D. Schmandt-Besserat, pp. 61–63. Malibu: UNDENA.

Reiner, E.
1985 The Uses of Astrology. *JAOS* 105: 589-95.

1970 *ana nalban*. *AfO* 23: 89-90.

Renger, J.
1990 Rivers, Watercourses and Irrigation Ditches. *Bulletin on Sumerian Agriculture* 5: 31–46.

Rich, B.
1973 *Modern Elementary Algebra*. Schaum's Outline Series. New York, etc.: McGraw Hill.

Rinaldi, G.
1949 Tablette sumeriche della III dinastia di Ur, serie III. *Aegyptus* 29: 102–17.

Robins, G., and Chute, C. C. D.
1985 Mathematical Bases of Ancient Egyptian Architecture and Graphic Art. *Historia Mathematica* 12: 107–22.

Roux, G.
1964 *Ancient Iraq*. London: Penguin Books.

Rowton, M. B.
1967 Watercourses and Water Rights in the Official Correspondence from Larsa and Isin. *JCS* 21: 267–74.

Sachs, A. J.
1944 Some Metrological Problems in Old-Babylonian Mathematical Texts. *BASOR* 96: 29–39.

1945 Rejoinder. *BASOR* 98: 26–27.

1946 Notes on Fractional Expressions in Old Babylonian Mathematical Texts. *JNES* 5: 203–14.

1947a Babylonian Mathematical Texts I. *JCS* 1: 219–40.

1947b Two Neo-Babylonian Metrological Tables from Nippur. *JCS* 1: 67–71.

1952 Babylonian Mathematical Texts II–III. *JCS* 6: 151–56.

1955 *Late Babylonian Astronomical and Related Texts*, copied by T. G. Pinches and J. N. Strassmaier, prepared for publication by A. J. Sachs with the cooperation of J. Schaumberger. Providence. [*LBAT*]

Safar, F.
1951 A Further Text of Shalmaneser III. From Assur. *Sumer* 7: 3–21 + 3 pl.

Saggs. H. W. F.
1960 A Babylonian Geometrical Text. *RA* 54: 131–45.

1962 *The Greatness That Was Babylon*. Toronto and New York, The New American Library.

Salonen, A.
1939 *Die Wasserfahrzeuge in Babylonien nach sumerisch-akkadischen Quellen (mit besonderer Berücksichtigung der 4. Tafel der Serie ḪAR-ra = ḫubullu)*. Eine lexikalische und kulturgeschichtliche Untersuchung (StOr 8/IV). Helsinki.

1965 *Die Waffen der alten Mesopotamier. Eine lexikalische und kulturgeschichtliche Untersuchung* (StOr 33). Helsinki.

1966 *Die Hausgeräte der Alten Mesopotamier nach sumerisch-akkadischen Quellen.* Part II, *Gefässe* (Annales Academiae Scientiarum Fennicae, Series B, No. 144). Helsinki.

1970 *Die Fischerei im Alten Mesopotamien nach sumerisch-akkadischen Quellen.* (Annales Academiae Scientiarum Fennicae, Series B, No. 166). Helsinki.

1972 *Die Ziegeleien im alten Mesopotamien* (Annales Academiae Scientiarum Fennicae, Series B, No. 171). Helsinki.

Sasson, J. M.

1969 *The Military Establishments at Mari* (Studia Pohl, Dissertationes Scientificae de Rebus Orientis Antiqui 3). Rome: Pontifical Biblical Institute.

Sayce, A. H.

1876 Babylonian Augury by Means of Geometrical Figures. *Transactions of the Society of Biblical Archaeology* 4: 302–14.

Scheil, V.

1895 Notes d'épigraphies et d'archéologie assyriennes. *Recueil de travaux relatifs à la philologie et à l'archéologie égyptiennes et assyriennes* 17: 27–41.

1900 *Textes élamites-sémitiques, première série.* Paris.

1902 *Une saison de fouilles à Sippar* (Mémoires publiés par les membres de l'Institut Français d'Archéologie Orientale du Caire I/1). Cairo.

1915 Le calcul des volumes dans un cas particulier à l'époque d'Ur. *RA* 12: 161–72.

1918 Sur le marché aux poissons de Larsa. *RA* 15: 183–94.

1919 Tablettes pictographiques. *RA* 26: 15–17.

1938 Tablettes susiennes: exercises scolaires, calcul des surfaces. *RA* 35: 92–103.

Scheil, V., and Dieulafoy, M.
1914 Esagil ou le temple de Bêl-Marduk à Babylone." *Memoires de l'Institut National de France, Académie des Inscriptions et Belles-Lettres* 39: 293–372.

Schmandt-Besserat, D.
1977 An Archaic Recording System and the Origin of Writing. *Syro-Mesopotamian Studies* 1: 1–32.

1979a An Archaic Recording System in the Uruk-Jemdet Nasr Period. *American Journal of Archaeology*: 19–48 + 1 pl.

1979b On the Origins of Writing in Early Technologies. In *Early Technologies* 3, ed. by D. Schmandt-Besserat, pp. 41–46. Malibu: UNDENA.

1980 The Envelopes That Bear the First Writing. *Technology and Culture* 21: 357–85.

1984 Before Numerals. *Visible Language* 18: 48–60.

1986 Tokens: Facts and Interpretation. *Visible Language* 20: 250–73.

1988 Tokens at Uruk. *BaM* 19 (1988) 1–175 + pls.

Schmidt, O.
1980 On Plimpton 322; Pythagorean Numbers in Babylonian Mathematics. *Centaurus* 24: 4–13.

Schneider, N.
1930 Die Geschäftsurkunden aus Drehem und Djoha in den Staatlichen Museen (VAT) zu Berlin in Autographie und mit systematischen Wörterindices herausgegeben.
Or 47–49.

Schott, A.
1933 Zur Terminologie der mathematischen Keilschrifttexte 1: *ki ma-ṣi*. *Quellen und Studien zur Geschichte der Mathematik* B 2: 364–68.

Schroeder, O.
1920 *Keilschrifttexte aus Assur verschiedenen Inhalts* (WVDOG 35). Leipzig.

Schuster, H-S.
1931 Quadratische Gleichungen der Seleukidenzeit aus Uruk. *Quellen und Studien zur Geschichte der Mathematik* B 1: 194–200.

Schwenzer, W.
1915 *Zum altbabylonischen Wirtschaftsleben. Studien über Wirtschfsbetrieb, Preise, Darlehen und Agrarverhältnisse* (MVAG 19/III). Leipzig.

Seidenberg, A.
1962a The Ritual Origin of Counting. *Archive for the History of Exact Sciences* 2: 1–40.

1962b The Ritual Origin of Geometry. *Archive for the History of Exact Sciences* 1: 488–527.

1965 The Sixty System of Sumer. *Archive for the History of Exact Sciences* 2: 436–40.

1972 On the Area of a Semi-Circle. *Archive for History of Exact Sciences* 9: 171–211.

1978 The Origin of Mathematics. *Archive for History of Exact Sciences* 18: 301–42.

1981 The Ritual Origin of the Circle and Square. *Archive for the History of Exact Sciences* 25: 269–327.

1984–85 On Ancient Mathematical Terminology. *Archive for History of Exact Sciences* 31: 1–13.

Shaffer, A.
1980 A New Musical Term in Ancient Mesopotamian Music. *Iraq* 42: 79–83.

Sidersky, D.
1929 La division de la circonférence en 360 parties. *RA* 26: 31–32.

Sjöberg, Å.
1972 In Praise of the Scribal Art. *JCS* 24: 126–31.

1975a Examenstext A. *ZA* 64: 137–46.

1975b The Old Babylonian **eduba**. In *Sumerological studies in honor of Thorkild Jacobsen* (AS 20). pp. 159–79.

Smith, S.
1876 The Temple of Belus. *The Athenaeum Journal of the Literature...*, (Feb. 12): 232–33.

1969 Babylonian Time Reckoning. *Iraq* 31: 74–81.

Smith, S., and Wiseman, D. J.
1956 *Cuneiform Texts from Cappadocian Tablets in the British Museum* 5. London.

Snell, D. C.
1982 *Ledgers and Prices: Early Mesopotamian Merchant Accounts* (YNER 8). New Haven: Yale University Press.

von Soden, W.
1937 Review of *Mathematische Keilschrifttexte*, 3 vols., by O. Neugebauer. *ZDMG* 91: 185–204.

1939 Review of *Textes Mathématiques Babyloniens* by F. Thureau-Dangin. *ZDMG* 93: 143–52.

1952 Zu den neuen mathematischen Aufgabentexten von Tel Harmal. *Sumer* 8: 49–56.

1964 Review of *Textes mathématiques de Suse* (MDP 34) by E. M. Bruins and M. Rutten. *BiOr* 21: 44–50.

1965 Leistung und Grenze sumerischer und babylonischer Wissenschaft (schluss). In *Die Welt als Geschichte* 2, pp. 411–64 and 509–57. Reprint. Darmstadt.

1967 Akkadisch *šukûšum*. In *Festschrift für Wilhelm Eilers. Ein Dokument der Internationalen Forschung zum 27 Sept., 1966*, p. 123. Weisbaden.

1971 Etemenanki vor Asarhaddon nach der Erzählung vom Turmbau zu Babel und dem Erra-Mythos. *UF* 3: 253–63.

1977 Mathematische Konstantenlisten als Zeugnisse für Arbeitsnormen in Babylonien. *XX. Deutscher Orientalistentag/Erlangen*: 107–09.

1982 Review of *Aspects of Cuneiform Writing* (1980), ed. by M. A. Powell. *ZA* 72: 291–93.

Sollberger, E.
1965 *Royal Inscriptions, Part II* (UET 8). London.

Soubeyran, D.
1984 Textes mathématiques de Mari. *RA* 78: 19–48.

Speleers, L.
1924 *Recueil des inscriptions de l'Asie Antérieure des Muséees Royaux du Cinquantenaire à Bruxelles. Textes sumériens, babyloniens et assyriens*. Paris.

Steele, F. R.
1951 Writing and History: the New Tablets from Nippur. *UMB* 16/2. [*UMB* 16/2]

Stolper, M.W.
1988 The *šaknu* of Nippur. *JCS* 40: 127–55.

Steinkeller, P.
1979 Alleged GUR.DA = **ugula-geš-da** and the Reading of the Sumerian Numeral 60. *ZA* 69: 176–87.

1981 The Renting of Fields in Early Mesopotamia and the Development of the Concept of "Interest" in Sumerian. *Journal of Economic and Social History of the Orient* 24: 113–45.

1988 Notes on the Irrigation System in Third Millennium Southern Babylonia. *Bulletin on Sumerian Agriculture* 4: 73-91.

Stephens, F. J.

1953 A Surveyor's Map of a Field. *JCS* 7: 1-4.

Stol, M.

1988 Old Babylonian Fields. *Bulletin on Sumerian Agriculture* 4: 173-81.

Talon, P.

1985 *Archives royales de Mari*. Vol. 24, *Textes administratifs des salles "y et z" du palais de Mari*. Paris.

Tanret, M.

1982 Les tablettes "scolaires" découvertes à Tell ed-Dēr. *Akkadica* 27: 46-49.

Thureau-Dangin, F.

1895 La comptabilité agricole en Chaldée. *RA* 3: 118-46.

1896 Quelques mots de métrologie. *ZA* 11: 428-32.

1897 Un cadastre chaldéen. *RA* 4: 13-27.

1903 *Recueil de tablettes chaldéennes*. Paris.

1909a Le rapport de valeur entre l'or et l'argent en Babylonie. *OLZ* 9: 382-84.

1909b L'U, le Qa et la Mine, leur mesures et leur rapport. *Journal asiatique* 13: 79-111.

1912 Notes assyriologiques. *RA* 9: 24-25.

1913 Distances entre étoiles fixes d'après une tablette de l'époque des Séleucides. *RA* 10: 215-25.

1921 Numération et métrologie sumériennes. *RA* 18: 123-42.

1922 *Tablettes d'Uruk à l'usage des prêtres du temple d'Anu au temps des Séleucides* (TCL 6). Paris.

1924a La correspondance de Ḫammurapi avec Šamaš-ḫâṣir. *RA* 21: 1-58.

1924b Notes assyriologiques. *RA* 21: 139–46.

1926 Notes assyriologiques. *RA* 23: 31–34.

1927 *Marru.* *RA* 24: 147–48.

1928 La division du cercle. *RA* 25: 187–88.

1930a Notes assyriologiques. *RA* 27: 53–55.

1930b Notes assyriologiques. *RA* 27: 115–19.

1931 Notes sur la terminologie des textes mathématiques. *RA* 38: 195–98.

1932a Mathématique babylonienne. *RA* 29: 59–66.

1932b La mesure du "**qa**." *RA* 29: 189–92.

1932c Notes assyriologiques. *RA* 29: 21–28.

1932d Notes assyriologiques. *RA* 29: 77–88.

1932e Notes assyriologiques. *RA* 29: 131–42.

1932f Le prisme mathématique AO 8862. *RA* 29: 1–10, pl. 1–4.

1932g Le prisme mathématique AO 8862: Un post-scriptum. *RA* 29: 89–90.

1932h La ville ennemie de Marduk. *RA* 29: 109–19.

1933a Clepsydre babylonienne et clepsydre égyptienne. *RA* 30: 51–52.

1933b Notes assyriologiques. *RA* 30: 144.

1933c Notes assyriologiques. *RA* 30: 183–88.

1933d Notes lexicographiques. *RA* 30: 50.

1933e Poliorcétique babylonienne. *RA* 30: 126.

1933f La tablette de Strasbourg n° 11. *RA* 30: 184–87.

1934a Encore une fois la tablette de Strasbourg No. 11. *RA* 31: 70.

1934b Notes assyriologiques. *RA* 31: 49–52.

1934c Une nouvelle tablette mathématique de Warka. *RA* 31: 61–69.

1934d La tablette de Strasbourg n° 11. *RA* 31: 30, 70.

1935a La mesure des volumes d'après une tablette inédite du British Museum. *RA* 32: 1–28.

1935b Terminologie mathématique babylonienne. *RA* 32: 188.

1936a L'équation du deuxième degré dans la mathématique babylonienne d'après une tablette inédite du British Museum. *RA* 33: 27–48.

1936b Notes sur la mathématique babylonienne. *RA* 33: 161–68.

1936c Notes sur la mathématique babylonienne. *RA* 33: 180–84.

1936d Textes mathématiques babyloniens. *RA* 33: 65–84.

1936e Review of *Mathematische Keilschrifttexte* 1–2, by O. Neugebauer. *RA* 33: 55–62.

1937a La clepsydre babylonienne. *RA* 34: 144.

1937b La mesure du "**qa**." *RA* 34: 80–86.

1937c Notes sur la mathématique babylonienne. *RA* 34: 9–28.

1937d Review of *Mathematische Keilschrifttexte* 3, by O. Neugebauer. *RA* 34: 87–92.

1938a Le "grain," mesure de surface. *RA* 35: 156–57.

1938b La méthode de fausse position et l'origine de l'algèbre. *RA* 35: 71–77.

1938c Notes assyriologiques. *RA* 35: 104–06.

1938d *Textes mathématiques babyloniens*. Leiden. [*TMB*]

1939 Sketch of a History of the Sexagesimal System. *Osiris* 7: 95–141.

1940 Notes sur la mathématique babylonienne. *RA* 37: 1–10.

Torraca, G.; Chiari, G.; and Gullini, G.
1972 Report on Mud Brick Preservation. *Mesopotamia* 7: 259–86.

Toulmin,. S.
1967 The Astrophysics of Berossos the Chaldean. *Isis* 58: 65–76.

Ungnad, A.
1916 Zur babylonischen Mathematik. *OLZ* 19: 363–68.

1917–18 Lexikalisches. *ZA* 31: 248–76.

Vaiman, A. A.
1955 Ermitaznaya klinopisnaya matematiceskaya tablicka N° 015189 (The Hermitage cuneiform mathematical tablet no. 015189). *Epigrafika Vostoka* 10: 71–83.

1958 Ermitaznaya klinopisnaya matematiceskaya tablicka N° 15188 (The Hermitage cuneiform mathematical tablet no. 15188). *Epigrafika Vostoka* 12: 89–93.

1959 O geometriceskoi figure *absamikku* klinopisnyh matematiceskih tekstov (On the geometric figure *absamikku* in cuneiform mathematical texts). *VDI*: 91–94.

1960 Vavilonskaya geometriceskie risunki prostranstvennyh figur (Babylonian geometric drawings of geometric figures). *Istoriko-matematiceskie issledovaniya* 13: 379–82.

1961 *Sumero-vavilonskaya matematika, III–I tysyaceletiya do n.e.* (*Sumero-Babylonian mathematics, third to first millennium,* B.C.). Moscow.

1963 Istolkovanie geometriceskih postoyannyh iz suzskogo klinopisnogo spiska I (Suzy) (Interpretation of the geometric constants in the cuneiform tablet I from Susa). *VDI* 83: 75–86.

Veenhof, K. R.

1981–82 Babylonian Expressions for "Over/At a distance of...," Notes on *bērum*, *zūzam* and Related Matters. *Jaarbericht Ex Oriente Lux* 27: 65–75.

Vetter, Q.

1936 Quatre notes sur les mathématiques babyloniens. *Osiris* 1: 692–702.

Vino, I., and Viola, T.

1981 Testo n. 73: u problema algebrica. *Materiali epigrafici di Ebla* 3: 278–85.

Vitale, R.

1982 La musique suméro-accadien—gamme et notation musicale. *UF* 14: 241–63.

Vogel, K.

1932–33 Eine Pyramidenstumpf-Aufgabe bei den Babyloniern. *AfO* 8: 220–21.

1936 Bemerkungen zu den quadratischen Gleichungen der babylonischen Mathematik. *Osiris* 1: 703–17.

1958 Ist die babylonische Mathematik sumerisch oder akkadisch? *Mathematische Nachrichten* 18: 377–82.

1960 Der «falsche Ansatz» in der babylonischen Mathematik. *Mathematische-Physikalische Semesterberichte* 7: 89–95.

Vygodskii, M. Ya.

1959 Proishozdenie anaka nulya v vavilonskoi numeracii (The origin of the zero in Babylonian numeration). *Istoriko-matematiceskie issledovaniya* 12: 393–420.

van der Waerden, B. L.

1961 *Science Awakening*. New York: Oxford University Press.

1980 On Pre-Babylonian Mathematics 1. *Archive for the History of Exact Sciences* 23: 1-46.

Waetzoldt, H.

1972 *Untersuchungen zur neusumerischen Textilindustrie*, *Studi economici e tecnologici* 1. Rome.

1990 Zu den Bewässerungseinrichtungen in der Provinz Umma. *Bulletin on Sumerian Agriculture* 5: 1-29.

Walker, C.B.F.

1987 *Cuneiform*. London: British Museum Publications.

Waschow, H.

1932-33 Angewandte Mathematik im alten Babylonien (um 2000 v. Chr.). Studien zu den Texten *CT* IX, 8-15. *AfO* 8: 127-31, 215-20.

1933 Verbesserungen zu den babylonischen Dreiecksaufgaben. *Quellen und Studien zur Geschichte der Mathematik* B 2: 211-14.

1936-37 Review of *Mathematische Keilschrifttexte* 1-2, by O. Neugebauer. *AfO* 11: 245-47.

1937 Review of *Mathematische Keilschrifttexte* 3, by O. Neugebauer. *AfO* 12: 277.

Waschow, H., and Neugebauer, O.

1932 Reihen in der babylonischen Mathematik. *Quellen und Studien zur Geschichte der Mathematik* B 2: 298-304.

Weidner, E. F.

1916 Die Berechnung rechtwinkliger Dreiecke bei den Akkadern um 2000 v. Chr. *OLZ* 19: 257-63.

1917 Zahlenspielereien in akkadischen Leberschautexten. *OLZ* 20: 257-66.

Weissbach, F. H.

1914 Zu den Massen des Tempels Esagila und des babylonischen Turmes. *OLZ* 17: 193-201.

Weitemeyer, M.
1962 *Some Aspects of the Hiring of Workers in the Sippar Region at the Time of Hammurabi*. With a chapter on Seal Impressions by Edith Porada and Paul Lampl. Copenhagen.

Westenholz, A.
1975 *Early Cuneiform Texts in Jena* (Det. Kongelige Danske Videnskabernes Selskab, Historisker-Filosofiske Skrifter 7).

Whiting, R. M.
1984 More Evidence for Sexagesimal Calculations in the Third Millennium B.C. *ZA* 74: 59–66.

Wiseman, D. J.
1972 A Babylonian Architect? *AnSt* 22: 141–47.

Wulstan, D.
1968 The Tuning of the Babylonian Harp. *Iraq* 30: 215–28.

1974 Music from Ancient Ugarit. *RA* 63: 125–28.

Yitzhaki, M.
1986 Citation Patterns of the Research Literature of Ancient Near Eastern Studies. *JCS* 38: 81–93.

Young, D.W.
1991 The Incredible Regnal Spans of Kish I in the Sumerian King List. *JNES* 50: 23–35.

1988 A Mathematical Approach to Certain Dynastic Spans in the Sumerian King List. *JNES* 47: 123–29.

Zaccagnini, C.
1986 The Dilmun Standard and Its Relationship with Indus and Near Eastern Weight Systems. *Iraq* 48: 19–23.

1975 The Yield of the Fields at Nuzi. *OrAnt* 14: 181–225.

Zimmern, H.
1916 Zu den altakkadischen geometrischen Berechnungs-aufgaben. *OLZ* 19: 321–25.